Geometric Modeling and Mesh Generation from Scanned Images

CHAPMAN & HALL/CRC MATHEMATICAL AND COMPUTATIONAL IMAGING SCIENCES

Series Editors

Chandrajit Bajaj
Center for Computational Visualization
The University of Texas at Austin

Guillermo Sapiro
Department of Electrical
and Computer Engineering
Duke University

Aims and Scope

This series aims to capture new developments and summarize what is known over the whole spectrum of mathematical and computational imaging sciences. It seeks to encourage the integration of mathematical, statistical and computational methods in image acquisition and processing by publishing a broad range of textbooks, reference works and handbooks. The titles included in the series are meant to appeal to students, researchers and professionals in the mathematical, statistical and computational sciences, application areas, as well as interdisciplinary researchers involved in the field. The inclusion of concrete examples and applications, and programming code and examples, is highly encouraged.

Published Titles

Proposals for the series should be submitted to the series editors above or directly to:
CRC Press, Taylor & Francis Group
3 Park Square, Milton Park, Abingdon, OX14 4RN, UK

CHAPMAN & HALL/CRC
MATHEMATICAL AND COMPUTATIONAL IMAGING SCIENCES

Geometric Modeling and Mesh Generation from Scanned Images

Yongjie Jessica Zhang

Carnegie Mellon University

Pittsburgh, Pennsylvania, USA

CRC Press
Taylor & Francis Group
Boca Raton London New York

CRC Press is an imprint of the
Taylor & Francis Group, an **informa** business
A CHAPMAN & HALL BOOK

CRC Press
Taylor & Francis Group
6000 Broken Sound Parkway NW, Suite 300
Boca Raton, FL 33487-2742

First issued in paperback 2020

© 2016 by Taylor & Francis Group, LLC
CRC Press is an imprint of Taylor & Francis Group, an Informa business

No claim to original U.S. Government works

ISBN-13: 978-1-4822-2776-5 (hbk)
ISBN-13: 978-0-367-65852-6 (pbk)

Visit the Taylor & Francis Web site at
http://www.taylorandfrancis.com

and the CRC Press Web site at
http://www.crcpress.com

Contents

Preface

With finite element method (FEM) and scanning technology seeing increased use in cyberinfrastructure-revolutionized research areas such as computational science and engineering, there is an emerging need for quality mesh generation from scanned images to enable the analysis, understanding and prediction of complex physical phenomena. It is well known that FEM is currently well developed and efficient, but mesh generation for complex geometries still takes 80% of the total analysis time and is the major obstacle to reducing the total computation time. High-fidelity geometric modeling for such sophisticated domains is critical for a lot of cyber-enabled innovative applications. As a newly emerging interdisciplinary research area, *Geometric Modeling and Mesh Generation from Scanned Images* aims to integrate image processing, geometric modeling and mesh generation with FEM to solve problems in computational biology, medicine, materials sciences and engineering.

This book, the first for such an interdisciplinary topic, is based on a course "Computational Biomodeling & Visualization" that I taught at Carnegie Mellon University over the past eight years and also some of our latest research. It aims at introducing the fundamentals of medical imaging, image processing, computational geometry, mesh generation, visualization and finite element analysis; and also to expose readers to novel and advanced applications in computational biology, medicine, material sciences and other engineering fields. The principal audience includes senior undergraduate students, graduate students and researchers in mechanical engineering, biomedical engineering, material sciences and other engineering departments. This book can be used directly for teaching, research and as a professional reference. It can also be useful to more general researchers in related industry companies, medical centers, government laboratories and other research institutions.

The subject integrates mechanical engineering, biomedical engineering, computer science and mathematics together. The covered topics include biomedical/material imaging, image processing, geometric modeling and visualization, finite element method and biomedical and engineering applications. This book starts with an introduction of pipelines from imaging to meshing and simulation as well as advances and challenges in this area (Chapter 1), followed by a review of numerical methods used in various modules of the pipelines (Chapter 2). Chapter 3 introduces several important scanning techniques, including computational tomography (CT), magnetic resonance imaging (MRI), nuclear medicine, ultrasound and cryo-electronic microscopy (cryo-EM), as well as techniques used to scan polycrystalline materials such as electron backscatter diffraction (EBSD) and high energy X-rays. In addition, this chapter also talks about some basic operators in image processing such as contrast enhancement, filtering, segmentation and registration. In Chapter 4, fundamentals of geometric modeling and computer graphics are first introduced, followed by geometric objects and transformations, as well as curves and surfaces. Since isocontouring is important for medical image visualization, we introduce two isocontouring methods in detail: marching cubes and dual contouring. The following two chapters (Chapters 5 and 6) talk about various triangular/tetrahedral and quadrilateral/hexahedral mesh generation techniques. These methods are also extended to mesh generation for multiple-material domains, resolving topology ambiguities and with guaranteed quality. After that, Chapter 7 discusses volumetric T-spline modeling for isogeometric analysis (IGA). Finally, Chapter 8 introduces FEM and its three new developments such as IGA, extended FEM (XFEM) and immersed FEM (IFEM); overviews the broad application areas; and explains how the above topics (or chapters) are used in various applications through several real projects.

This book touches on many related topics and has many unique features compared to other books, including

- Being the first textbook for this new and interdisciplinary area;

- Covering all the related topics from image to simulation and application;

- Integrating mechanical engineering, biomedical engineering, computer science and mathematics together;

- Being a textbook suitable for both senior undergraduate students and graduate students in mechanical engineering, biomedical engineering and material sciences;

- The latest state-of-the-art techniques; and

- For each topic, discussions of the latest research advances, challenges and related applications.

These unique features can be beneficial to many people working on various interdisciplinary research areas.

This book introduces our latest research projects. Therefore, I thank my former and current students for their great effort in our research projects, and also their help in generating pictures for this book: Joshua Chen, Kangkang Hu, Arjun Kumar, Yicong Lai, Xinghua Liang, Tao Liao, Lei Liu, Aishwarya Pawar, Jin Qian, Wenyan Wang, Xiaodong Wei; visiting PhD students Yue Jia and Juelin Leng; as well as many collaborators who have made contributions: Drs. Chandrajit L. Bajaj, Yuri Bazilevs, Lei Dong, Andrew B. Geltmacher, Shaolie S. Hossain, Thomas J.R. Hughes, Pete M. Kekenes-Huskey, Trond Kvamsdal, Shawn Litster, Wing Kam Liu, Jim Lua, Andrew McCammon,

Andrew McCulloch, Timon Rabczuk, Michael A. Scott, Thomas W. Sederberg, Kenji Shimada and Guoliang Xu. Our projects were supported by ONR-PECASE, NSF-CAREER, ONR-YIP, ONR, NRL, NAVAIR, AFOSR, NSF-MRSEC, NIH-UCSD Subaward, SINTEF in Norway, UPMC-HTI, Winters Foundation, Berkman Faculty Development Foundation and Chinese NSF. Their support is greatly appreciated.

Chapter 2 was written based on my lecture notes from two teaching courses on numerical methods (one at the graduate level and the other one at the undergraduate level). I thank my colleagues Professors Alan J. H. McGaughey and Metin Sitti who kindly shared their lecture notes with me. Their help is greatly appreciated! I also thank my course teaching assistants, Xinghua Liang, Tao Liao, Kangkang Hu and Onder Erin, for organizing the solution manual for all the homework and project assignments.

Jessica Zhang
Pittsburgh, USA

Chapter 1

Introduction and Pipelines

An introduction is provided to explain the pipelines from imaging to meshing and simulation. The pipelines involve five main topics: imaging, image processing, geometric modeling and computer graphics, mesh generation and finite element simulations. These five topics can be applied to many applications in computational biology, medicine, material sciences and engineering.

1.1 Introduction

Over recent years, finite element method (FEM) and scanning technology have been developed and advanced rapidly, especially in cyberinfrastructure-revolutionized research areas such as computational science and engineering. They have very broad applications in computational biology, medicine, materials sciences and engineering. There is an emerging need for quality mesh generation of the spatially realistic domains to enable the analysis, understanding and prediction of complex physical or biological phenomena. In images obtained from computer tomography (CT), magnetic resonance imaging (MRI) or microscopy scanning, the domain of focus often possesses homogeneous, heterogeneous materials and/or functionally different properties. For each of the partitioned materials, high-fidelity (correct topology and accurate geometry) geometric models and quality meshes are needed, with meshes conforming at the manifold or non-manifold boundaries. However, quality meshes for such complex geometries are typically generated manually and thus usually take months to be created. It is well known that FEM is currently well developed and efficient, but mesh generation for complex geometries (e.g., the human body) still takes 80% of the total analysis time and is the major obstacle to reduce the total computation time. High-fidelity geometric modeling for such sophisticated domains is critical for a lot of cyber-enabled innovative applications. As a new emerging interdisciplinary research area, "geometric modeling and mesh generation from scanned images" integrates image processing, geometric modeling and mesh generation with FEM to solve problems in computational biology, medicine, material sciences and engineering.

Image-based mesh generation is a relatively new field. Normally researchers first extract boundary surfaces using isocontouring [189, 256] which usually involves manual interaction, and then construct tetrahedral or hexahedral (hex) meshes. The research on mesh generation is dominated by tetrahedral meshing algorithms, which can be grouped into Delaunay triangulation [109, 396], advancing front [251, 252], or grid-based methods [338, 377]. Fewer algorithms exist for automatic all-hex mesh generation due to its intrinsic complexities, and all these existing methods have limitations. For example, the frequently used, easy to implement block-structured method [71, 280] produces non-conforming boundaries and a large number of elements. The grid-based method [325, 326], which puts structured grids inside the volume while adding elements at the boundaries afterwards, cannot be extended to all-hex mesh generation for heterogeneous domains with conformal non-manifold boundaries. Today, the key barriers scientists face are:

- A lack of automatic and robust meshing techniques for multiscale modeling and heterogeneous domains;

- Robust unstructured all-hex mesh generation with sharp feature preservation for complicated geometry and topology is still a challenge;

- The inability of existing methods to effectively improve the quality of non-manifold meshes with feature preservation and topology validation; and

- A lack of volumetric parameterization (e.g., NURBS and T-spline) techniques for complicated domains to support a new development of FEM called isogeometric analysis.

Many simulations cannot hereby be effectively carried out due to the lack of analysis-suitable finite element meshes. In this book, we introduce the fundamentals of imaging, image processing, computational geometry, mesh generation, visualization, finite element analysis as well as their novel and advanced applications in computational biology, medicine, material sciences and other engineering fields.

1.2   Pipelines and Five Topics

As a newly emerging interdisciplinary research area, "geometric modeling and mesh generation from scanned images" integrates mechanical engineering, biomedical engineering, computer science, and mathematics together. From image scanning to meshing and then to simulation applications, the entire procedure forms comprehensive image-mesh-simulation pipelines [426]. As shown in Figure 1.1, there are mainly five modules in the : imaging, image processing, geometry processing, finite element methods, and application.

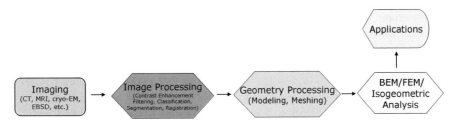

FIGURE 1.1: Image-mesh-simulation pipeline with five basic topics.

There are many techniques and processes developed to create images, such as CT, MRI, nuclear medicine, ultrasound, and fluorescence for biomedical imaging. CT utilizes X-rays while MRI utilizes radio wave and magnetic field to scan objects. For both CT and MRI, the resulting images provide anatomy structure. Unlike CT and MRI, Nuclear Medicine injects radioactive source into the body, and then measures the radiation emitted from the body. The resulting images provide physiological functions of the object. Ultrasound utilizes high frequency sound waves in real time, generally 3~10 megahertz. Fluoroscopy uses a constant of X-rays, and it is also in real time. Cryo-EM (cryo-electron microscopy) is a popular scanning technique in structural biology to study small-scale objects like viruses at the level of Angstroms. In addition, people also scan metal materials. For example, electron backscatter diffraction (EBSD) is a microstructural-crystallographic technique used to examine the crystallographic orientation and elucidate texture or preferred orientation of polycrystalline materials. Instead of destroying the metal material like EBSD, high-energy X-ray is a nondestructive scanning technique to study polycrystalline materials. These various scanning techniques produce images for different types of objects. In some research areas like computer-aided design (CAD), people also use computational ways like signed distance function to compute volumetric imaging data. For biomolecules or proteins, people use the atomic resolution information to build electron density map and solve partial differential equations to obtain electron static potential distribution on regular grids.

The scanned image data V is a scalar field over sampled rectilinear grids, $V = \{F(i, j, k)| \ i, j, k$ are indices of x, y, z coordinates in a rectilinear grid$\}$. Due to the limitation of the scanning resolution, object motion and other issues, the obtained images may have low contrast, noises and unclear boundary for materials of interest. Therefore, people develop different kinds of image processing techniques to improve the quality of images and try to extract useful information to enable a better understanding. More specifically, people use computational methods to enhance the contrast, remove noises, classify and segment materials or regions of interest, or register images from difference time phases or modalities. After that, the processed images are fed into the third module in the pipeline to perform geometry processing. It includes geometric modeling, mesh generation and quality improvement. Various quality finite element meshes can be created in this module, which are ready for the following finite element simulations.

The fourth module includes various mechanics simulation techniques, such as boundary element methods, finite element methods and isogeometric analysis [179]. The last module is for applications, it can be very broad ranging from mechanical engineering, biomedical engineering, material sciences, petroleum engineering, electrical engineering to civil engineering and architecture. Generally, various applications have different research aims and they need different techniques in the first four modules to handle the entire process. For example, in cardiovascular blood flow simulation and brain biomechanics applications, image segmentation and quality piecewise-linear elements like tetrahedral/hexahedral meshes or solid NURBS (non-uniform rational B-splines)/T-spline models [332, 334] are required. In dynamic lung modeling, image registration techniques are needed to track the motion of tumor in the lung. In cardiac mechanical property study, high order elements like tricubic Hermite models are preferable. These five topics provide various fundamental technologies, based on which users can develop their own detailed pipelines.

Figure 1.2 shows a comprehensive image-mesh-simulation computational

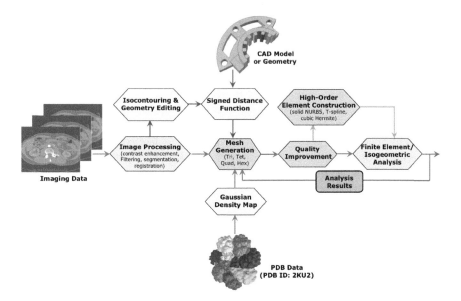

FIGURE 1.2: A comprehensive computational framework of image-mesh-simulation with several pipelines.

framework that includes several pipelines. Following the red arrows, we start with scanned images, use image processing techniques to extract and segment the region of interest, and then we generate piecewise-linear elements (triangle, quadrilateral, tetrahedra or hexahedra) and improve the mesh quality. The resulting meshes can be used in finite element analysis and the analysis results are then fed to the meshing process to fine tune the finite element meshes. Following the green arrows, the obtained piecewise-linear meshes can be used as control meshes to build high-order elements like solid NURBS, T-splines or cubic Hermite models, which are then used in isogeometric analysis. Another pipeline handles CAD models or input geometry (blue arrows), which can be designed from CAD software, extracted and edited from segmented images, or obtained from points cloud in reverse engineering. They can be converted to volumetric gridded data using signed distance function. The last pipeline (magenta arrows) takes atomic resolution data of biomolecules or proteins from the protein data bank (PDB data) as input, and builds Gaussian density map. The molecular surface can be approximated as an isocontour in this density map. This comprehensive framework can be broadly applied to many applications in computational biology, medicine, materials sciences and engineering.

1.3   Challenges and Advances

There are many challenges on image-based geometric modeling and mesh generation. Many times, the scanned images are not in good quality. For example, nano-CT using Zernike phase contrast sometimes produces images with artifacts like halo and shade-off, which makes the following image segmentation a big challenge, especially for fuel cell electrode and other porous materials. Image restoration techniques [211] are needed to resolve this issue. There are many uncertainties and difficulties in image processing. The scanned images usually contain noises and we need to develop filtering techniques to remove them and try to preserve important image features simultaneously. Although many techniques have been developed for image segmentation, it is still challenging to classify and segment what we need exactly due to the fuzzy boundary definition in images. There is still a lack of stable, efficient, automatic and robust segmentation techniques especially for large complex domains. To align two images, we need efficient, robust and accurate registration algorithms and we also need to handle large deformations.

For geometry processing and mesh generation, we are still lacking automatic meshing techniques for multiscale modeling and heterogeneous domains. Robust unstructured mesh generation with topology ambiguity resolved [436, 437] is still a difficulty. The inability of existing methods to effectively improve the mesh quality and provide quality guarantee yields a lot of limitations to us. Moreover, we need volumetric parameterization (e.g., solid NURBS or T-spline) techniques to support isogeometric analysis applications to integrate design with analysis.

Various applications often provide us different challenges on imaging, image processing, geometry processing and simulation. For example, in biomedical applications, vascular blood flow simulation using isogeometric analysis needs high-order elements like solid NURBS [430] or T-splines [391, 393, 394, 438, 439], it is difficult to estimate wall-thickness and anisotropic material property for fluid-structure interaction, and it is hard to characterize complex geometries like aneurysm dome using machine learning techniques. In engineering applications, critical feature determination of polycrystalline microstructure materials [306] needs high-fidelity quality finite element meshes; we also need to handle a huge amount of data and it is hard to effectively and efficiently obtain the information we need. All these challenges require us to develop new techniques to facilitate various applications.

In each module of the computational modeling pipeline, numerical methods are the key to develop algorithms for various research problems. In the following, we will first review the fundamentals of various numerical methods.

Chapter 2

Review of Numerical Methods

This chapter aims at reviewing the fundamentals of the theory behind the numerical techniques accessible in programs such as Matlab® or Mathematica, and exposing readers to novel applications in engineering. The topics to be covered include introduction, algebraic equations, curve fitting, ordinary differential equations, eigenvalue problems, partial differential equations, integration, Fourier analysis, and optimization. Based on these fundamentals, readers can develop computer algorithms and employ them in a variety of engineering applications. There are some homework problems provided at the end of this chapter. Readers are free to pick a program language in working on them, but C/C++ is preferred especially for graduate students.

This chapter was written based on my lecture notes of both graduate-level and undergraduate-level courses of "Numerical Methods." I thank my colleague Professors Alan J. H. McGaughey and Metin Sitti for kindly sharing their lecture notes with me. Their help is greatly appreciated!

2.1   Introduction

Numerical methods [81, 152, 301] are techniques by which real world problems are simplified into mathematical models and then mathematical problems are formulated so that they can be solved using arithmetic and logic operations. Numerical methods are also referred to as "computer mathematics." Currently, traditional engineering fields have been extended a lot to cutting-edge areas, and engineering students need a solid background not only in math, physics and so on, but also in understanding of numerical methods. Numerical methods reinforce our understanding of math, learning to use computers, and also enable us to use software packages to design our own programs. Numerical methods are essential problem-solving tools in the modern society because we often need to solve large-scale, nonlinear, multiphysics and multiscale problems.

Numerical methods did not start with the advent of computers. In old days, people used abacuses, slide rules (a mechanical analog computer) and log tables. Gradually, digital computation received more attention as computers became more powerful. A computer can speed up the computation significantly and reduce the risk of error. In general, a numerical method is a powerful tool that solves problems that are not tractable analytically, especially for nonlinear equations, partial differential questions and complex engineering problems. See Table 2.1 for the detailed comparison between analytical solution and numerical solution.

TABLE 2.1: Analytical Solution vs Numerical Solution

Analytical Solution	Numerical Solution
exact (if solvable)	approximate with an error bound
manual or symbolic computing	algorithms, programming & numerical computing
gives direct physical intuition	gives numbers & indirect intuition
not applicable for complex systems	applicable for complex systems

In numerical methods, the solution is always an approximation and better accuracy requires more computation. There is also a trade-off between speed and accuracy in numerical methods. From a computational standpoint, we want a method with least amount of computation for a desired accuracy. In the following, we will discuss a pair of numerical errors and also review the basic finite difference method.

2.1.1 Round-Off and Truncation Errors

Numerical computation introduces numerical errors, which are generally defined based on accuracy and precision. *Accuracy* refers to how closely a computed or measured value agrees with the true value, while *precision* refers to how closely individual computed or measured values agree with each other. Given the true value or analytical solution, we can compute the true relative error

$$\epsilon_t = \frac{true\ value - approximation}{true\ value} \times 100\%. \tag{2.1}$$

If you do not have the true value, we can compute the approximation relative error

$$\epsilon_a = \frac{current\ approximation - previous\ approximation}{current\ approximation} \times 100\%. \tag{2.2}$$

There are two kinds of numerical errors, namely, the round-off error and the truncation error. The *round-off error* is introduced because digital computers cannot represent some quantities exactly, for example, π, e and $\sqrt{7}$. In addition, some numerical manipulations are highly sensitive to round-off errors. Differently, the *truncation error* is introduced by using an approximation to replace an exact mathematical procedure such as the Taylor series expansion. For example, given $f(x_i)$, we can use the Taylor series expansion to approximate $f(x_{i+1})$ where $x_{i+1} = x_i + h$. We have

$$\begin{aligned} f(x_{i+1}) &= f(x_i) + f'(x_i)h + \frac{f''(x_i)}{2!}h^2 + \cdots \\ &+ \frac{f^{(n)}(x_i)}{n!}h^n + R_n, \end{aligned}$$

where R_n is the remainder or the truncation error. Assume $f(x)$, $f'(x)$, $\cdots$, $f^{(n)}(x)$ exist for $x_i < x < x_{i+1}$, then we have

$$R_n = \frac{f^{(n+1)}(\xi)}{(n+1)!}h^{n+1} = O(h^{n+1}), \qquad x_i < \xi < x_{i+1}. \tag{2.3}$$

In other words, the truncation error is in the order of h^{n+1} or $\propto h^{n+1}$. It is obvious that a smaller h gives a smaller truncation error, and higher order approximation (larger n) reduces the truncation error provided $h < 1$.

The total numerical error consists of the round-off and truncation errors. There is a trade-off between them. Choosing a small h gives us a small truncation error, but it needs more computation or iterations, leading to large accumulated round-off error. For a specific numerical scheme, we need to study and choose the optimal step size h_{opt} in order to minimize the total numerical error for the computation.

2.1.2 Finite Difference Methods

The finite difference method is a numerical method to approximate derivatives using finite difference equations. For the first derivative approximation, we have the forward difference, the backward difference and the centered difference methods, which can be derived using the Taylor series expansion. The centered difference is the most accurate one in terms of the truncation error. Here, we define four operators to explain the finite different methods in a more general sense, which can also be applied to a set of discrete data.

1. Forward difference: $\Delta F(x_i) = \Delta F_i = F_{i+1} - F_i$;

2. Backward difference: $\nabla F(x_i) = \nabla F_i = F_i - F_{i-1}$;

3. Stepping: $EF_i = F_{i+1}$; and

4. Derivative: $DF_i = \frac{dF_i}{dx}\big|_{x_i}$.

Note that Δ, ∇, E and D are operators; they can be applied together. For example,

$$
\begin{aligned}
\Delta^2 F_i &= \Delta(\Delta F_i) \\
&= \Delta(F_{i+1} - F_i) \\
&= \Delta F_{i+1} - \Delta F_i \\
&= F_{i+2} - F_{i+1} - F_{i+1} + F_i \\
&= F_{i+2} - 2F_{i+1} + F_i.
\end{aligned}
\tag{2.4}
$$

Another example

$$
\begin{aligned}
(1 + \Delta)F_i &= F_i + \Delta F_i \\
&= F_i + F_{i+1} - F_i \\
&= F_{i+1} \\
&= EF_i,
\end{aligned}
\tag{2.5}
$$

therefore,

$$
1 + \Delta = E.
\tag{2.6}
$$

We want to develop expressions for DF_i and $D^2 F_i$ that are easy to evaluate. Considering EF_i and expanding it as a Taylor series about F_i (we replace Δx with h here to avoid the confusion with the operators), we obtain

$$
\begin{aligned}
EF_i = F_{i+1} &= F_i + hDF_i + \frac{h^2}{2}D^2 F_i + \frac{h^3}{6}D^3 F_i + \cdots \\
&= (1 + hD + \frac{h^2}{2}D^2 + \frac{h^3}{6}D^3 + \cdots)F_i \\
&= e^{hD} F_i.
\end{aligned}
$$

Then, we have

$$
E = e^{hD}, \tag{2.7}
$$

$$
lnE = hD,
$$

$$
D = \frac{1}{h}lnE = \frac{1}{h}ln(1 + \Delta). \tag{2.8}
$$

Expanding $ln(1 + \Delta)$ as a Taylor series gives

$$
ln(1 + \Delta) = \Delta - \frac{\Delta^2}{2} + \frac{\Delta^3}{3} - \frac{\Delta^4}{4} + \cdots
$$

so that

$$
\begin{aligned}
DF_i &= \frac{1}{h}ln(1 + \Delta)F_i \\
&= \frac{1}{h}(\Delta - \frac{\Delta^2}{2} + \frac{\Delta^3}{3} - \frac{\Delta^4}{4} + \cdots)F_i.
\end{aligned}
$$

We have

$$
DF_i = \frac{1}{h}(\Delta F_i - \frac{\Delta^2}{2}F_i + \frac{\Delta^3}{3}F_i - \frac{\Delta^4}{4}F_i + \cdots). \tag{2.9}
$$

In the following, we use Eq. (2.9) to derive the forward difference formula. Considering the first term only, we have

$$
\begin{aligned}
D^{(1)}F_i &= \frac{1}{h}\Delta F_i + O(h) \\
&= \frac{F_{i+1} - F_i}{h} + O(h).
\end{aligned}
$$

Considering the first two terms, we have

$$
\begin{aligned}
D^{(2)}F_i &= \frac{1}{h}(\Delta F_i - \frac{1}{2}\Delta^2 F_i) + O(h^2) \\
&= \frac{1}{h}(F_{i+1} - F_i - \frac{1}{2}F_{i+2} + F_{i+1} - \frac{1}{2}F_i) + O(h^2) \\
&= \frac{1}{h}(-\frac{1}{2}F_{i+2} + 2F_{i+1} - \frac{3}{2}F_i) + O(h^2).
\end{aligned}
$$

Note that when one more extra term is considered, the order of error increases by one. In the following, let us find the second derivative.

$$
\begin{aligned}
D^2 F_i &= D(DF_i) \\
&= D\left[\frac{1}{h}(\Delta - \frac{1}{2}\Delta^2 + \frac{1}{3}\Delta^3 - \cdots)F_i\right] \\
&= \frac{1}{h^2}(\Delta - \frac{1}{2}\Delta^2 + \frac{1}{3}\Delta^3 - \cdots)(\Delta - \frac{1}{2}\Delta^2 + \frac{1}{3}\Delta^3 - \cdots)F_i \\
&= \frac{1}{h^2}(\Delta^2 - \Delta^3 + \frac{11}{12}\Delta^4 + \cdots)F_i. \tag{2.10}
\end{aligned}
$$

The first term of the mth derivative will always look like

$$
D^m F_i = \frac{1}{h^m}\Delta^m F_i.
$$

If we truncate at the first term for $D^2 F_i$ in Eq. (2.10), we obtain

$$
\begin{aligned}
D^{2(1)} F_i &= \frac{1}{h^2}\Delta^2 F_i + O(h) \\
&= \frac{1}{h^2}(F_{i+2} - 2F_{i+1} + F_i) + O(h).
\end{aligned}
$$

Similarly using the backward difference operator, ∇, we can show that DF_i can also be expressed as

$$
DF_i = \frac{1}{h}(\nabla + \frac{1}{2}\nabla^2 + \frac{1}{3}\nabla^3 + \cdots)F_i. \tag{2.11}
$$

Eq. (2.11) can be used to derive the backward difference formula. For higher order D, the first term always looks like

$$
D^m F_i = \frac{1}{h^m}\nabla^m F_i.
$$

Then, we average the two expressions for DF_i in Eqns. (2.9) and (2.11) and obtain

$$
D = \frac{1}{2h}\left[(\Delta + \nabla) - \frac{1}{2}(\Delta^2 - \nabla^2) + \frac{1}{3}(\Delta^3 + \nabla^3) - \frac{1}{4}(\Delta^4 - \nabla^4) + \cdots\right]. \tag{2.12}
$$

Eq. (2.12) is used to derive the central difference formula. Considering only the first term, we have

$$
\begin{aligned}
DF_i &= \frac{1}{2h}(\Delta + \nabla)F_i + O(h^2) \\
&= \frac{1}{2h}(\Delta F_i + \nabla F_i) + O(h^2) \\
&= \frac{1}{2h}(F_{i+1} - F_i + F_i - F_{i-1}) + O(h^2) \\
&= \frac{1}{2h}(F_{i+1} - F_{i-1}) + O(h^2). \tag{2.13}
\end{aligned}
$$

It is obvious that the central difference method is more accurate than the forward and backward difference methods. When we use the central difference approximation for the second derivative, we obtain

$$D^2 F_i = \frac{1}{2h}(F_{i+1} - 2F_i + F_{i-1}) + O(h^2). \tag{2.14}$$

Following the same way, we can derive the finite difference formula for any degree of accuracy you require. The error is a function of the step size h. When h decreases, we obtain better accuracy, but the computational cost also increases.

2.2 Linear and Nonlinear Algebraic Equations

In this section, we will first review matrix notations and computation, and then discuss how to use finite difference methods for solving boundary value problems, sets of linear and nonlinear algebraic equations.

2.2.1 Matrix Operations and LU Decomposition

Matrix computation is important in solving linear systems. Here let us first review the basic notations and matrix computation. A matrix A with n rows and m columns is written as

$$A = \begin{bmatrix} a_{11} & a_{12} & \cdots & a_{1m} \\ a_{21} & a_{22} & \cdots & a_{2m} \\ \vdots & & & \\ a_{n1} & a_{n2} & \cdots & a_{nm} \end{bmatrix},$$

where a_{ij} is an element of A. A vector is a column matrix defined as $\vec{x} = \{x_1, x_2, \cdots, x_n\}^T$. The transpose of A_{nm} is A_{nm}^T, and we have $a_{ji}^T = a_{ij}$. The inverse of A is A^{-1} such that $A^{-1}A = I$, where I is the identify matrix with $I_{ij} = \delta_{ij}$. The trace of a square matrix A is $tr(A) = \sum_i a_{ii}$ and its determinate is denoted as $det(A)$.

The "matrix norm" is used to assess the sensitivity of a system of equations to errors introduced by solution techniques or uncertainty in the input data; we use $||A||$ to denote it. The norm can be defined in various ways, but it must satisfy the following properties:

1. $||A|| \geq 0$;

2. $||kA|| = |k|||A||$, where k is a scalar;

3. $||A + B|| \leq ||A|| + ||B||$; and

4. $||AB|| \le ||A||||B||$.

There are three popular norm definitions:

$$||A||_1 = \max_{1 \le j \le n} \sum_{i=1}^{m} |a_{ij}|;$$

$$||A||_\infty = \max_{1 \le i \le m} \sum_{j=1}^{n} |a_{ij}|;$$

$$||A||_e = \left(\sum_{i=1}^{m} \sum_{j=1}^{n} a_{ij}^2 \right)^{1/2}.$$

We define the condition number of A as $co(A) = ||A||||A^{-1}||$. For an ill-conditioned matrix A, a small variation in the matrix may yield a large variation in the solution. Suppose we want to solve $A\vec{x} = \vec{b}$, but due to some error in the initial data, we end up with solving $(A + E)\vec{x}' = \vec{b}$, where E is an error term. We next tend to estimate the error $\vec{x} - \vec{x}'$. Starting from $A\vec{x} = \vec{b}$, we can obtain

$$\begin{aligned} \vec{x} &= A^{-1}\vec{b} = A^{-1}A'\vec{x}' = A^{-1}(A + E)\vec{x}' \\ &= A^{-1}A\vec{x}' + A^{-1}E\vec{x}' = \vec{x}' + A^{-1}E\vec{x}'. \end{aligned}$$

Therefore, $\vec{x} - \vec{x}' = A^{-1}E\vec{x}'$. We do not know E, but we have an estimate of its size (i.e., its norm). Using the norm property (4), we can obtain a relationship between the relative error in solution and the relative error in the input:

$$||\vec{x} - \vec{x}'|| \le ||A^{-1}||||E||||\vec{x}'||$$

$$\Rightarrow \frac{||\vec{x} - \vec{x}'||}{||\vec{x}'||} \le ||A^{-1}||\frac{||A||||E||}{||A||}$$

$$\Rightarrow \frac{||\vec{x} - \vec{x}'||}{||\vec{x}'||} \le ||co(A)||\frac{||E||}{||A||}.$$

Suppose we solve $A\vec{x} = \vec{b}$ and obtain a solution $\vec{x}' \ne \vec{x}$ due to the error introduced in the solution procedure such as the rounding error. If we evaluate the residual $\vec{r} = \vec{b} - A\vec{x}$, will $\vec{r}$, $\Delta\vec{r}$ or $||\vec{r}||$ be a good measure of whether A is ill-conditioned? The answer is "NO" because for an ill-conditioned matrix, a very wrong $\vec{x}'$ may yield a small $\vec{r}$. If we define $\vec{e} = \vec{x} - \vec{x}'$, then it can be proved that

$$\frac{1}{co(A)} \frac{||\vec{r}||}{||\vec{b}||} \le \frac{||\vec{e}||}{||\vec{x}||} \le co(A) \frac{||\vec{r}||}{||\vec{b}||}. \tag{2.15}$$

Eq. (2.15) gives the range of possible relative error in the solution. As shown in Figure 2.1, the range of the possible error increases as the condition number

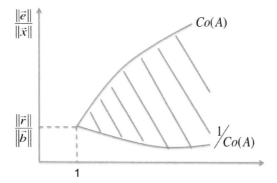

FIGURE 2.1: The schematic plot of the range of possible error vs the condition number of a matrix.

increases, and the residual is only a good estimate of the error when $co(A)$ is small.

LU decomposition is a popular means to solve a system of linear equations $(A\vec{x} = \vec{b})$. In the following, we will review the LU decomposition in detail. The basic idea is to decompose A into a product of a lower-triangular matrix (L) and an upper-triangular matrix (U), and then solve the system using a forward substitution and a backward substitution. The algorithm can be summarized in the following six steps.

1. For $A\vec{x} = \vec{b}$, let $A = LU$;

2. Then $LU\vec{x} = \vec{b}$;

3. Let $U\vec{x} = \vec{y}$;

4. So that $L\vec{y} = \vec{b}$;

5. Solve for $\vec{y}$ in Step (4) using forward substitution; and

6. Solve for $\vec{x}$ in Step (3) using backward substitution.

In a general $n \times n$ matrix, there are n^2 elements. In an upper or lower diagonal matrix, there are $\sum_{i=1}^{n} = \frac{n(n+1)}{2}$ nonzero elements. Therefore, we have $\frac{2n(n+1)}{2} - n^2 = n^2 + n - n^2 = n$ free parameters to specify. For a 3×3 matrix, $A = LU$ can be written as

$$\begin{bmatrix} a_{11} & a_{12} & a_{13} \\ a_{21} & a_{22} & a_{23} \\ a_{31} & a_{32} & a_{33} \end{bmatrix} = \begin{bmatrix} l_{11} & 0 & 0 \\ l_{21} & l_{22} & 0 \\ l_{31} & l_{32} & l_{33} \end{bmatrix} \begin{bmatrix} 1 & u_{12} & u_{13} \\ 0 & 1 & u_{23} \\ 0 & 0 & 1 \end{bmatrix}. \qquad (2.16)$$

Here we adopt the Crout reduction, where we set $u_{ii} = 1$. In this way, the values of l_{ij} and u_{ij} will fall out directly. In addition, $det(A) = det(L)det(U) = det(L) = \prod_{i=1}^{n} l_{ii}$. We use the following five steps to solve alternative columns and rows of L and U.

1. L column 1 (l_{i1}, where $i = 1, \cdots, n$);

2. U row 1 (u_{1j}, where $j = 2, \cdots, n$);

3. L column 2 (l_{i2}, where $i = 2, \cdots, n$);

4. U row 2 (u_{2j}, where $j = 3, \cdots, n$); and

5. L column 3 (l_{i3}, where $i = 3, \cdots, n$).

In the following, we go step by step (a total of five steps) through the solution for the 3×3 case.

1. L column 1, using entries in A column 1 (a_{i1}, where $i = 1, \cdots, n$);
   $a_{11} = l_{11}(1) + 0(0) + 0(0) \Rightarrow l_{11} = a_{11}$.
   Similarly, $l_{21} = a_{21}$ and $l_{31} = a_{31}$.

2. U row 1 (u_{1j}, using the remaining entries in A row 1 or a_{1j}, where $j = 2, \cdots, n$);
   $a_{12} = l_{11}u_{12} + 0(1) + 0(0) \Rightarrow u_{12} = \frac{a_{12}}{l_{11}}$.
   Similarly, $u_{13} = \frac{a_{31}}{l_{11}}$.

3. L column 2 (l_{i2}, using the remaining entries in A column 2 or a_{i2}, where $i = 2, \cdots, n$);
   $a_{22} = l_{21}u_{12} + l_{22}(1) + 0(6) \Rightarrow u_{22} = a_{22} - l_{21}u_{12}$.
   Similarly, $l_{32} = a_{32} - l_{31}u_{12}$.

4. U row 2 (u_{2j}, where $j = 3, \cdots, n$);
   We can obtain $u_{23} = \frac{a_{23} - l_{21}u_{13}}{l_{22}}$.

5. L column 3 (l_{i3}, where $i = 3, \cdots, n$).
   We can obtain $l_{33} = a_{33} - l_{31}u_{13} - l_{32}u_{23}$.

In general, we can show that (under the Crout reduction, where $u_{ii} = 1$)

$$l_{ij} = a_{ij} - \sum_{k=1}^{j-1} l_{ik}u_{kj} \quad for \ \ i = 1, \cdots, n \ \ and \ \ j \leq i; \tag{2.17}$$

$$u_{ij} = \frac{a_{ij} - \sum_{k=1}^{i-1} l_{ik}u_{kj}}{l_{ii}} \quad for \ \ j = 2, \cdots, n \ \ and \ \ i < j. \tag{2.18}$$

To account for $l_{ii} = 0$, we might need to introduce the concept of pivoting (row operations). By now, we have obtained L and U. Then we consider $L\vec{y} = \vec{b}$ and use the forward substitution to solve for $\vec{y}$. We have

$$\begin{bmatrix} l_{11} & 0 & 0 \\ l_{21} & l_{22} & 0 \\ l_{31} & l_{32} & l_{33} \end{bmatrix} \begin{Bmatrix} y_1 \\ y_2 \\ y_3 \end{Bmatrix} = \begin{Bmatrix} b_1 \\ b_2 \\ b_3 \end{Bmatrix}. \tag{2.19}$$

We can solve for $\vec{y}$ terms directly starting from y_1. In general, we have

$$y_i = \frac{b_i - \sum_{k=1}^{i-1} l_{ik} y_k}{l_{ii}}, \quad i = 1, \cdots, n. \tag{2.20}$$

Similarly, we use the backward substitution to solve for $\vec{x}$ from $U\vec{x} = \vec{y}$.

$$\begin{bmatrix} 1 & u_{12} & u_{13} \\ 0 & 1 & u_{23} \\ 0 & 0 & 1 \end{bmatrix} \left\{ \begin{array}{c} x_1 \\ x_2 \\ x_3 \end{array} \right\} = \left\{ \begin{array}{c} y_1 \\ y_2 \\ y_3 \end{array} \right\}. \tag{2.21}$$

The solution $\vec{x}$ can be written as

$$x_i = y_i - \sum_{k=i+1}^{n} u_{ik} x_k, \quad i = n, \cdots, 1. \tag{2.22}$$

Compared to Gaussian elimination, LU decomposition separates the time-consuming forward elimination and only needs to decompose $A = LU$ once when $\vec{b}$ varies, leading to an efficient calculation of the inverse matrix A^{-1}.

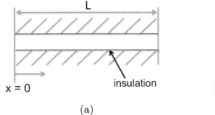
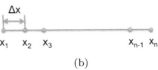

(a) (b)

FIGURE 2.2: The steady 1D conduction problem (a boundary value problem) and its domain discretization.

2.2.2 Linear Boundary Value Problems

A boundary value problem is a differential equation with a set of boundary condition constraints. In the following, let us take a look at a simple boundary value problem as shown in Figure 2.2, the steady 1D conduction equation with constant properties. The governing equation can be written as

$$\frac{d}{dx}\left(-k\frac{dT}{dx}\right) = q(x) \tag{2.23}$$

with boundary conditions $T(0) = T_A$ and $T(1) = T_B$, where k is the thermal conductivity constant and $q(x)$ is the energy generation term. For a constant k, Eq. (2.23) can be rewritten as

$$\frac{d^2T}{dx} = -\frac{q(x)}{k}. \tag{2.24}$$

This problem can be solved analytically to obtain the continuous temperature profile $T(x)$ when $q(x)$ is in some special forms. For complex $q(x)$, we have to discretize the domain and solve the problem numerically. The most straightforward way is to use the finite difference method. First, we discretize the 1D domain into n data points, and we have $T(x) = \{T_1, T_2, \cdots, T_n\}$, where $T_i = T(x_i)$ and $i = 1, 2, \cdots, n$. From the boundary conditions, we have $T_1 = T(0) = T_A$ and $T_n = T(L) = T_B$. Therefore, there are $n - 2$ unknowns. Note that $(n - 1)\Delta x = L$, we have $n = \frac{L}{\Delta x} + 1$, where Δx is chosen to obtain the desired resolution on $T(x)$. Next, we need to write the governing equation at each node, and then linearize it. At node i, we have

$$\left.\frac{d^2 T}{dx^2}\right|_{x_i} = -\frac{q(x_i)}{k} = -\frac{q_i}{k}. \tag{2.25}$$

The left second derivative term can be approximated using the finite difference method, and then we have

$$\left.\frac{d^2 T}{dx^2}\right|_{x_i} \approx \frac{T_{i+1} - 2T_i + T_{i-1}}{(\Delta x)^2}. \tag{2.26}$$

The energy equation for node i is then written as

$$T_{i-1} - 2T_i + T_{i+1} = -\frac{q_i(\Delta x)^2}{k} \tag{2.27}$$

for all internal nodes $i = 2, 3, \cdots, n - 1$. In this way, we convert an ordinary differential equation into a system of linear algebraic equations. The finite difference equations can be written in the matrix form, and the boundary conditions give us equations for T_1 and T_n. We have

$$\begin{bmatrix} 1 & 0 & 0 & 0 & 0 & \cdots & 0 \\ 1 & -2 & 1 & 0 & 0 & \cdots & 0 \\ 0 & 1 & -2 & 1 & 0 & \cdots & 0 \\ 0 & 0 & 1 & -2 & 1 & \cdots & 0 \\ \vdots & & & & & & \\ 0 & 0 & & \cdots & 1 & -2 & 1 \\ 0 & 0 & & \cdots & 0 & 0 & 1 \end{bmatrix} \begin{Bmatrix} T_1 \\ T_2 \\ T_3 \\ T_4 \\ \vdots \\ T_{n-1} \\ T_n \end{Bmatrix} = \begin{Bmatrix} T_A \\ -q_2(\Delta x)^2/k \\ -q_3(\Delta x)^2/k \\ -q_4(\Delta x)^2/k \\ \vdots \\ -q_{n-1}(\Delta x)^2/k \\ T_B \end{Bmatrix}.$$

By now, we have converted a second order boundary value problem into a set of linear equations ($Ax = b$). Boundary conditions are included in the first and last equations. We then can use LU decomposition to solve this problem.

2.2.3 Nonlinear Boundary Value Problems

In the following, we will talk about how to solve nonlinear ordinary differential equation (ODE) boundary value problems (BVPs). First, let us compare it with a linear ODE BVP. Both linear and nonlinear ODE BVPs have one independent variable. In a linear ODE BVP, every term only has one "variable" part, i.e., x, $f(x)$, $\frac{df}{dx}$, $\frac{d^2f}{dx^2}$, etc. In such cases, LU decomposition can be applied to obtain a solution directly. Differently, a nonlinear ODE BVP has some nonlinear terms like x^2, $sinx$, $xf(x)$, $x\frac{df(x)}{dx}$ and so on. Solving a nonlinear ODE BVP is much more complicated. In the following, let us first consider a 1D heat conduction problem with $k = k(T)$. The conduction equation becomes

$$\frac{d}{dx}\left(-k\frac{dT}{dx}\right) = q(x).$$

The term on the left-hand side can be rewritten into two terms. We have

$$k\frac{d^2T}{dx^2} + \frac{dk}{dx}\frac{dT}{dx} = -q(x);$$

furthermore we obtain

$$k\frac{d^2T}{dx^2} + \frac{dk}{dT}\frac{dT}{dx}\frac{dT}{dx} = -q(x),$$

and finally we have the governing equation

$$k(T)\frac{d^2T}{dx^2} + k'(T)\left(\frac{dT}{dx}\right)^2 = -q(x). \tag{2.28}$$

Here we consider a crystal at high temperature with $k = \frac{B}{T}$. Recall that we have derived the central difference equations at node i as

$$\frac{d^2T}{dx^2}\bigg|_i = \frac{T_{i+1} - 2T_i + T_{i-1}}{(\Delta x)^2}$$

and

$$\frac{dT}{dx}\bigg|_i = \frac{T_{i+1} - T_{i-1}}{2\Delta x}.$$

We substitute them into the governing equation and obtain

$$\frac{B}{T_i}\frac{T_{i+1} - 2T_i + T_{i-1}}{(\Delta x)^2} - \frac{B}{T_i^2}\left(\frac{T_{i+1} - T_{i-1}}{2\Delta x}\right)^2 = -q_i. \tag{2.29}$$

Rewrite this finite difference equation in the form

$$F_1(T_{i-1}, T_i, T_{i+1}, x)T_{i-1} + F_2(T_{i-1}, T_i, T_{i+1}, x)T_i + F_3(T_{i-1}, T_i, T_{i+1}, x)T_{i+1} = F_4(x). \tag{2.30}$$

Here, the F_i ($i = 1, \cdots, 4$) functions are variable coefficients for the temperature and there is no unique way to solve this equation. Considering $q(x) = 0$, we have

$$\frac{B}{T_i}\frac{T_{i+1} - 2T_i + T_{i-1}}{(\Delta x)^2} - \frac{B}{T_i^2}\left(\frac{T_{i+1} - T_{i-1}}{2\Delta x}\right)^2 = 0,$$

which can be reorganized into

$$\underbrace{\left(1 - \frac{T_{i-1}}{4T_i}\right) T_{i-1}}_{F_1} \underbrace{-2\,T_i}_{F_2} + \underbrace{\left(1 - \frac{T_{i+1}}{4T_i} + \frac{T_{i-1}}{2T_i}\right) T_{i+1}}_{F_3} = 0.$$

We will guess a solution, evaluate the coefficients, solve the system and compare the new solution with the old one. If they agree within a given tolerance, then we stop. Otherwise, we take the new solution, evaluate the coefficients and iterate until a convergent result is achieved.

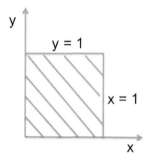

FIGURE 2.3: The steady conduction in two dimensions.

2.2.4 Nonlinear Equations in One Unknown

Given a nonlinear function $f(x)$ with one unknown variable, our task is to find its root x_r such that $f(x_r) = 0$. Let us first consider an example as shown in Figure 2.3, the 2D steady conduction problem. Its governing equation can be written as

$$\frac{\partial^2 T(x,y)}{\partial x^2} + \frac{\partial^2 T(x,y)}{\partial y^2} = 0 \qquad (2.31)$$

with the boundary conditions $T(0,y) = 0$, $T(x,0) = 0$, $T(1,y) + T'(1,y) = 0$ and $T(x,1) = 0$.

Before studying the numerical solution of this partial differential equation, let us first derive its analytical solution by using separation of variables. Let $T(x,y) = R(x)S(y)$, and then we have

$$R''(x)S(y) + R(x)S''(y) = 0$$

and furthermore we obtain

$$\frac{R''(x)}{R(x)} = -\frac{S''(y)}{S(y)} = -\lambda^2.$$

Consider the x-equation: $R''(x) + \lambda^2 R(x) = 0$ with $R(0) = 0$ and $R(1) + R'(1) = 0$. For $\lambda^2 > 0$ (i.e., real x), it has a solution in the form $R(x) = A\cos\lambda x + B\sin\lambda x$. Applying the boundary conditions, we have

$$R(0) = 0 \quad \Rightarrow \quad A = 0;$$
$$R(1) + R'(1) = 0 \quad \Rightarrow \quad B\sin\lambda + B\lambda\cos\lambda = 0.$$

Note that $B \neq 0$ for nontrivial solution, so that $tan\lambda = -\lambda$. We plot the left- and right-hand sides of this equation; see Figure 2.4(a). It is obvious that there are an infinite number of solutions. Note that $\lambda < 0$ will not give any additional information. To cast the problem as $f(x) = 0$, we rewrite $tan\lambda = -\lambda$ into $\frac{\sin\lambda}{\lambda} + \cos\lambda = 0$.

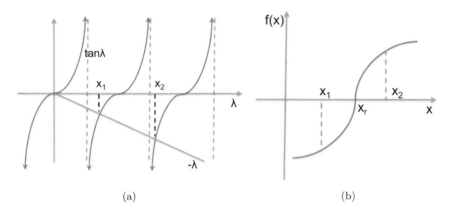

(a)   (b)

FIGURE 2.4: (a) Plot the left- and right-hand sides of $tan\lambda = -\lambda$ to get a sense of the solution; and (b) the root x_r is bracketed within $[x_1, x_2]$.

The first step for a general problem is to plot the function and find out if there are real roots. The plot will give us an initial guess, and then we can iterate to a solution using root-finding schemes until a desired tolerance is met. An ideal method would not have any human judgment involvement. It can characterize for example by $\epsilon_{n+1} = k(\epsilon_n)^m$, where ϵ_n and ϵ_{n+1} are the error at step n and $n+1$, respectively, and m is the order of convergence.

To find the roots x_r for $f(x) = 0$, we identify two initial values x_1 and x_2 such that $x_1 < x_r < x_2$ and $f(x_1)f(x_2) < 0$; see Figure 2.4(b). In other words, we are "bracketing" the root. Note that there are two types of exceptions for the bracketing method as shown in Figure 2.5: asymptote, and repeating/multiple roots or very close roots. These exceptions can be avoided by plotting the function and observation.

If we can bracket the root, the most robust method for root finding is the *bisection method*, while it is relatively slow compared to other techniques. It requires two initial values (x_1 and x_2) to define a bracket with function sign change over it. As shown in Figure 2.6(a), there are three steps:

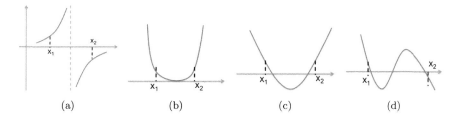

(a)   (b)   (c)   (d)

FIGURE 2.5: Exceptions for the bracketing method. (a) Asymptote; and (b-d) multipe roots or very close roots.

1. Find the middle point of the bracket $x_3 = \frac{x_1+x_2}{2}$;

2. Evaluate $f(x_3)$ and compare with the desired tolerance τ. If $|f(x_3)| < \tau$, stop; otherwise

3. Pick (x_1 or x_2) and x_3 as the new bracket and continue.

The maximum size of error after n iterations is $\left|\frac{x_1-x_2}{2^n}\right|$, leading to a linear rate of convergence with $k = 1/2$ and $m = 1$.

The *false position* method also starts with two initial points and a bracket defined by them. It estimates the root using a linear approximation instead of choosing the middle point as in the bisection method. False position works better than the bisection method mostly, except for a few special cases like functions with significant curvature. As shown in Figure 2.6(b), this method also consists of four main steps:

1. Connect $[x_1, f(x_1)]$ and $[x_2, f(x_2)]$ with a line;

2. Determine where this line intersects the x-axis. Call this point x_3;

3. Evaluate $f(x_3)$ and compare with τ. If $|f(x_3)| < \tau$, stop; otherwise

4. Choose (x_1 or x_2) and x_3 to bracket the root and continue.

According to similar triangles, we have $\frac{-f(x_1)}{x_3-x_1} = \frac{f(x_2)}{x_2-x_3}$ and further obtain $x_3 = \frac{x_2 f(x_1)-x_1 f(x_2)}{f(x_1)-f(x_2)}$. This method has a "super linear" convergence with $1 < m < 2$, and it guarantees to work if the root lies inside the bracket.

In the limit of $x_1 = x_2$ in the false position method, we get the Newton's method or Newton–Raphson method, where we use the slope of the function to get the next guess. The Newton–Raphson method is one kind of open method that needs only one initial point in general instead of two in the bracketing methods. Table 2.2 shows a comparison of these two kinds of methods. As shown in Figure 2.6(c), this method consists of four steps:

1. Find $f'(x_1)$;

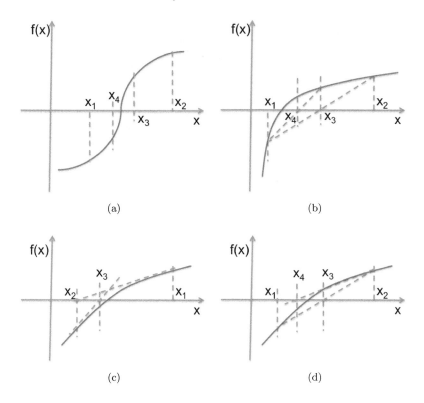

FIGURE 2.6: Four root-finding methods for 1D nonlinear function. (a) Bisection; (b) false position; (c) Newton–Raphson; and (4) secant.

2. Determine where the tangential line intersects the x-axis to give x_2;

3. Evaluate $f(x_2)$ and compare with τ. If $|f(x_2)| < \tau$, stop; otherwise

4. Choose x_2 as the new study point and continue.

Using the Taylor series expansion about x_i, we have $f(x_2) \approx f(x_1) + f'(x_1)(x_2 - x_1) = 0$ and further obtain $x_2 = x_1 - \frac{f(x_1)}{f'(x_1)}$. We need to make sure that the initial guess is "good" (not near a local extremum). The implementation requires that $f'(x)$ is easy to evaluate. This method is one of the most popular methods for root finding due to its quadratic convergence rate with $m = 2$. The Newton–Raphson method has pitfalls. It may oscillate around an inflection, local min or max point, it may jump to another root when multiple roots exist. We also need to avoid choosing study points with zero-slope.

Similar to the false position method, the *secant method* shown in Figure 2.6(d) starts with two initial points but no bracketing is required. It may diverge depending on the input points and the function. Instead of using two initial guesses, the *modified secant method* starts with one initial guess and

TABLE 2.2: Bracketing Methods vs Open Methods

Bracketing Methods	Open Methods
$(-)$ two bracketing initial values x_1, x_2, require $f(x_1)f(x_2) < 0$	$(+)$ one (or two) starting point, not require $f(x_1)f(x_2) < 0$
$(+)$ converges	$(-)$ may diverge
$(-)$ can be very slow (linear)	$(+)$ fast if converges (quadratic)
$(+)$ can find one root if $f(x_1)f(x_2) < 0$	$(+)$ can find one root

introduces a small perturbation to compute the second data point. There are also other methods of varying degrees of complexity such as the modified New−Raphson method for multiple roots. We do not discuss them here but please refer to the related textbooks [81].

2.2.5 Systems of Nonlinear Equations

In this subsection, we consider a multidimensional application of the Newton−Raphson method to solve a system of nonlinear equations. Given $F_1(\vec{x})$, $F_2(\vec{x})$, $\cdots$, $F_n(\vec{x})$ where $\vec{x} = (x_1, x_2, \cdots, x_n)$, we aim to find $\vec{x_r}$ such that $F_1(\vec{x_r}) = F_2(\vec{x_r}) = \cdots = F_n(\vec{x_r}) = 0$. For example,

$$\begin{cases} F_1(x_1, x_2) = x_1^2 + x_2^2 - 4, \\ F_2(x_1, x_2) = e^{x_1} + \sin x_2 - 2. \end{cases}$$

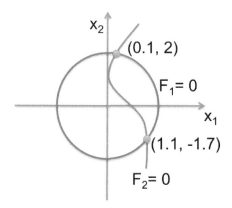

FIGURE 2.7: An example of 2D nonlinear systems. Note that $(0.1, 2)$ and $(1.1, -1.7)$ can be used as two initial guesses in finding the two roots using Newton's method.

If we plot these two equations as shown in Figure 2.7, we can observe that there are two solutions. In solving this problem using the Newton's method, it is important to pick good initial guesses. If there are n unknowns, n equations are required. It is possible to have a unique solution, multiple solutions or no solution. The general formulation is similar to the one-variable case: given an initial guess $\vec{x} = (x_1, x_2, \cdots, x_n)$, solve $F_i(x_1, x_2, \cdots, x_n)$, where $i = 1, \cdots, n$. Here we adopt the Newton–Raphson method for nonlinear equations. First, let us expand each function as a Taylor series and obtain

$$F_i(\vec{x} + \delta\vec{x}) = F_i(\vec{x}) + \sum_{j=1}^{n} \frac{\partial F_i}{\partial x_j}\bigg|_{\vec{x}} \delta x_j + \cdots.$$

Ignoring higher order terms (only keeping the first two terms), we can linearized the problem. We want $F_i(\vec{x} + \delta\vec{x}) = F_i(\vec{x}) + \sum_{j=1}^{n} \frac{\partial F_i}{\partial x_j}\big|_{\vec{x}} \delta x_j = 0$, where $\delta\vec{x} = (\delta x_1, \delta x_2, \cdots, \delta x_n)$ are unknowns. Consider the following series of equations:

$$\begin{cases} i = 1: \quad \sum_{j=1}^{n} \frac{\partial F_1}{\partial x_j}\big|_{\vec{x}} \delta x_j = -F_1(\vec{x}), \\[2mm] i = 2: \quad \sum_{j=1}^{n} \frac{\partial F_2}{\partial x_j}\big|_{\vec{x}} \delta x_j = -F_2(\vec{x}), \\[2mm] \vdots \\[2mm] i = n: \quad \sum_{j=1}^{n} \frac{\partial F_n}{\partial x_j}\big|_{\vec{x}} \delta x_j = -F_n(\vec{x}). \end{cases}$$

They can be rewritten as

$$\begin{cases} \frac{\partial F_1}{\partial x_1}\big|_{\vec{x}} \delta x_1 + \frac{\partial F_1}{\partial x_2}\big|_{\vec{x}} \delta x_2 + \cdots + \frac{\partial F_1}{\partial x_n}\big|_{\vec{x}} \delta x_n = -F_1(\vec{x}), \\[2mm] \frac{\partial F_2}{\partial x_1}\big|_{\vec{x}} \delta x_1 + \frac{\partial F_2}{\partial x_2}\big|_{\vec{x}} \delta x_2 + \cdots + \frac{\partial F_2}{\partial x_n}\big|_{\vec{x}} \delta x_n = -F_2(\vec{x}), \\[2mm] \vdots \\[2mm] \frac{\partial F_n}{\partial x_1}\big|_{\vec{x}} \delta x_1 + \frac{\partial F_n}{\partial x_2}\big|_{\vec{x}} \delta x_2 + \cdots + \frac{\partial F_n}{\partial x_n}\big|_{\vec{x}} \delta x_n = -F_n(\vec{x}), \end{cases}$$

or in matrix form,

$$\begin{bmatrix} \frac{\partial F_1}{\partial x_1} & \frac{\partial F_1}{\partial x_2} & \cdots & \frac{\partial F_1}{\partial x_n} \\[2mm] \frac{\partial F_2}{\partial x_1} & \frac{\partial F_2}{\partial x_2} & \cdots & \frac{\partial F_2}{\partial x_n} \\[2mm] \vdots & & & \vdots \\[2mm] \frac{\partial F_n}{\partial x_1} & \frac{\partial F_n}{\partial x_2} & \cdots & \frac{\partial F_n}{\partial x_n} \end{bmatrix}_{\vec{x}} \begin{Bmatrix} \delta x_1 \\ \delta x_2 \\ \vdots \\ \delta x_n \end{Bmatrix} = - \begin{Bmatrix} F_1(\vec{x}) \\ F_2(\vec{x}) \\ \vdots \\ F_n(\vec{x}) \end{Bmatrix}. \qquad (2.32)$$

Eq. (2.32) is a set of n linear equations in n unknowns, which can be solved using Gaussian elimination or LU decomposition. We write this matrix equation in a compact form as $J(\vec{x})\delta\vec{x} = -\vec{F}(\vec{x})$, and then obtain $\delta\vec{x} = -J^{-1}(\vec{x})\vec{F}(\vec{x})$. Solve for $\delta\vec{x}$, and then we have $\vec{x}_{new} = \vec{x}_{old} + \delta\vec{x}$. We compare $\vec{F}(\vec{x}_{new})$ with a given tolerance; continue to iterate until the error tolerance is reached.

2.3   Curve Fitting

Curve fitting is an important application problem in engineering, for example, estimating the linear spring constant based on a set of measured data points. There are two kinds of approaches for curve fitting: regression and interpolation. Given data exhibiting significant error or noise, *regression* derives a single curve that represents the general trend of the data, which may not pass through the given points. Differently, *interpolation* fits a curve or a series of curves that passes through all the given points when the data are precise. In the following, we will review one regression method (the linear regression) and two interpolation methods (Newton's interpolating polynomials, quadratic and cubic splines).

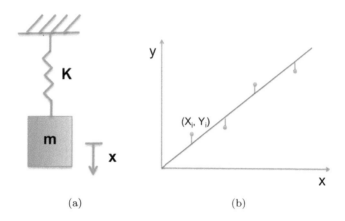

(a)   (b)

FIGURE 2.8: (a) A linear spring system; and (b) curve fitting of a set of data points (X_i, Y_i) $(i = 1, \cdots, N)$ to estimate the linear spring constant using linear regression.

2.3.1 Linear Regression

We take a linear spring as the example and aim to estimate the spring constant k based on a set of measured data points. Given (X_i, Y_i) $(i = 1, ..., N)$ as shown in Figure 2.8, we tend to find a linear fit $y = a_0 + a_1 x$. In the linear regression, the least square method is used to minimize the total distance from each data point to the line. We have

$$S = \sum_{i=1}^{N} e_i^2 = \sum_{i=1}^{N} (Y_i - a_0 - a_1 X_i)^2,$$

which is a 2D optimization problem with respect to (a_0, a_1). By setting its first derivatives to be zero, we can obtain a set of two equations

$$\begin{cases} \frac{\partial S}{\partial a_0} = \sum_{i=1}^{N} 2(Y_i - a_1 X_i - a_0)(-1) = 0 \\ \frac{\partial S}{\partial a_1} = \sum_{i=1}^{N} 2(Y_i - a_1 X_i - a_0)(-X_i) = 0 \end{cases}$$

and rewrite them in the matrix form

$$\begin{bmatrix} N & \sum X_i \\ \sum X_i & \sum X_i^2 \end{bmatrix} \begin{Bmatrix} a_0 \\ a_1 \end{Bmatrix} = \begin{Bmatrix} \sum Y_i \\ \sum X_i Y_i \end{Bmatrix}.$$

a_0 and a_1 can be obtained easily by solving the above linear system. Note that for a linear spring, if we enforce it to pass through the origin (assume no deformation when there is no force acted on it), we can simplify the system by setting $a_0 = 0$. Then we have only one equation in the system. To measure how good the fit is, we define a metric R such as

$$R = 1 - \frac{\sum (Y_i - y_i)^2}{\sum (Y_i - \bar{Y})^2}, \tag{2.33}$$

where $\bar{Y}$ is the mean of Y_i for $i = 1, \cdots, n$.

This concept can be easily extended to a polynomial fit of order n: $y = a_0 + a_1 x + a_2 x^2 + \cdots + a_n x^n$. Then the error is $e_i = Y_i - y_i = Y_i - a_0 - a_1 x - a_2 x^2 - \cdots - a_n x^n$ and $S = \sum_{i=1}^{N} \left(Y_i - a_0 - a_1 x - a_2 x^2 - \cdots - a_n x^n \right)^2$. We need to solve the set of equations defined by $\frac{\partial S}{\partial a_i} = 0$. You will end up with an $(n+1) \times (n+1)$ matrix equation

$$\begin{bmatrix} N & \sum X_i & \cdots & \sum X_i^n \\ \sum X_i & \sum X_i^2 & \cdots & \sum X_i^{n+1} \\ \vdots & & & \\ \sum X_i^n & \sum X_i^{n+1} & \cdots & \sum X_i^{2n} \end{bmatrix} \begin{Bmatrix} a_0 \\ a_1 \\ \vdots \\ a_n \end{Bmatrix} = \begin{Bmatrix} \sum Y_i \\ \sum X_i Y_i \\ \vdots \\ \sum X_i^n Y_i \end{Bmatrix}. \tag{2.34}$$

Note that this is a set of equations linear in $(a_0, a_1, \cdots, a_n)$. We can then use Gaussian elimination or LU decomposition to solve them.

2.3.2 Newton's Interpolating Polynomials

Different from regression which approximates the general trend of the input data, interpolation passes through all the given data points. *Newton's interpolation* finds the polynomial $f(x)$ that passes through all data points (x_i, f_i) where $i = 0, 1, \cdots, n$. In other words, given $(n+1)$ data points, there exists only one n^{th}-order polynomial passing through all the data points

$$f(x) = a_0 + a_1 x + \cdots + a_n x^n,$$

where $a_0, a_1, \cdots, a_n$ are unknown coefficients we need to compute. For example, given two points x_0 and x_1 ($n = 1$), we can obtain a linear (first-order) interpolation

$$
\begin{aligned}
f_1(x) &= f(x_0) + \frac{f(x_1) - f(x_0)}{x_1 - x_0}(x - x_0) \\
&= f(x_0) + f[x_1, x_0](x - x_0) \\
&= b_0 + b_1(x - x_0) = a_0 + a_1 x,
\end{aligned}
$$

where $b_0 = f(x_0)$ and $b_1 = f[x_1, x_0]$ is the finite divided difference (FDD) approximation of the first derivative. We can also obtain $a_0 = b_0 - b_1 x_0$ and $a_1 = b_1$. Given three points x_0, x_1 and x_2 ($n = 2$), we can obtain a quadratic (second-order) interpolation

$$
\begin{aligned}
f_2(x) &= f(x_0) + f[x_1, x_0](x - x_0) + f[x_2, x_1, x_0](x - x_0)(x - x_1) \\
&= b_0 + b_1(x - x_0) + b_2(x - x_0)(x - x_1) = a_0 + a_1 x + a_2 x^2,
\end{aligned}
$$

where $b_2 = f[x_2, x_1, x_0]$ is the FDD approximation of the second derivative. We can also obtain $a_0 = b_0 - b_1 x_0 + b_2 x_0 x_1$, $a_1 = b_1 - b_2(x_0 + x_1)$ and $a_2 = b_2$. Similarly, given four points x_0, x_1, x_2 and x_3 ($n = 3$), we can obtain a cubic (third-order) interpolation

$$
f_3(x) = b_0 + b_1(x - x_0) + b_2(x - x_0)(x - x_1) + b_3(x - x_0)(x - x_1)(x - x_2),
$$

where $b_3 = f[x_3, x_2, x_1, x_0]$ is the FDD approximation of the third derivative. We can observe that the FDD approximations have a very nice recursive definition. Using the same manner, we can extend to the nth-order interpolation for given $(n + 1)$ points.

2.3.3 Quadratic and Cubic Splines

Given $(n + 1)$ points (x_i, y_i) ($i = 0, \cdots, n$), Newton's interpolation yields one high-order (n^{th}-order) polynomial which may have large oscillations. Instead, here we can fit the input points with "n" piecewise low-order (quadratic or cubic) polynomials, which are called quadratic and cubic splines.

In *quadratic spline*, each piece of the parabola can be represented as $f_i(x) = a_i x^2 + b_i x + c_i$ with three unknowns. Therefore, we have a total of $3n$ unknowns and we need to construct $3n$ equations or conditions. First, for each piece we have two endpoints (x_{i-1}, y_{i-1}) and (x_i, y_i); the two endpoint locations should match with the input data. We have $f_i(x_{i-1}) = y_{i-1}$ and $f_i(x_i) = y_i$, and also obtain a total of $2n$ conditions. Second, for each internal node shared by two pieces, the slope continuity should match. We have $\frac{df_i}{dx}(x_i) = \frac{df_{i+1}}{dx}(x_i)$, and obtain a total of $(n - 1)$ conditions because we have $(n - 1)$ internal nodes. Finally, we need one more condition to construct $3n$ equations. We can set the first or the last segment to be linear (natural spline), for example $f_0''(x_0) = 0$ or $f_n''(x_n) = 0$. All these $3n$ equations can be organized into a matrix form

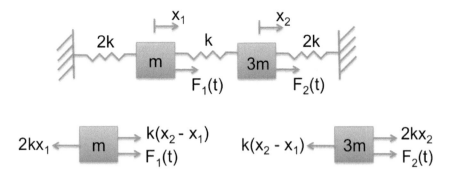

FIGURE 2.9: The mass-spring system and its free body diagrams.

$A\vec{x} = \vec{b}$, and the unknown coefficients can be obtained by solving this linear system.

Similarly, we can extend this concept to *cubic splines*. For each piece, we have a cubic polynomial $f_i(x) = a_i x^3 + b_i x^2 + c_i x + d_i$ with four unknowns. We have a total of $4n$ unknowns, and we need to construct $4n$ equations. We can still use the endpoint locations to obtain $2n$ condition. For the internal node continuity, we can enforce both the first and the second derivatives to match, then we obtain $2(n-1)$ conditions. Finally, we set both the first and the last segment to be linear to obtain the last two conditions.

In summary, each curve-fitting scheme has its own properties. Users can pick the proper one based on the detailed requirements from various application projects.

2.4 Ordinary Differential Equations

Ordinary differential equations (ODEs) contain only one independent variable, which can be divided into two categories: boundary value problems (BVPs) with a spatial independent variable and initial value problems (IVPs) with a temporal independent variable. We have discussed about BVPs in Chapter 2.2. Here we will talk about IVPs and the stiffness of ODEs.

2.4.1 Initial Value Problems

Figure 2.9 shows an example of the IVPs, the mass-spring system. The equations of motion can be derived using free-body diagrams. We have

$$\begin{cases} \sum F_1 = k(x_2 - x_1) + F_1(t) - 2kx_1 = m\ddot{x}_1, \\ \sum F_2 = F_2(t) - 2kx_2 - k(x_2 - x_1) = 3m\ddot{x}_2. \end{cases}$$

They can be reorganized into

$$\begin{cases} m\ddot{x}_1 = -3kx_1 + kx_2 + F_1(t), \\ 3m\ddot{x}_2 = kx_1 - 3kx_2 + F_2(t). \end{cases}$$

We have now obtained a set of two coupled linear second-order ODEs. We need four initial conditions such as $x_1(0)$, $\dot{x}_1(0)$, $x_2(0)$ and $\dot{x}_2(0)$.

The general problem of a nonlinear m^{th}-order ODE can be written as

$$\sum_{i=1}^{m} g_i(t, x) \frac{d^i x(t)}{dt^i} = g_0(t, x) \tag{2.35}$$

with initial conditions $x(0)$, $x'(0)$, $\cdots$, $x^{m-1}(0)$. Our task is to find $x(t)$. In principle, any m^{th}-order ODE can be rewritten as a set of m first-order ODEs. Eq. (2.35) can be rewritten as

$$\begin{cases} \frac{dx_1(t)}{dt} = F_1(t, x_1, x_2, \cdots, x_m), \\ \frac{dx_2(t)}{dt} = F_2(t, x_1, x_2, \cdots, x_m), \\ \vdots \\ \frac{dx_m(t)}{dt} = F_m(t, x_1, x_2, \cdots, x_m) \end{cases} \tag{2.36}$$

with the initial conditions $x_1(0)$, $x_2(0)$, $\cdots$, $x_m(0)$.

When $m = 1$, the problem becomes how to find $x(t)$ given $\frac{dx(t)}{dt} = F(t, x)$ with $x(0) = x_0$. For a discretized solution with the step Δt and $x(t_n) = x_n$, we have

$$\begin{aligned} x_{n+1} = x(t_{n+1}) = x(t_n + \Delta t) &= x(t_n) + x'(t_n)\Delta t + \frac{x''(t_n)}{2}(\Delta t)^2 \\ &+ \cdots + \frac{x^{(p)}(t_n)}{p!}(\Delta t)^p + \cdots. \end{aligned}$$

Note that $x'(t) = F(t, x)$, and then we have

$$x_{n+1} = x(t_n) + F(t_n, x_n)\Delta t + \frac{1}{2}\frac{dF(t, x)}{dt}\bigg|_{t_n, x_n}(\Delta t)^2 + \cdots.$$

Note that $\frac{dF(t,x)}{dt}\big|_{t_n, x_n} = \left(\frac{\partial F}{\partial t} + \frac{\partial F}{\partial x}\frac{dx}{dt}\right)_{t_n, x_n} = (F_t + F_x F)_n$, so that

$$x_{n+1} = x(t_n) + F_n \Delta t + \frac{1}{2}(F_t + F_x F)_n(\Delta t)^2 + \cdots. \tag{2.37}$$

In the Euler method, the Taylor series is truncated after the first derivative term and we can obtain

$$x_{n+1} = x_n + F(t_n, x_n)\Delta t.$$

Recall that F is the slope of $x(t)$. We use the slope at the known data point to go to the next point. This is a first-order method with poor accuracy and stability (a very small Δt is required to match the performance of other methods), but it gives us some motivation for the Runge–Kutta methods.

In Runge–Kutta methods, we try to match the Taylor series to higher order by using data between (t_n, x_n) and (t_{n+1}, x_{n+1}). Consider the second-order formulation (RK2)

$$x_{n+1} = x_n + ak_1 + bk_2, \tag{2.38}$$

where

$$\begin{cases} k_1 = \Delta t \cdot F(t_n, x_n), \\ k_2 = \Delta t \cdot F(t_n + \alpha\Delta t, x_n + \beta k_1), \end{cases}$$

so that

$$x_{n+1} = x_n + a\Delta t F(t_n, x_n) + b\Delta t F[t_n + \alpha\Delta t, x_n + \beta\Delta t F(t_n, x_n)].$$

From the Taylor series expansion, we know $F[t_n + \alpha\Delta t, x_n + \beta\Delta t F(t_n, x_n)] \approx F_n + F_{t,n}\alpha\Delta t + F_{x,n}\beta\Delta T F_n$. Plugging it into the above equation, we can obtain

$$x_{n+1} = x_n + (a + b)\Delta t F_n + b(\alpha F_{t,n} + \beta F_{x,n}F)(\Delta t)^2. \tag{2.39}$$

Comparing Eqns. (2.37) and (2.39), we can obtain $a + b = 1$, $b\alpha = \frac{1}{2}$, and $b\beta = \frac{1}{2}$. We have three equations in four unknowns. There is one free choice to guarantee that the Taylor series is matched to the second-order and the error is $O(\Delta t)^3$. Some physical sense should be used here, for example $0 \le a, b, \alpha, \beta \le 1$. The popular choices are the modified Euler method with $a = b = \frac{1}{2}$ and $\alpha = \beta = 1$. Some people also choose $a = 0$, $b = 1$, and $\alpha = \beta = \frac{1}{2}$.

For one equation $\frac{dx(t)}{dt} = F(t, x)$ with $x(0) = x_0$, the fourth-order RK method (RK4) is

$$x_{n+1} = x_n + \frac{1}{6}(k_1 + 2k_2 + 2k_3 + k_4), \tag{2.40}$$

where

$$\begin{cases} k_1 = \Delta t F(t_n, x_n), \\ k_2 = \Delta t F(t_n + \frac{\Delta t}{2}, x_n + \frac{k_1}{2}), \\ k_3 = \Delta t F(t_n + \frac{\Delta t}{2}, x_n + \frac{k_2}{2}), \\ k_4 = \Delta t F(t_n + \Delta t, x_n + k_3). \end{cases}$$

We can extend the RK4 algorithm to calculate $x_1(t)$, $x_2(t)$, $\cdots$, and $x_n(t)$ for a system of IVPs as defined in Eq. (2.36). There will be different k_1, k_2, k_3 and k_4 for each x_i.

2.4.2 Stiffness of ODEs

Stiffness occurs in a problem where there are two or more very different scales of the independent variables on which the dependent variables are changing. Let us first try to understand what is stiffness of ODEs by considering

$$\begin{cases} \dot{x}_1 = 1195x_1 - 1995x_2 \\ \dot{x}_2 = 1197x_1 - 1997x_2 \end{cases}$$

with initial conditions $x_1(0) = 2$ and $x_2(0) = -2$. The exact solution is

$$\begin{cases} x_1(t) = 10e^{-t/0.5} - 8e^{-t/0.00125} = 10e^{-2t} - 8e^{-800t}, \\ x_2(t) = 6e^{-t/0.5} - 8e^{-t/0.00125}. \end{cases}$$

In the solution, the first term will dominate because these two terms are at very different time scales. You might think to pick a $\Delta t < 0.00125$ to accurately model this system, but such a small Δt may not be practical.

Let us try the Euler scheme, $x_{n+1} = x_n + \Delta t \dot{x}_n$ with $\Delta t = 0.1$. We find $x_1(0.1) = 640$ and $x_2(0.1) = 636$, whereas the exact solutions are $x_1(0.1) = 8.187$ and $x_2(0.1) = 4.912$. It is obvious that there is something terribly wrong. This is because the initial slope was a terrible guess as to what would happen.

To understand why, let us consider the single equation $\dot{x} = -cx$, where $c > 0$ and $x(0) = A$. The analytical solution is $x(t) = Ae^{-ct}$. Note that $x(t \to \infty) = 0$. According to the Euler scheme, we have $x_{n+1} = x_n + \Delta t \dot{x}_n$ and also $x_{n+1} = x_n - \Delta t c x_n = (1 - c\Delta t)x_n$, so that $x_{n+1}/x_n = 1 - c\Delta t$. From the exact solution we know that $|x_{n+1}/x_n| < 1$, therefore $-1 < 1 - c\Delta t < 1 \Rightarrow \Delta t < 2/c$ and $c\Delta t > 0$ (this is OK). To obtain a stable solution, we have to satisfy this restriction on the time step $\Delta t < 2/c$.

$x_{n+1} = x_n + \Delta t \dot{x}_n$ is explicit. Now let us consider an implicit solution $x_{n+1} = x_n + \Delta t \dot{x}_{n+1}$ using information at the final point for the last term. Then we obtain $x_{n+1} = x_n - \Delta t c x_{n+1}$ and also $x_{n+1} = x_n/(1 + c\Delta t)$. This is stable for all Δt and as $t \to \infty$, $x_{n+1} \to 0$. If we apply such an implicit scheme to the earlier system, we obtain

$$\begin{cases} x_{1,n+1} = x_{1,n} + 0.1(1195x_{1,n+1} - 1995x_{2,n+1}), \\ x_{2,n+1} = x_{2,n} + 0.1(1197x_{1,n+1} - 1997x_{2,n+1}). \end{cases}$$

The solution can be easily obtained. We have $x_1(0.1) = 8.23$ and $x_2(0.1) = 4.90$. They are much closer to the exact solution. From the above derivation, we can observe that a good implicit scheme helps handle the stiffness issues.

2.5 Eigenvalue Problems

A general eigenvalue problem is defined as: given a square matrix A, find the constants λ and vectors $\vec{x}$ such that

$$A\vec{x} = \lambda\vec{x} \quad \Rightarrow \quad (A - \lambda I)\vec{x} = 0, \tag{2.41}$$

where λ are eigenvalues and $\vec{x}$ are corresponding eigenvectors. Recall the vibration of a mass-spring system as shown in Figure 2.9. When $F_1(t) = F_2(t) = 0$, the equations of motion are

$$\begin{cases} m\ddot{x}_1 + 3kx_1 - kx_2 = 0, \\ 3m\ddot{x}_2 - kx_1 + 3kx_2 = 0. \end{cases}$$

Write these two equations in matrix form to obtain

$$\begin{bmatrix} m & 0 \\ 0 & 3m \end{bmatrix} \begin{Bmatrix} \ddot{x}_1 \\ \ddot{x}_2 \end{Bmatrix} + \begin{bmatrix} 3k & -k \\ -k & 3k \end{bmatrix} \begin{Bmatrix} x_1 \\ x_2 \end{Bmatrix} = \vec{0},$$

or in compact form

$$M\ddot{\vec{x}} + k\vec{x} = 0.$$

Then, we have $-\omega^2 M\vec{x} + k\vec{x} = \vec{0}$ or $(\omega^2 M - k)\vec{x} = \vec{0}$. Divide through by ω^2 and multiply through by k^{-1} to obtain $\left(k^{-1}M - \frac{1}{\omega^2}I\right)\vec{x} = \vec{0}$. $K^{-1}M$ is called the dynamical matrix D such that $(D - \lambda I)\vec{x} = \vec{0}$ with $\lambda = \frac{1}{\omega^2}$. We can find the eigenvalues from $|D - \lambda I| = 0$.

2.5.1 Power Method

In some cases, we may only need the biggest or smallest eigenvalues, e.g., the lowest natural frequency of a beam. We can apply **the Power method** to do this. Given a dynamical matrix

$$D = \begin{bmatrix} 3/8 & 3/8 \\ 1/8 & 9/8 \end{bmatrix},$$

we use the following steps to find the biggest eigenvalue and corresponding eigenvector:

1. Guess eigenvector $\vec{e}_1 = \begin{pmatrix} 1 \\ 1 \end{pmatrix}$;

2. Generate $\vec{e}_2 = D\vec{e}_1 = \begin{bmatrix} 3/8 & 3/8 \\ 1/8 & 9/8 \end{bmatrix} \begin{pmatrix} 1 \\ 1 \end{pmatrix} = \begin{pmatrix} 3/4 \\ 5/4 \end{pmatrix}$;

3. Normalize $\vec{e}_2$ to obtain $\vec{e}_2 = \vec{e}_2/|\vec{e}_2| = \begin{pmatrix} 0.5145 \\ 0.8575 \end{pmatrix}$;

4. Generate $\vec{e}_3 = D\vec{e}_2 = \begin{bmatrix} 3/8 & 3/8 \\ 1/8 & 9/8 \end{bmatrix} \begin{pmatrix} 0.5145 \\ 0.8575 \end{pmatrix} = \begin{pmatrix} 0.5145 \\ 1.0290 \end{pmatrix}$;

5. Normalize $\vec{e}_3$;

$\vdots$

Continue until $\vec{e}_n \approx \vec{e}_{n-1}$. $\vec{e}_n$ is the eigenvector and the scaling factor $|\vec{e}_n|$ is the eigenvalue.

The Power method can also be used to find the smallest eigenvalue using A^{-1}. Given $A\vec{x} = \lambda\vec{x}$, we have $A^{-1}A\vec{x} = A^{-1}\lambda\vec{x}$ then $\vec{x} = \lambda A^{-1}\vec{x}$ and finally $A^{-1}\vec{x} = \frac{1}{\lambda}\vec{x}$. If λ is an eigenvalue of A, then $\frac{1}{\lambda}$ is an eigenvalue of A^{-1}. Therefore, the largest $\left|\frac{1}{\lambda}\right|$ will give the smallest $|\lambda|$. The disadvantage is that we need to calculate A^{-1}, which can be computationally expensive.

Why does the Power method work? Recall that the eigenvectors form basis. For any vector $\vec{x}$ in that space, we have $\vec{x} = \sum_i C_i\vec{e}_i$ and $A\vec{x} = \sum_i C_i A\vec{e}_i$. The eigenvalue problem is $A\vec{e}_i = \lambda\vec{e}_i$, so that $A\vec{x} = \sum_i C_i\lambda_i\vec{e}_i$. Multiple through by A again to obtain $AA\vec{x} = A^2\vec{x} = \sum_i C_i\lambda_i A\vec{e}_i = \sum_i C_i\lambda_i^2\vec{e}_i$; continue this process until $A^n\vec{x} = \sum_i C_i\lambda_i^n\vec{e}_i = C_1\lambda_1^n\vec{e}_1 + C_2\lambda_2^n\vec{e}_2 + \cdots$. Here λ_1 is the largest eigenvalue and as n increases, the first term will start to dominate. The rate of convergence depends on how close λ_1 and λ_2 are. Note that the initial guess $\vec{x}$ cannot be perpendicular to $\vec{e}_1$.

2.5.2 Example of Eigenvalue Problems

Let us take a look at the standard 1D wave equation $\frac{\partial^2 u}{\partial t^2} = C^2\frac{\partial^2 u}{\partial x^2}$ with two fixed ends. Using separation of variables, we have $u(x, t) = y(x)T(t)$. The spatial part of the solution gives $\frac{d^2 y}{dx^2} + k^2 y = 0$ with boundary conditions $y(0) = y(1) = 0$. For real k, the analytical solution has the form

$$y(x) = A\sin kx + B\cos kx,$$

where

$$\begin{cases} y(0) = 0 \Rightarrow B = 0, \\ y(1) = 0 \Rightarrow A\sin k = 0. \end{cases} \tag{2.42}$$

For nontrivial solution $A \neq 0$, we have $\sin k = 0$. This gives $k = n\pi$ where $n = \pm 1, \pm 2, \cdots$. $k = 2\pi/l$ and l is the wavelength. Therefore, $l = 2\pi/k = \pm 2, \pm 1, \pm\frac{2}{3}, \pm\frac{1}{2}, \cdots$. These are the wavelengths of the allowed vibration modes in the string.

To solve the problem numerically, we use the finite difference method to discretize the problem as shown in Figure 2.10. We get

$$\frac{y_{i+1} - 2y_i + y_{i-1}}{(\Delta x)^2} + k^2 y_i = 0,$$

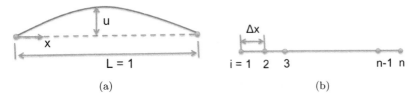

FIGURE 2.10: 1D wave problem and the domain discretization.

which can be written as

$$-y_{i-1} + [2 - k^2(\Delta x)^2]y_i - y_{i+1} = 0.$$

Explicitly for the internal nodes, we have

$$i = 2, \qquad -y_1 + [2 - k^2(\Delta x)^2]y_2 - y_3 = 0,$$

$$i = 3, \qquad -y_2 + [2 - k^2(\Delta x)^2]y_3 - y_4 = 0,$$

$$\vdots$$

$$i = n - 1, \quad -y_{n-2} + [2 - k^2(\Delta x)^2]y_{n-1} - y_n = 0.$$

Apply the boundary conditions $y_1 = y_n = 0$ to get $(n-2)$ equations with $(n-2)$ unknowns. These equations can be written in matrix form

$$\begin{bmatrix} 2 - k^2(\Delta x)^2 & -1 & 0 & \cdots & 0 \\ -1 & 2 - k^2(\Delta x)^2 & -1 & \cdots & 0 \\ \vdots & & & & \\ 0 & 0 & \cdots & -1 & 2 - k^2(\Delta x)^2 \end{bmatrix} \begin{Bmatrix} y_2 \\ y_3 \\ \vdots \\ y_{n-1} \end{Bmatrix} = \vec{0}.$$

For convenience, let $\lambda = k^2(\Delta x)^2$ and rewrite the above equation as

$$\left(\begin{bmatrix} 2 & -1 & 0 & \cdots & 0 \\ -1 & 2 & -1 & \cdots & 0 \\ \vdots & & & & \\ 0 & 0 & \cdots & -1 & 2 \end{bmatrix} - \lambda \begin{bmatrix} 1 & & & \\ & 1 & & \\ & & \vdots & \\ & & & 1 \end{bmatrix} \right) \begin{Bmatrix} y_2 \\ y_3 \\ \vdots \\ y_{n-1} \end{Bmatrix} = \vec{0}$$

or $(A - \lambda I)\vec{y} = \vec{0}$. Our coefficient matrix is $A - \lambda I$, meaning that we require

$$\begin{vmatrix} 2 - \lambda & -1 & 0 & \cdots & 0 \\ -1 & 2 - \lambda & -1 & \cdots & 0 \\ \vdots & & & & \\ 0 & 0 & \cdots & -1 & 2 - \lambda \end{vmatrix} = 0.$$

Expanding this out will give an $(n-2)^{th}$-order polynomial whose roots are the eigenvalues. In practice, this may not be the best way to find the eigenvalues if you recall that finding the determinant of a matrix is expensive (needs $n!$). The Power method can be a good choice if you only need to find the largest or smallest eigenvalue.

2.6   Partial Differential Equations

Most real engineering problems involve more than one independent variable, for example, multiple space directions and time. The general linear form with two independent variables is

$$A\frac{\partial^2 u}{\partial x^2} + B\frac{\partial^2 u}{\partial x \partial y} + C\frac{\partial^2 u}{\partial y^2} + D\left(x, y, \frac{\partial u}{\partial x}, \frac{\partial u}{\partial y}\right) = 0, \tag{2.43}$$

where A, B and C are coefficients. We aim to find $u(x, y)$ given appropriate boundary or initial conditions.

2.6.1 Classification of PDEs

We can classify partial differential equations (PDEs) in terms of A, B and C values following

1. Elliptical: $B^2 - 4AC < 0$;

2. Parabolic: $B^2 - 4AC = 0$; and

3. Hyperbolic: $B^2 - 4AC > 0$.

Various PDEs need different approaches to solve analytically and numerically. In the following, let us take a look at an example for each type of the PDEs.

Static heat conduction in 2D. This is a boundary value problem. The governing equation is

$$\frac{\partial^2 T(x, y)}{\partial x^2} + \frac{\partial^2 T(x, y)}{\partial y^2} = -\frac{q(x, y)}{k}, \tag{2.44}$$

where $A = C = 1$, $B = 0$, and $D = q(x, y)/k$. We have $B^2 - 4AC = -4 < 0$, therefore this is an elliptical PDE. This PDE can also be written in the compact form: $\nabla^2 T = -q(x, y)/k$ which is the Poisson equation. If $q(x, y) = 0$, then $\nabla^2 T = 0$ which is the Laplace equation.

Time varying heat conduction in 1D. This is a boundary and initial value problem, and the governing equation is

$$\frac{\partial^2 T(x, t)}{\partial x^2} = \frac{\rho C_p}{k}\frac{\partial T(x, t)}{\partial t} = \frac{1}{\alpha}\frac{\partial T(x, t)}{\partial t}, \tag{2.45}$$

where $A = 1$, $B = C = 0$, and $D = \frac{\rho C_p}{k}\frac{\partial T}{\partial t}$. We have $B^2 - 4AC = 0$, therefore this is a parabolic PDE. This PDE can be extended to multiple spatial dimensions.

Wave equation. This is a boundary and initial value problem, and the governing equation is

$$\frac{\partial^2 u(x,t)}{\partial t^2} = C^2 \frac{\partial^2 u(x,t)}{\partial x^2}, \tag{2.46}$$

where $A = 1$, $B = 0$, and $C = -c^2$. We have $B^2 - 4AC = 4c^2 > 0$, which is a hyperbolic PDE. This PDE can also be extended to multiple spatial domains.

2.6.2 Solving PDEs Using Finite Difference Techniques

In this subsection, we discuss how to use the finite difference techniques to solve the elliptical, parabolic and hyperbolic PDEs through an example for each of them.

Elliptical PDE. Consider a 2D heat transfer problem as shown in Figure 2.11(a). The governing equation is

$$\frac{\partial^2 T}{\partial x^2} + \frac{\partial^2 T}{\partial y^2} = -\frac{q(x,y)}{k}, \tag{2.47}$$

with the boundary conditions applied to the edges, $T(0,y) = T_A$, $T(L_x, y) = T_B$, $T(x,0) = T_C$, and $T(x, L_y) = T_D$. We discretize the solution domain to obtain a 2D grid of nodes; see Figure 2.11(b). Taking $\Delta x = \Delta y = h$, the central difference technique gives us

$$\left. \frac{\partial^2 T}{\partial x^2} \right|_{i,j} \approx \frac{T_{i+1,j} - 2T_{i,j} + T_{i-1,j}}{h^2}$$

and

$$\left. \frac{\partial^2 T}{\partial y^2} \right|_{i,j} \approx \frac{T_{i,j+1} - 2T_{i,j} + T_{i,j-1}}{h^2}.$$

Plugging the central difference expressions into the PDE gives

$$\frac{T_{i+1,j} - 2T_{i,j} + T_{i-1,j}}{h^2} + \frac{T_{i,j+1} - 2T_{i,j} + T_{i,j-1}}{h^2} = -\frac{q_{i,j}}{k},$$

and then we have

$$T_{i+1,j} + T_{i-1,j} - 4T_{i,j} + T_{i,j+1} + T_{i,j-1} = -\frac{q_{i,j} h^2}{k}$$

for all internal nodes in the discretized solution domain. Thus we can transform the differential equation into a set of linear algebraic equations, which can be solved using LU decomposition. For irregular grid, finite difference methods are no longer suitable and the finite element method can be a good choice.

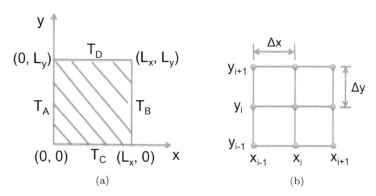

(a)   (b)

FIGURE 2.11: 2D elliptical problem and the domain discretization.

Parabolic PDE. Let us consider a 1D time-varying heat conduction problem

$$\frac{\partial^2 T(x,t)}{\partial x^2} = \frac{1}{\alpha}\frac{\partial T(x,t)}{\partial t}, \tag{2.48}$$

with boundary conditions $T(0,t) = T_A$ and $T(L,t) = T_B$, and initial conditions $T(x,0) = T^0(x)$. We discretize the problem in time with Δt and in space with Δx. We still use the central difference technique for the spatial domain, but adopt the forward difference technique for the time domain. We have

$$\left.\frac{\partial^2 T}{\partial x^2}\right|_{x_i,t_j} \approx \frac{T_{i+1}^j - 2T_i^j + T_{i-1}^j}{(\Delta x)^2} + O((\Delta x)^2)$$

and

$$\left.\frac{\partial T}{\partial t}\right|_{x_i,t_j} \approx \frac{T_i^{j+1} - T_i^j}{\Delta t} + O(\Delta t).$$

Plug them into the governing equation to obtain

$$\frac{T_{i+1}^j - 2T_i^j + T_{i-1}^j}{(\Delta x)^2} = \frac{1}{\alpha}\frac{T_i^{j+1} - T_i^j}{\Delta t}.$$

We aim to solve for T_i^{j+1},

$$T_i^{j+1} = \frac{\alpha\Delta t}{(\Delta x)^2}(T_{i+1}^j + T_{i-1}^j) + \left(1 - \frac{2\alpha\Delta t}{(\Delta x)^2}\right)T_i^j.$$

For convenience, let $\frac{\alpha\Delta t}{(\Delta x)^2} = r$ and we have

$$T_i^{j+1} = r(T_{i+1}^j + T_{i-1}^j) + (1 - 2r)T_i^j.$$

Note that this method is explicit; it may not be stable when Δt is big. An implicit method like the Crank–Nicolson method can perform better.

Go back to the governing equation in Eq. (2.48) and use the central difference technique in time and space at $t = t_{j+1}$. Then we can obtain

$$\frac{1}{2}\left[\frac{T_{i+1}^{j} - 2T_i^j + T_{i-1}^j}{(\Delta x)^2} + \frac{T_{i+1}^{j+1} - 2T_i^{j+1} + T_{i-1}^{j+1}}{(\Delta x)^2}\right] = \frac{1}{\alpha}\frac{T_i^{j+1} - T_i^j}{\Delta t}.$$

Note that we now have three unknowns $(T_{i+1}^{j+1}, T_i^{j+1}, T_{i-1}^{j+1})$ instead of the single unknown we had in the explicit scheme before. With $r = \frac{\alpha \Delta t}{(\Delta x)^2}$, we have

$$-rT_{i-1}^{j+1} + 2(1+r)T_i^{j+1} - rT_{i+1}^{j+1} = rT_{i-1}^j + 2(1-r)T_i^j + rT_{i+1}^j,$$

which turns out to be stable for all r, but not necessarily accurate and will need more effort to solve.

Hyperbolic PDE. Consider the 1D wave equation

$$\frac{\partial^2 u(x,t)}{\partial t^2} = c^2\frac{\partial^2 u(x,t)}{\partial x^2}. \tag{2.49}$$

Applying the central difference technique leads to

$$\frac{u_i^{j+1} - 2u_i^j + u_i^{j-1}}{(\Delta t)^2} = c^2\frac{u_{i+1}^j - 2u_i^j + u_{i-1}^j}{(\Delta x)^2},$$

which can be reorganized into

$$u_i^{j+1} = \frac{c^2(\Delta t)^2}{(\Delta x)^2}(u_{i+1}^j + u_{i-1}^j) - u_i^{j-1} + 2\left(1 - \frac{c^2(\Delta t)^2}{(\Delta x)^2}\right)u_i^j.$$

This is an implicit scheme, and it needs u_i^{j-1} in the initial condition.

2.7 Numerical Integration

A simple example of integral is to calculate the area or volume of a 2D or 3D domain. When we derive the analytical solution of a parabolic PDE, we use separation of variables and integration is also needed in the derivation process. For a continuous function whose integral cannot be evaluated using analytical techniques, we have to use numerical integration. Another motivation for numerical integration is that sometimes we need to integrate discrete data from measurements.

2.7.1 Deterministic and Stochastic Approaches

The basic problem is to find I, where $I = \int_a^b f(x)dx$ in two different cases: (1) $f(x)$ is continuous; and (2) $f(x)$ is discrete data. We can treat these two cases by applying the same techniques using the deterministic approach or the stochastic approach.

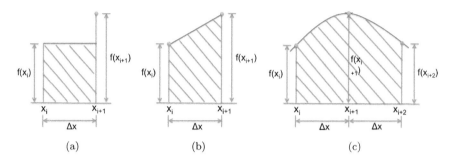

FIGURE 2.12: Deterministic approaches for numerical integration. (a) 0th-order approach; (b) trapezoidal rule; and (c) Simplson's 1/3 rule.

Deterministic approaches. In this method, we use Newton Cotes formulas and replace a complicated function or tabulated data with piecewise polynomials that are easy to integrate. For a continuous function, how to discretize the data or how to pick Δx is important. For discrete data, the sampling rate is set. There is a trade-off between accuracy and computation time.

1. 0th-order approach - fit a constant function between $(x_i, f(x_i))$ and $(x_{i+1}, f(x_{i+1}))$ as shown in Figure 2.12(a). We have

$$\int_{x_i}^{x_{i+1}} f(x)dx = f_i \Delta x, \tag{2.50}$$

therefore $\int_a^b f(x)dx = \Delta x(f_1 + f_2 + \cdots + f_{n-1})$ with error $O(\Delta x)$. This method is only exact for a constant function.

2. Trapezoidal rule - fit a linear function between $(x_i, f(x_i))$ and $(x_{i+1}, f(x_{i+1}))$ as shown in Figure 2.12(b). We have

$$\int_{x_i}^{x_{i+1}} f(x)dx = \frac{\Delta x}{2}(f_i + f_{i+1}), \tag{2.51}$$

and $\int_a^b f(x)dx = \frac{\Delta x}{2}(f_1 + 2f_2 + 2f_3 + \cdots + 2f_{n-1} + f_n)$ with error $O((\Delta x)^2)$. This method is exact for linear functions.

3. Simpson's rule - fit a parabola to (x_i, f_i), (x_{i+1}, f_{i+1}), and (x_{i+2}, f_{i+2}) as shown in Figure 2.12(c). We can show that

$$\int_{x_i}^{x_{i+2}} f(x)dx = \frac{\Delta x}{3}(f_i + 4f_{i+1} + f_{i+2}), \qquad (2.52)$$

and $\int_a^b f(x)dx = \frac{\Delta x}{3}(f_1 + 4f_2 + 2f_3 + 4f_4 + \cdots + 4f_{n-1} + f_n)$ with error $O((\Delta x)^4)$. This method is exact for a second-order function.

Stochastic approaches. In 1D, if we divide the data into n discrete points, we need to perform n function evaluations. In 2D, dividing each dimension into n discrete points will lead to n^2 function evaluations; and multi-dimensional integrals over hundreds or thousands of independent variables come up in statistical mechanics all the time. The purpose of stochastic approaches like the Monte Carlo simulation is to evaluate such integrals in an efficient way. Considering the integral $\int_a^b f(x)dx$, we can relate it with the average value of $f(x)$ between a and b. Then we have $\int_a^b f(x)dx = < f(x) >_{a-b} (b-a)$. How do we sample $f(x)$ between a and b, regularly or randomly? Such a situation leads to the idea of importance sampling, which is a key feature of Monte Carlo simulations. If we know something about a function, we can bias the random sampling to specific regions.

2.7.2 Gaussian Quadrature

To study Gaussian quadrature, let us start with the integration methods with equally spaced data. For a two-point $(n = 2)$ method, we have

$$\int_{x_i}^{x_{i+1}} f(x)dx = \Delta x[w_i f(x_i) + w_{i+1}f(x_{i+1})],$$

where the weights w_i and w_{i+1} are two unknowns. In the trapezoidal rule, we select $w_i = w_{i+1} = 1/2$. For a three-point $(n = 3)$ method, we have

$$\int_{x_i}^{x_{i+2}} f(x)dx = \Delta x[w_i f(x_i) + w_{i+1}f(x_{i+1}) + w_{i+2}f(x_{i+2})],$$

where the three unknown weights give a second-order fit. In Simpson's rule, we choose $w_i = 1/3$, $w_{i+1} = 4/3$, and $w_{i+2} = 1/3$.

Let us remove the requirement that we need to evaluate the function at the x_i point, meaning that we have an explicit expression for $f(x)$ and this approach will not work for discrete data. Then, for an n-point method, there will be $2n$ unknowns: where the function is evaluated (n unknowns) and the weights at each point (n unknowns). Based on what we found above, this approach will give us a fit accurate to order $(2n - 1)$. So the 2-point method is now postulated as

$$\int_{x_i}^{x_{i+1}} f(x)dx \approx w_1 f(x_1) + w_2 f(x_2),$$

where $x_i \leq x_1, x_2 \leq x_{i+1}$. The four unknowns will give us a result exact to the third-order. Considering the integral $\int_{-1}^{1} f(t)dt \approx w_1 f(t_1) + w_2 f(t_2)$, we have four unknowns w_1, w_2, t_1 and t_2. The integral should be exact to the third order, i.e., it should give the exact solution for $f(t) = 1, t, t^2$ and t^3.

$$f(t) = t^3 \quad \Rightarrow \quad \int_{-1}^{1} t^3 dt = 0 = w_1 t_1^3 + w_2 t_2^3, \tag{2.53}$$

$$f(t) = t^2 \quad \Rightarrow \quad \int_{-1}^{1} t^2 dt = \frac{2}{3} = w_1 t_1^2 + w_2 t_2^2, \tag{2.54}$$

$$f(t) = t \quad \Rightarrow \quad \int_{-1}^{1} t dt = 0 = w_1 t_1 + w_2 t_2, \tag{2.55}$$

$$f(t) = 1 \quad \Rightarrow \quad \int_{-1}^{1} 1 dt = 2 = w_1 + w_2. \tag{2.56}$$

(2.53)-t_1^2(2.55) gives $w_1 t_1^3 + w_2 t_2^3 - w_1 t_1^3 + w_2 t_1^2 t_2 = w_2 t_2 (t_2^2 - t_1^2) = w_2 t_2 (t_2 - t_1)(t_2 + t_1) = 0$, which leaves us with a number of possibilities: (1) $w_2 = 0$; (2) $t_2 = 0$; (3) $t_2 = t_1$; and (4) $t_2 = -t_1$. The first three possibilities reduce to a one-point formula, and only the last one makes sense. If $t_2 = -t_1$, then

$$\begin{aligned} (2.55) \quad &\Rightarrow \quad w_1 t_1 - w_2 t_1 = 0 \Rightarrow w_1 = w_2; \\ (2.56) \quad &\Rightarrow \quad w_1 + w_2 = 2 \Rightarrow w_1 = w_2 = 1; \\ (2.54) \quad &\Rightarrow \quad t_1^2 + t_2^2 = \frac{2}{3} \Rightarrow t_1 = \frac{1}{\sqrt{3}}, t_2 = -\frac{1}{\sqrt{3}}. \end{aligned}$$

Therefore $w_1 = w_2 = 1$, $t_1 = \frac{1}{\sqrt{3}}$, $t_2 = -\frac{1}{\sqrt{3}}$, and $\int_{-1}^{1} f(t)dt = w_1 f(t_1) + w_2 f(t_2) = f(t = \frac{1}{\sqrt{3}}) + f(t = -\frac{1}{\sqrt{3}})$. This is exact for $f(t)$ to the third order for only two function evaluations, which results in a significant improvement to what we had before with the trapezoidal or Simpson's rule.

For a general integral $\int_a^b f(x)dx$, we can transform it to an integral in t from -1 to 1. We have $x = \frac{b-a}{2}t + \frac{a+b}{2}$ and $dx = \frac{b-a}{2}dt$; plug them into the integral to obtain

$$\begin{aligned} \int_a^b f(x)dx \quad &= \quad \frac{b-a}{2} \int_{-1}^{1} f \left[\frac{(b-a)t + a + b}{2} \right] dt \\ &= \quad \frac{b-a}{2} \left[f \left(\frac{(b-a)\frac{-1}{\sqrt{3}} + a + b}{2} \right) + f \left(\frac{(b-a)\frac{1}{\sqrt{3}} + a + b}{2} \right) \right]. \end{aligned}$$

This is a powerful technique and can be extended to higher orders like $\int_{-1}^{1} f(t)dt \approx \sum_{i=1}^{n} w_i f(t_i)$.

2.8 Fourier Analysis

In mathematics, Fourier analysis is to represent or approximate a general function using a sum of trigonometric functions. Let us first take a look at an example. Consider the 1D heat conduction (parabolic PDE) as shown in Figure 2.2,

$$\frac{\partial^2 T(x,t)}{\partial x^2} = \frac{1}{\alpha}\frac{\partial T(x,t)}{\partial t}$$

with boundary conditions $T(0,t) = 0$, $T(L,t) = 0$ and initial conditions $T(x,0) = f(x)$. By separation of variables, let $T(x,t) = P(x)Q(t)$, then the PDE can be rewritten as

$$QP'' = \frac{1}{\alpha}PQ'$$

and

$$\frac{P''}{P} = \frac{1}{\alpha}\frac{Q'}{Q} = -\lambda^2.$$

Then we obtain a spatial equation $P'' + \lambda^2 P = 0$ with $P(0) = P(L) = 0$, and a temporal equation $Q'(t) = -\alpha\lambda^2 Q(t)$. For the spatial equation, it has a general solution in the form of $P(x) = A\cos\lambda x + B\sin\lambda x$ with $P(0) = 0$. We can obtain $A = 0$ and $P(L) = 0$, or $B\sin\lambda L = 0$. Then we have $B = 0$ (trivial solution) or $\sin\lambda L = 0$, yielding $\lambda = \frac{n\pi}{L}(n = 0, \pm1, \pm2, \cdots)$. Therefore we have $P(x) = B_n \sin\left(\frac{n\pi x}{L}\right)$. For the temporal equation, the solution is $Q(t) = Ce^{-\lambda^2\alpha t}$. The final solution can be written as

$$T(x,t) = \sum_{n=1}^{\infty} B_n \sin\left(\frac{n\pi x}{L}\right) e^{-\lambda^2\alpha t}.$$

To find the B_n values, we apply the initial condition

$$T(x,0) = f(x) = \sum_{n=1}^{\infty} B_n \sin\left(\frac{n\pi x}{L}\right).$$

In this way, we have decomposed a function into a basis set, but how do we calculate the coefficients B_n? Fourier analysis can be the solution.

2.8.1 Fourier Series and Transforms

First, let us consider some general properties of integrating odd and even functions over the interval $(-P/2, P/2)$. As shown in Figure 2.13, an even

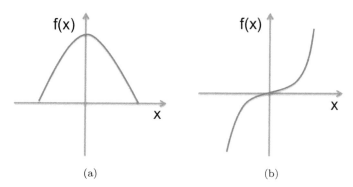

(a)   (b)

FIGURE 2.13: (a) An even function; and (b) an odd function.

function like $cosx$ and x^2 satisfies $f(x) = f(-x)$. Note that $\int_{-P/2}^{P/2} f_e(x)dx = 2\int_0^{P/2} f_e(x)dx$. Differently, an odd function like $sinx$ and x^3 satisfies $f(x) = -f(-x)$. Note that $\int_{-P/2}^{P/2} f_o(x)dx = 0$. Symbolically, we have the following three properties:

$$f_{e,1}(x) \cdot f_{e,2}(x) = f_{e,3}(x); \qquad (2.57)$$
$$f_{o,1}(x) \cdot f_{o,2}(x) = f_{e,1}(x); \qquad (2.58)$$
$$f_{e,1}(x) \cdot f_{o,1}(x) = f_{o,2}(x). \qquad (2.59)$$

Fourier analysis represents a function using a basis set of *sine* and *cosine* functions, $\{1, sinx, cosx, sin2x, cos2x, \cdots\}$. Note that all these functions are linearly independent. If we use periodic terms for the basis, the approximation will also be periodic. If the period is P, this will mean that $f(x) = f(x+P) = f(x+2P) = \cdots$ and $f(x) = f(x-P) = f(x-2P) = \cdots$. Therefore, say that we want a representation of $f(x)$ that is periodic on $[-\pi, \pi]$, then

$$f(x) = \frac{A_0}{2} + \sum_{n=1}^{\infty}[A_n cos(nx) + B_n sin(nx)]. \qquad (2.60)$$

The next problem is to find A_0, A_n and B_n. For A_0, we integrate both sides of Eq. (2.60) over $-\pi$ to π:

$$\int_{-\pi}^{\pi} f(x)dx = \int_{-\pi}^{\pi} \frac{A_0}{2}dx + \sum_{n=1}^{\infty}\int_{-\pi}^{\pi} A_n cos(nx)dx + \sum_{n=1}^{\infty}\int_{-\pi}^{\pi} B_n sin(nx)dx.$$

As we all know, integrating a periodic function over an integer number of periods gives zero. Therefore the last two terms on the right-hand side are zero. We have

$$\int_{-\pi}^{\pi} f(x)dx = A_0\pi \quad \Rightarrow \quad A_0 = \frac{1}{\pi}\int_{-\pi}^{\pi} f(x)dx.$$

To find A_m, we multiply through by $cos(mx)$ and integrate over $-\pi$ to π.

$$\int_{-\pi}^{\pi} f(x)cos(mx)dx = \int_{-\pi}^{\pi} \frac{A_0}{2}cos(mx)dx + \sum_{n=1}^{\infty} \int_{-\pi}^{\pi} A_n cos(nx)cos(mx)dx$$
$$+ \sum_{n=1}^{\infty} \int_{-\pi}^{\pi} B_n sin(nx)cos(mx)dx.$$

We can easily obtain that the first and last terms on the right-hand side are zero. For the second term, it should be $\delta_{nm}\pi A_n$ due to the orthogonality of *sine* and *cosine* functions. We have

$$\int_{-\pi}^{\pi} f(x)cos(mx)dx = \pi A_m \quad \Rightarrow \quad A_m = \frac{1}{\pi}\int_{-\pi}^{\pi} f(x)cos(mx)dx.$$

Similarly, we can obtain

$$B_m = \frac{1}{\pi}\int_{-\pi}^{\pi} f(x)sin(mx)dx.$$

Therefore, the Fourier series of a function $f(x)$ can be written as

$$f(x) = \frac{A_0}{2} + \sum_{n=1}^{\infty}[A_n sin(nx) + B_n cos(nx)] \tag{2.61}$$

for $(-\pi, \pi)$, where

$$A_0 = \frac{1}{\pi}\int_{-\pi}^{\pi} f(x)dx, \tag{2.62}$$

$$A_n = \frac{1}{\pi}\int_{-\pi}^{\pi} f(x)cos(nx)dx, \tag{2.63}$$

$$B_n = \frac{1}{\pi}\int_{-\pi}^{\pi} f(x)sin(nx)dx. \tag{2.64}$$

Generally, if a function is periodic over $-P/2$ to $P/2$ (i.e., with period P), you will use functions of the form $cos\left(\frac{n\pi x}{P/2}\right)$, $sin\left(\frac{n\pi x}{P/2}\right)$ in the Fourier series and

$$A_n = \frac{2}{P}\int_{-P/2}^{P/2} f(x)cos\left(\frac{n\pi x}{P/2}\right)dx, \tag{2.65}$$

$$B_n = \frac{2}{P}\int_{-P/2}^{P/2} f(x)sin\left(\frac{n\pi x}{P/2}\right)dx, \tag{2.66}$$

where $n = 0, 1, 2, \cdots$. Note that numerical methods are important for performing these integrations.

Now suppose that we have a function only defined on $(0, L)$ that we wish to characterize (i.e., find its Fourier series representation). The Fourier series require a function periodic on $(-L, L)$. We are free to define the function in even or odd by choosing either $A_n = 0$ for odd or $B_n = 0$ for even. These two situations are known as the *cosine* and *sine* expansions. We have

$$f(x)_{cos} = \sum_{n=0}^{\infty} A_n cos \left(\frac{n\pi x}{L}\right), \tag{2.67}$$

$$f(x)_{sin} = \sum_{n=0}^{\infty} B_n sin \left(\frac{n\pi x}{L}\right), \tag{2.68}$$

where

$$A_n = \frac{2}{L} \int_0^L f(x) cos \left(\frac{n\pi x}{L}\right) dx, \quad n = 0, 1, 2, \cdots, \tag{2.69}$$

$$B_n = \frac{2}{L} \int_0^L f(x) sin \left(\frac{n\pi x}{L}\right) dx, \quad n = 1, 2, 3, \cdots. \tag{2.70}$$

These formulas can be applied to continuous and discrete data using numerical integrations. There are two approximations here: the number of terms in summation, and how integral is performed.

The Fourier series can also be represented in the form of complex exponential. Note that $e^{inx} = cos(nx) + isin(nx)$ and $e^{-inx} = cos(nx) - isin(nx)$. This fact suggests that we can express $f(x)$ using a basis set of e^{inx} and e^{-inx} terms. Suppose that

$$
\begin{aligned}
f(x) &= C_0 + \sum_{n=1}^{\infty} \left(C_n e^{inx} + D_n e^{-inx}\right) \\
&= C_0 + \sum_{n=1}^{\infty} \left(C_n [cos(nx) + isin(nx)] + D_n [cos(nx) - isin(nx)]\right) \\
&= C_0 + \sum_{n=1}^{\infty} \left[(C_n + D_n)cos(nx) + i(C_n - D_n)sin(nx)\right]. \tag{2.71}
\end{aligned}
$$

We compare Eqns. (2.61) and (2.71) to obtain

$$C_0 = \frac{A_0}{2}, \tag{2.72}$$

$$C_n = \frac{A_n - iB_n}{2}, \tag{2.73}$$

$$D_n = \frac{A_n + iB_n}{2}. \tag{2.74}$$

Note that C_n and D_n are complex conjugates; we have $D_n = C_{-n}$. Plugging

$D_n = C_{-n}$ into Eq. (2.71), we can obtain

$$
\begin{aligned}
f(x) &= C_0 + \sum_{n=1}^{\infty} \left(C_n e^{inx} + C_{-n} e^{-inx} \right) \\
&= C_0 + \sum_{n=1}^{\infty} C_n e^{inx} + \sum_{n=-\infty}^{-1} C_n e^{inx} \\
&= \sum_{n=-\infty}^{\infty} C_n e^{inx}.
\end{aligned}
\tag{2.75}
$$

Similar to the *sine* and *cosine* terms, the e^{inx} terms form an orthogonal basis. Therefore, we multiply Eq. (2.75) through by e^{-imx} and integrate over 0 to 2π to find C_m,

$$
C_m = \frac{1}{2\pi} \int_0^{2\pi} f(x) e^{-imx} dx, \quad m = 0, \pm 1, \pm 2, \cdots .
\tag{2.76}
$$

2.8.2 Discrete Fourier Transform

The Fourier coefficients cannot typically be evaluated analytically, and the process of the numerical integration is the discrete Fourier transform (DFT). Given a set of discrete data on $[0, 2\pi]$, evenly spaced at $x = \frac{2\pi k}{N}$ where $k = 0, 1, 2, \cdots, N$. By definition, we have

$$
C_n = \frac{1}{2\pi} \int_0^{2\pi} f(x) e^{-inx} dx.
\tag{2.77}
$$

Numerically integrating by the rectangle method, we can obtain

$$
\begin{aligned}
C_n &= \frac{1}{2\pi} \sum_{k=0}^{N-1} f(x_k) e^{-in\frac{2\pi k}{N}} \cdot \frac{2\pi}{N} \\
&= \frac{1}{N} \sum_{k=0}^{N-1} f(x_k) \left[e^{-in\frac{2\pi k}{N}} \right]^{nk} .
\end{aligned}
$$

Let $w = e^{-i2\pi/N}$, then we have

$$
C_n = \frac{1}{N} \sum_{k=0}^{N-1} f(x_k) w^{nk}.
\tag{2.78}
$$

For $N = 4$, we have

$$
\begin{aligned}
C_0 &= \frac{1}{4}\left[w^0 f(x_0) + w^0 f(x_1) + w^0 f(x_2) + w^0 f(x_3)\right], \\
C_1 &= \frac{1}{4}\left[w^0 f(x_0) + w^1 f(x_1) + w^2 f(x_2) + w^3 f(x_3)\right], \\
C_2 &= \frac{1}{4}\left[w^0 f(x_0) + w^2 f(x_1) + w^4 f(x_2) + w^6 f(x_3)\right], \\
C_3 &= \frac{1}{4}\left[w^0 f(x_0) + w^3 f(x_1) + w^6 f(x_2) + w^9 f(x_3)\right].
\end{aligned}
$$

In matrix form, we have

$$
\vec{C} = \frac{1}{4}
\begin{bmatrix}
w^0 & w^0 & w^0 & w^0 \\
w^0 & w^1 & w^2 & w^3 \\
w^0 & w^2 & w^4 & w^6 \\
w^0 & w^3 & w^6 & w^9
\end{bmatrix}
f(\vec{x}).
\tag{2.79}
$$

To get the DFT, multiply an $N \times N$ matrix by an N-row column vector. In the fast Fourier transform (FFT), we explore properties of the $\vec{C}$ matrix. Consider for $N = 4$, we have $w^1 = exp(-i2\pi/4) = -i$ and $w^5 = w^9 = -i$. In general $w^k = w^{k \bmod N}$. This allows the development of FFT, which is $O(Nlog_2N)$ and much faster than $O(N^2)$.

2.9   Optimization

There are many optimization problems in engineering. Curve fitting is an example as we discussed earlier. The general problem definition is as follows: given $f(\vec{x})$ where $\vec{x} = (x_1, x_2, \cdots)$, find a set of points x_0 such that $f(\vec{x})$ is minimized or maximized. x_0 is called a stationary point.

In general, $\vec{x}_0$ is a global minimum if $f(\vec{x}_0) \leq f(\vec{x})$ for all $\vec{x}$, and a local minimum if $f(\vec{x}_0) \leq f(\vec{x})$ for $|\vec{x} - \vec{x}_0| < r$. Optimization problems can be unconstrained or constrained. The extrema points can be obtained by setting the first derivative to be zero. In other words, the optimization problem is basically a root-finding problem. We can check the second derivative to decide if it is minimum or maximum. A minimum has $f''(\vec{x}_0) > 0$ and a maximum has $f''(\vec{x}_0) < 0$. $f''(\vec{x}_0) = 0$ gives a saddle point. Similar arguments can be obtained for higher dimensions using the Hessian matrix H; see Table 2.3.

In the following, let us take a look at two gradient-based approaches: steepest descent and conjugate gradient, as well as Newton's method in 2D or Hessian-based method.

Steepest Descent. Consider $f(x_1, x_2) = x_1^2 + 3x_2^2$, which we know has a minimum at $(0, 0)$ and an initial guess at $(6, 3)$. In the steepest descent

TABLE 2.3: Optimality Condition and Extrema Points

	Optimality Condition	**Min or Max**
1D	$\frac{df}{dx} = 0$	$f''(x) \begin{cases} < 0: & max \\ > 0: & min \\ = 0: & inflection\ point \end{cases}$
2D	$\vec{\nabla} = \begin{bmatrix} \frac{\partial f}{\partial x_1} \\ \frac{\partial f}{\partial x_2} \end{bmatrix} = 0$ or $\frac{\partial f}{\partial x_1} = \frac{\partial f}{\partial x_2} = 0$	$\|H\| > 0 \begin{cases} \frac{\partial^2 f}{\partial x^2} < 0: & max \\ \frac{\partial^2 f}{\partial x^2} > 0: & min \end{cases}$
	$H = \begin{bmatrix} \frac{\partial^2 f}{\partial x_1^2} & \frac{\partial^2 f}{\partial x_1 \partial x_2} \\ \frac{\partial^2 f}{\partial x_1 \partial x_2} & \frac{\partial^2 f}{\partial x_2^2} \end{bmatrix}$	$\|H\| < 0 :$ saddle point
	$\Rightarrow \|H\| = \frac{\partial^2 f}{\partial x_1^2}\frac{\partial^2 f}{\partial x_2^2} - \left(\frac{\partial^2 f}{\partial x_1 \partial x_2}\right)^2$	$\|H\| = 0 :$ degenerated critical point

method, we choose to move from a point in the direction in which the function is changing the fastest (i.e., in the direction of the gradient vector). We find the place along that 1D line where the function is minimized, and take that as the next point in the iteration. In 2D, the gradient is perpendicular to the contour lines. Now let us take a look at the five steps in each iteration.

1. $\nabla f = \left(\frac{\partial f}{\partial x_1}, \frac{\partial f}{\partial x_2}\right) = (2x_1, 6x_2) = 2x_1\hat{i} + 6x_2\hat{j}$;

2. The parametric equation of a line along the gradient vector through (x_1, x_2) is $\vec{z}(t) = \vec{x} - t\nabla f = x_1\hat{i} + x_2\hat{j} - t2x_1\hat{i} - t6x_2\hat{j} = (1 - 2t)x_1\hat{i} + (1 - 6t)x_2\hat{j}$. We want to find the point along $\vec{z}(t)$ where $f(x_1, x_2)$ is minimized;

3. The 1D optimization is to build a function $g(t)$, satisfying $g(t) = f(\vec{z}(t)) = (1 - 2t)^2 x_1^2 + 3(1 - 6t)^2 x_2^2$;

4. Minimize $g(t)$ via $g'(t) = 0$, and we have $t = \frac{x_1^2 + 9x_2^2}{2x_1^2 + 54x_2^2}$. For complex equations, we can use numerical methods such as bisection, golden search or Newton's method; and

5. $\vec{x}_{n+1} = \vec{x}_n - t\nabla f(\vec{x}_n)$.

Conjugate Gradient. This method works in a similar way as the steepest descent except that we choose the directions for the line search in a much more efficient manner. Again, consider $f(x_1, x_2) = x_1^2 + 3x_2^2$ with $\vec{x}_0 = (6, 3)$. Generating $\vec{x}_1$ using the steepest descent method, let us start the iteration from $\vec{x}_n$.

1. Find steepest descent direction, $\nabla f(\vec{x}_n) = \nabla f_n$;

2. Evaluate β, which will be used to obtain new directions, $\beta_n = \frac{(\nabla f_n - \nabla f_{n-1}) \cdot \nabla f_n}{\nabla f_{n-1} \cdot \nabla f_{n-1}}$;

3. The conjugate direction is $\vec{\Lambda}_n = -\nabla f_n - \beta_n \nabla f_{n-1}$;

4. Perform a line search to minimize $f(x)$ on line through $\vec{x}_n$ with direction $\vec{\Lambda}_n$ (gives t); and

5. $\vec{x}_{n+1} = \vec{x}_n - t\vec{\Lambda}_n$, iterate.

The conjugate gradient method is derived based on a quadratic assumption, which ensures that any quadratic function will converge in two steps. Clearly, it is a superior approach to the steepest descent.

Newton's Method in 2D or Hessian-Based Method. Suppose $g(\vec{x}) = f'(\vec{x})$. From the Newton's method in 1D, we can obtain $\vec{x}_{n+1} = \vec{x}_n - \frac{g(\vec{x}_n)}{g'(\vec{x}_n)} = \vec{x}_n - \frac{f'(\vec{x}_n)}{f''(\vec{x}_n)}$. Similarly in 2D, we have

$$
\begin{aligned}
\vec{x}_{n+1} &= \vec{x}_n - \left[\nabla^2 f(\vec{x}_n)\right]^{-1} \vec{\nabla} f(\vec{x}_n) \\
&= \vec{x}_n - H^{-1} \vec{\nabla} f(\vec{x}_n),
\end{aligned}
\tag{2.80}
$$

where

$$
H = \begin{bmatrix} \frac{\partial^2 f}{\partial x_1^2} & \frac{\partial^2 f}{\partial x_1 \partial x_2} \\ \frac{\partial^2 f}{\partial x_1 \partial x_2} & \frac{\partial^2 f}{\partial x_2^2} \end{bmatrix}.
\tag{2.81}
$$

This method works well if $\vec{x}_0$ is close to the local minimum; otherwise it may blow up.

For constrained optimization, we define a new objective function by including each constraint via a Lagrange multiplier (for equality constraints) or a "slack" variable (for inequality constraints). In this way, we convert constrained optimization problems into unconstrained optimization. Then we can use the above steepest descent, conjugate gradient or Hessian-based method to solve it. Note that the Lagrange multipliers and "slack" variables are treated as unknowns, and we may need to give some discussions on the "slack" variables when multiple possibilities happen.

In summary, we have reviewed several commonly used topics of numerical methods in this chapter. For other topics, the readers can refer to related textbooks [81, 152, 301].

Homework

Problem 2.1. Describe a real world mechanical engineering problem example for each of the following four complex systems where we need to use numerical methods: large-scale systems (large numbers of parameters and equations), nonlinear systems, multiphysics systems and multiscale systems.

Problem 2.2. In this chapter, we defined the following operators:

- Forward difference: $\Delta f_i = f_{i+1} - f$;

- Backward difference: $\nabla f_i = f_i - f_{i-1}$;

- Stepping: $E f_i = f_{i+1}$; and

- Derivative: $D f_i = \left. \frac{df}{df} \right|_i$.

(a) Prove that

- $\nabla = 1 - E^{-1}$;

- $E\nabla = \Delta$; and

- $D = \frac{1}{h} ln(\frac{1}{1-\nabla})$.

(b) Expand $D = \frac{1}{h} ln(\frac{1}{1-\nabla})$ as a Taylor series to prove that

$$D = \frac{1}{h} \left(\nabla + \frac{1}{2}\nabla^2 + \frac{1}{3}\nabla^3 + \cdots \right).$$

(c) Use the result of (b) to develop the backward difference formulas for $D f_i$ to the first and second order.

(d) Use the result of (b) to find an expression for D^2 that depends on ∇ and its higher powers. Use the result to find the first- and second-order backward difference formulas for $D^2 f_i$.

Problem 2.3. Prove that the number of multiplications/divisions needed to find the Crout LU decomposition of an $n \times n$ matrix is $n^3/3 - n/3$ and that the number of multiplications/divisions needed to solve an associated system of equations is n^2.

Problem 2.4. Answer the following questions briefly:

1. Compare analytical and numerical methods in the respect of exactness, type of computing and applicable systems.

2. Compare bracketing and open methods in the respect of required initial points, convergence and rate of convergence.

3. Choosing smaller step size h can reduce the truncation error. Explain why smaller h increases the round-off error.

4. For a set of numerical data points, which numerical methods (include every possible method) can we use to find its root?

5. For $f(x) = (x+1)^4$, which numerical methods can we use (include every possible method) to find its root? Which method would converge to the root the fastest? Discuss the above two questions again for $f(x) = (x+1)^3$.

Problem 2.5. Write a program to find the roots of a single nonlinear equation. For a given function, your program should be able to:

- Identify bracketed regions that enclose a root so as to generate initial guesses;

- Iterate using Newton's method to the root to within a given tolerance; and

- Find all roots in one execution.

Use your program to find the real roots of the following two equations to within 10^{-6}. Submit your code and the generated output.

1. $sinx + lnx = 1$;

2. $e^x + x^2 = 3x + 2$.

Problem 2.6. Use the multidimensional Newton's method to find the roots for the following system:

$$\begin{cases} f_1(x_1, x_2) &= x_1^2 + x_2^2 - 4, \\ f_2(x_1, x_2) &= e^{x_1} + sinx_2 - 2. \end{cases}$$

1. Plot $f_1(x_1, x_2)$ and $f_2(x_1, x_2)$ to get initial guesses for the roots.

2. Determine the Jacobian matrix J,

$$J = \begin{pmatrix} \frac{\partial f_1}{\partial x_1} & \frac{\partial f_1}{\partial x_2} \\ \frac{\partial f_2}{\partial x_1} & \frac{\partial f_2}{\partial x_2} \end{pmatrix}.$$

3. Set up the system of equations required to find δx,

$$J\delta x = -f(x),$$

and find an analytical expression for δx by inverting J.

4. Iteratively find one solution at a time using Newton's method and an initial guess. Be sure to find all the solutions. Use a tolerance of 10^{-6} on $f_1(x_1, x_2)$ and $f_2(x_1, x_2)$.

Problem 2.7. For given n data points (x_i, y_i), write down the cost function to minimize to find the best fitting quadratic curve (quadratic regression) to these data points with possible error or noise. Then, minimize this cost function manually and find the resulting linear equations. Do not solve these equations analytically. Use five data points given as $(0, 0)$, $(1, 2)$, $(2, 3)$, $(3, 1.5)$ and $(4, 0.5)$, and then numerically solve the resulting quadratic regression curve and write its equation.

Problem 2.8. For the mass-spring system in Figure 2.9, solve the equations of motion analytically. Use the fourth-order Runge–Kutta method (RK4) to obtain the numerical solution for $m = k = 1$, $x_1(0) = 1$, $x_2(0) = 0$, $\dot{x}_1(0) = 0$, $\dot{x}_2(0) = 0$, and $F_1(t) = F_2(t) = 0$ for $0 < t < 20$. Be sure to choose an appropriate time step. Compare your numerical solution with the analytical solution.

Problem 2.9. Manually evaluate the integral

$$\int_0^3 x^2 e^x \, dx \tag{2.82}$$

with the methods below. For each numerical method, calculate the percent relative error based on the analytical result.

1. analytically;

2. single application of the trapezoid rule; and

3. single application of Simpsons 1/3 rule (two evenly spaced intervals).

Problem 2.10. The n-point Gaussian quadrature formula is

$$\int_{-1}^1 f(t)dt \approx \sum_{i=1}^n w_i f(t_i),$$

which is exact for a polynomial of order $2n - 1$. For the 2-point $(n=2)$ case, we have $w_1 = w_2 = 1$, $t_1 = -\frac{1}{\sqrt{3}}$, and $t_2 = \frac{1}{\sqrt{3}}$. In the following, work through an approach to find the w_i and t_i values for a given n.

1. The required t_i values for the n-point formula are the roots of the nth-order Legendre polynomial. The Legendre polynomials are defined recursively as

$$(n + 1)L_{n+1}(t) - (2n + 1)tL_n(t) + nL_{n-1}(t) = 0,$$

with $L_0(t) = 1$ and $L_1(t) = t$. Find $L_n(t)$ for all n values between 2 and 6.

2. Use the Newton's method to find the roots of $L_n(t)$ for all n values between 1 and 6.

3. In Chapter 2.6, we generated a system of four nonlinear equations for w_1, w_2, t_1 and t_2 for $n=2$. Generalize that approach to prove that the $2n$ equations for the n-point quadrature are

$$w_1 t_1^k + \cdots + w_n t_n^k = \begin{cases} 0 & for \ k = 1, 3, \cdots, 2n - 1 \\ \frac{2}{k+1} & for \ k = 0, 2, \cdots, 2n - 2 \end{cases}.$$

Using the specified t_i values, the system of equations becomes linear in the weights w_i. Solve it for all n values between 1 and 6.

Problem 2.11. The time-dependent temperature profile in a 1D beam is described by the parabolic PDE

$$\frac{\partial^2 T}{\partial x^2} = \frac{1}{\alpha} \frac{\partial T}{\partial t},$$

with the boundary conditions $T(0, t) = T(L, t) = 0$ and the initial condition $T(x, 0) = 100 \left[1 - 4(x - 0.5)^2\right]$. Take $\alpha = 0.01$ and $L = 1$, and solve the problem using Fourier series analysis. The analytical solution for $L = 1$ is

$$T(x, t) = \sum_{n=1}^{\infty} B_n sin(n\pi x) e^{-(n\pi)^2 \alpha t}.$$

By applying the initial condition, the B_n values are obtained from

$$f(x) = \sum_{n=1}^{\infty} B_n sin(n\pi x),$$

which is a Fourier sine series over $(0, 1)$.

1. Evaluate the B_n coefficients analytically. Plot B_n as a function of n for $1 \le n \le 10$.

2. Using the above results, plot $T(x, t)$ for $t = 0.1$, 1 and 10 for an upper limit on the solution summation of 3 and 5.

3. Take $\Delta x = 0.05$ and obtain all the B_n values up to $n = 10$ using the rectangle rule for integration. Plot the results with those from (1); how does the error introduced by the numerical integration change as n is increased?

4. Plot $T(x, t)$ for $t = 0.1$, 1 and 10 and upper limits on the summation of 3 and 5 based on the results of (3). Compare the solutions to the results from (2).

Problem 2.12. Minimize the cost function

$$f(x, y) = (x - 2)^2 + (2y - 1)^2 - x^2 y^2 + 2xy \qquad (2.83)$$

with the initial point $(x_0,\ y_0) = (0,\ 0)$ using (1) steepest descent; and (2) Newton–Raphson's (Hessian) method. For each method, manually perform only two steps in the algorithm with a termination condition of $\epsilon = 10^{-6}$.

Chapter 3

Scanning Techniques and Image Processing

This chapter introduces several important scanning techniques in biomedical imaging, including computational tomography (CT), magnetic resonance imaging (MRI), nuclear medicine, ultrasound, fluoroscopy and cryo-electronic microscopy (cryo-EM). The retrospective on 3D medical imaging is first reviewed, followed by the mechanism of each scanning technique. In addition, this chapter also introduces the techniques used to scan the microstructure of polycrystalline materials like electron backscatter diffraction (EBSD) and high-energy X-rays. Various scanning techniques are featured and compared.

This chapter also talks about some basic operators in image processing. Contrast enhancement, filtering, segmentation and registration are discussed in detail. The latest start-of-the-art techniques are discussed and compared for each of them. In filtering, we cover linear and nonlinear filtering. In segmentation, we discuss watershed segmentation, statistical pattern recognition, region merging, Markov random field, level-set methods and Centroid Voronoi Tessellation (CVT)-based segmentation. In registration, both affine and deformable image registration techniques are included. In the end, we also talk about the optical flow and multiscale methods with applications.

3.1   Scanning Techniques

Biomedical imaging [415] refers to the techniques and processes used to create images of the human body and biological structures for clinical, biological or medical purposes. In the clinical context, medical imaging is generally also called radiology or "clinical imaging." The commonly used biomedical imaging techniques include CT, MRI, nuclear medicine, ultrasound, fluoroscopy and electron microscopy.

CT uses X-rays to scan the objects like the human body, and the resulting images provide anatomical structure. MRI is another popular scanning technique used in clinical studies, which uses radio wave and magnetic field and the resulting images also provide anatomical structure. In both CT and MRI, the radioactive sources are outside the human body. In nuclear medicine, radioactive sources are injected into the body and then the radiation emitted from the body is measured. The resulting images give physiological functions of the object. Ultrasound imaging is a very safe scanning technique that is being used a lot on patients, for example, for checking the fetus growth in the womb of pregnant women. Fluoroscopy also uses X-rays, which uses a constant of X-rays and can monitor the object motion in real time. Electron microscopy is generally used to scan small-scale objects like the virus at the molecular or cellular levels. From various scanning techniques, the resulting imaging data are a scalar, vector or tensor field over a 2D/3D rectilinear grid.

3.1.1 History Review on Medical Imaging

As reviewed in [415], medical imaging is a relatively young field. It requires multidisciplinary knowledge, including physics, chemistry, medicine, mathematics, computer sciences and so on. In 1895, Wilhelm C. Röntgen discovered X-rays during his experiments with a Crookes tube, the precursor to the cathode ray tube which is commonly used today in video applications. At the beginning, he did not realize they were electromagnetic radiation. He called them "X"-rays because "X" represents the unknown and nobody knew their source. What was the most surprising is that X-rays were able to go through materials like muscles, soft tissues and the entire human body. It is important that Röntgen immediately recognized the potential application of X-ray radiation in diagnostic imaging. He took an image of his wife's hand;

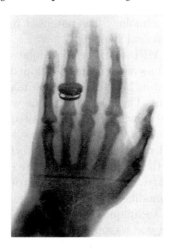

FIGURE 3.1: The famous X-ray image of the hand of Röntgen's wife [3], courtesy of public domain in the United States (PD-1923) because its copyright has expired.

you can see the bones and the metal wedding ring clearly in Figure 3.1. Due to such a noteworthy and revolutionary discovery, Röntgen received a Nobel Prize in Physics in 1901 within only six years after announcing the initial work.

Medical imaging has been aimed at providing clinicians and physicians with the ability to see through the body, and to diagnose the patient's condition. In the following years, the primary focus of this development has been to improve the quality of images for us to evaluate, and to extract useful information for diagnose and treatment. The entire image-based diagnose process is so sophisticated that it could not be possible until the computer technology became mature enough. It is obvious that there is a natural partnership between computer sciences, radiology and medicine.

This development needs mathematics and computation with the assistant of mature-enough computer facilities. Back to the early 20th century, a Czech mathematician named Johann Radon derived a mathematical transform to reconstruct cross-sectional information using planar projections around the object. Although this powerful theory had been developed for over 50 years, computing the transform on real, large data was not possible until digital computers became mature enough in the 1970s.

Godfrey Hounsfield and Alan Cormack developed X-ray computed tomography and 3D imaging emerged in 1972. Seven years later in 1979, they shared a Nobel Prize in Medicine. This achievement is so noteworthy because it was largely based on many underlying sciences developed decades earlier, such as engineering and theoretical mathematics. It combined mechanical engineering and computer sciences to complete such a sophisticated task that was impossible previously. This technology was patented in 1975, and immediately thereafter was applied in clinical systems.

Compared to CT, the MRI scanning technique was developed relatively later. While X-ray techniques were being developed and improved, organic chemists had been investigating how to use nuclear magnetic resonance (NMR) to analyze chemical samples. The NMR technique was studied by two scientists named Feliz Bloch and Edward Purcell in 1940s, and in 1952 they shared the Nobel Prize in Physics. Back to 1970s, people started to develop imaging applications from NMR phenomena. Paul Lauterbur, Peter Mansfield and Raymond Damadian were the first among them. In 1972, Lauterbur was able to create tomographic images of a physical phantom[1] constructed of capillary tubes and water in a modified spectrometer. In 1975, Damadian created animal images. Later, the technique of creating medical images using magnetic fields and radio waves was renamed magnetic resonance imaging (MRI). Lauterbur and Mansfield won the Nobel Prize in Medicine in 2003. Unfortunately Damadian was not honored, but his contribution is equally significant and should not be overlooked.

In the early 1970s, tomographic images in the axial plane from both X-ray CT and MRI became possible. CT matured first due to the earlier and long development of X-ray technology. Whereas for MRI, the difficulties and expensive cost of generating strong enough magnetic fields remained a major obstacle for clinical MRI scanner development until producing superconducting magnets became practical. MRI is now available worldwide although it is still very expensive. Different from CT, it creates precise images of the human body, outlining the anatomy information and physiology of internal structures other than bones. Table 3.1 shows a summary of four Nobel Prizes on medical imaging. For both CT and MRI, we can see that Physics played a fundamental role and the first Nobel Prize always went to Physics and then followed by the second Nobel Prized in Medicine. CT was developed 40 to 50 years earlier than MRI due to the long development history of X-rays.

In addition to the CT and MRI scanning techniques, nuclear imaging and volumetric ultrasound were also developed. For images scanned from different modalities, we need to align or register them in order to study multiple types of information together. Medical visualization handles the display of patient data. The radiologist needs to read and understand presentations of the raw data. Improving a diagnosis requires us to carefully craft the acquisition, maximize the contrast among the surrounding tissues, as well as suppress noise, fog and scatter related to regions of interest. It is obvious that improving the scanned image quality directly improves the resulting visualization, leading to more accurate understanding and diagnosis.

---

[1] Phantoms are used to measure doses in radiation protections and evaluate radiotherapy treatment plans in the Monte Carlo model. There are two kinds of phantoms in medical imaging: physical phantoms (can be measured) and computational phantoms.

TABLE 3.1: Four Nobel Prizes in Medical Imaging

Modality	Year	Event
CT	1901	Wilhelm Röntgen received the Nobel Prize in Physics due to his discovery of X-rays.
CT	1979	X-ray computed tomography was developed by Godfrey Hounsfield and Alan Cormack; 3D imaging emerged in 1972. They shared a Nobel Prize in Medicine in 1979.
MRI	1952	Felix Bloch and Edward Purcell studied nuclear magnetic resonance (NMR) in the 1940s, and they shared the Nobel Prize in Physics in 1952.
MRI	2003	Paul Lauterbur and Peter Mansfield developed imaging applications from NMR phenomena. The means of creating medical images using magnetic fields and radio waves was renamed Magnetic Resonance Imaging (MRI). Lauterbur and Mansfield won the 2003 Nobel Prize in Medicine.

3.1.2 Computational Tomography (CT)

X-ray CT is one of the most popular techniques of 3D medical imaging. CT has been formally referred to various names over the years, including computer-assisted tomography (CAT), computerized axial tomography (CAT scanning), computerized transaxial tomography (CTAT), computerized reconstruction tomography (CRT), and digital axial tomography (DAT).

Since early in the 20th century, the mathematics for tomographic reconstruction from multiple views have been developed. It took almost 50 years before all the components required for X-ray CT became mature enough, and then the instrument manufacture became economically feasible. As a room-sized X-ray instrument, a CT scanner requires a well-shielded environment to protect the clinic technologists from exposure during routine use. In early days, there were many manufacturers of CT scanners and the competition was so serious sometimes. The number of manufacturers has been steadily decreasing with only several vendors existing now, such as the famous Siemens and GE. The cost of a CT scanner is very expensive, ranging from $400,000 to $1,000,000.

There are three essential system components for a CT machine in general, including an accurately calibrated moving bed to translate the patient through the scanner, an X-ray tube that can revolve about the patient, and an array of X-ray detectors that can be gas-filled detectors or crystal scintillation detectors. In a gantry, the X-ray tube and detector array are mounted in such a way that the detector assembly is located directly across from the X-ray source, which is collimated by lead jaws. The X-rays form a flat fan beam with a pre-defined thickness. During the image acquisition, the source-detector ring rotates around the body and a cross-sectional transaxial image is reconstructed from the back-projected output of the detector array. By moving the patient along the bed direction, a series of slices can be obtained and they form a 3D representation of the human body anatomy.

Within the last 20 to 30 years, there have been significant advances and developments in CT technology, including faster spiral acquisition and reduced dose to patients, multislice detector arrays permitting acquisition of several slices simultaneously, scanners designed with fixed X-ray detectors, and only X-ray tubes needed to revolve. These significant advances in CT technology result in much faster image acquisition, improved patient throughput in a CT scanner, reduced artifacts from patient motion, reduced dose absorption, as well as multiple sensor layers in the detection ring. The new technologies are producing a huge amount of datasets with increasingly larger sizes, which need us to study and explore.

CT employs digital geometry processing to generate a 3D image of the internal structure of an object from a series of 2D X-ray images, which are taken around a single axis of rotation (along the movement of the bed). *In CT, X-ray radiation is absorbed at different rates in different types of tissues such as bone, muscle and fat.* Dense objects absorb more photons, therefore dense objects like bones have higher or brighter intensity values in the image, and the intensity in CT reflects the material density; see the 2D CT slice image of the human abdomen in Figure 3.2(a).

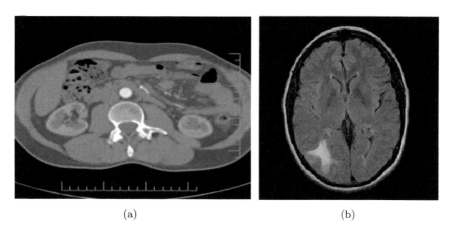

(a)   (b)

FIGURE 3.2: Example images of CT and MRI scanning techniques. (a) A CT image of the human abdomen; and (b) an MRI image of the brain.

Generally a CT scanner can acquire a transaxial slice image in seconds, like 1 to 5 seconds or even faster. The sampling resolution is either 256×256 or 512×512 voxels square. In the longitudinal direction, the sampling resolution is bounded because collimating the photons into thin planes is physically limited (e.g., 1 millimeter). The spatial resolution is the size of a voxel or a pixel, like 0.5 to 2 millimeters. Too high spatial resolution leads to low signal-to-noise ratios, creating voxels with too little signal which cannot accurately measure the X-ray absorption.

Aliasing artifacts are introduced by the reconstruction and sampling procedure. Partial voluming is an aliasing artifact when the contents of a pixel are distributed across multiple tissue types, blending the absorption characteristics of different materials. Patient motion also generates a variety of blurring and ring artifacts during reconstruction. In addition, artifacts are also introduced by the X-ray modality, for example, embedded dense objects like dental fixtures, fillings and bullets yield beam shadows and streak artifacts.

In summary, CT's primary benefit is the ability to separate anatomical structures at different depths within the body. CT uses X-rays to create images. The source or detector makes a complete 360 degree rotation about the subject obtaining a complete set of data from which images are reconstructed. CT employs digital geometry processing to generate volumetric images from a stack of 2D images. In CT, X-ray radiation is absorbed at different rates in different tissue types, and dense objects absorb more photons. Multiple detector rings with increasing rotation speeds are adopted in multislice CT, e.g., 4, 8, 16, 32, 40 and 64 detector rings. Figure 3.3(a) shows a human heart image scanned by CT 64 slices.

The scanning resolution of CT has been improved a lot over the years, supporting pixel sizes of the cross sections in the micrometer and even nanometer range. Micro-CT machines are used to scan many microstructures of biomedical samples such as trabecular bone in Figure 3.3(b) and foam materials in Figure 3.3(c). The imaging system can be either fan-beam or cone-beam. Nano-CT scanning supports the nanometer resolution, and some machines like the Xradia UltraXRM-L200 nanoXCT can reach about 50 nanometers per pixel for a field of view of 16 micrometers; see the polymer electrolyte fuel cell electrode image in Figure 3.3(d). Carnegie Mellon University has such a facility led by Professor Shawn Litster's laboratory in Mechanical Engineering [11].

3.1.3 Magnetic Resonance Imaging (MRI)

Different from CT, magnetic resonance imaging (MRI) does not use ionizing radiation to scan images. An MRI scanner is a sophisticated facility consisting of a large magnet, a microwave transmitter, a microwave antenna, and several electronic components to decode the signal and reconstruct cross-sectional images. The cost of an MRI system is more expensive than a CT system, ranging from $500,000 to $2,000,000 or even $3,000,000. In an MRI scanner, the bore is often 6 to 8 feet or 2 meters long. During the image acquisition, patients are positioned in the middle of the magnetic field. Due to large Faraday cages and substantial iron masses surrounding the magnet, the environments for MRI scanners must be well shielded like CT facilities to avoid possible magnetic and radiofrequency interference.

During image scanning, the patient is placed inside a very high inten-

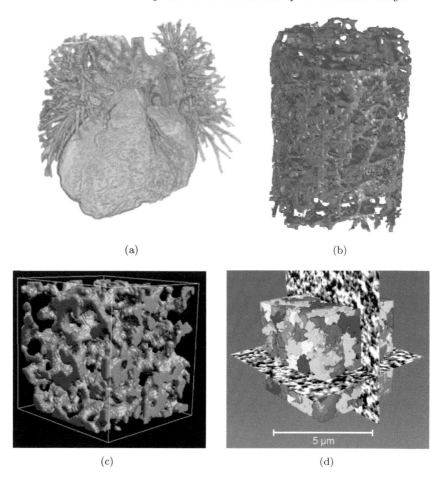

(a)   (b)

(c)   (d)

FIGURE 3.3: (a) A CT 64-slice image of the human heart visualized by volume rendering; (b) the micro-CT trabecular bone structure; (c) a micro-CT foam structure; and (d) a nano-CT polymer electrolyte fuel cell electrode image courtesy of Professor Shawn Litster at Carnegie Mellon University.

sity magnetic field. The field strengths can be 0.35 or even 1.5 Tesla, about 7,000~30,000 times of the Earth's magnetic field. Note that 1 Tesla = 10,000 Gauss and the Earth's magnetic field is about 0.5 Gauss only. This high intensity magnetic field forces the hydrogen atoms inside the body to align along the principal direction of the superconducting magnet. After that, low-level radio waves are transmitted through the patient in the microwave frequencies (15 to 60 MHz), which causes the magnetic moments of the hydrogen nuclei to resonate and re-emit microwaves. Then, a radiofrequency antenna is used to record the emitted microwaves, which are further filtered, amplified and reconstructed into tomographic slice MRI images.

While all the hydrogen nuclei resonate at a fixed frequency, *different tissues resonate longer than others*. This allows us to discriminate different tissues according to the magnitude of the signal. Figure 3.2(b) shows an MRI image of the human brain with a clear contrast of wrinkle boundaries. Unlike CT which always produces transaxial slices, the slices from MRI can be oriented in any plane because the orientations of the magnetic field can be easily adjusted, while the moving direction of the bed is fixed in CT machines.

In MRI images, the resulting intensity values at each pixel are not calibrated to any particular scale. Various values are obtained for different scan parameters, patient size, weights and magnetic characteristics. The inhomogeneity property of the magnetic field is a weakness of the MRI scanning technique. In an MRI image, pixels with different intensity values may belong to the same tissue, and pixels belonging to different tissues may have the same intensity value. Such inhomogeneity property makes image segmentation much more difficult for an MRI dataset.

Unlike X-ray-based scanning techniques, MRI intensity values have no physical analogs or no physical meaning. Whereas in CT, the intensity value reflects the material density. An MRI device measures the radio signals emitted by water drops over time. Due to this reason, skin and fat are generally much brighter than bone. Segmentation and classification are much more difficult for MRI data. In general, the sampling resolution for MRI images are still 256×256 or 512×512. The spatial resolution is mainly dependent on the strength of the gradient magnets and their ability to separate the slices along the gradient directions. Since MRI utilizes non-ionizing radio frequency (RF) signals to acquire its images, it has excellent soft-tissue contrast.

Like CT, MRI also has artifacts introduced by partial voluming, patient motion, and aliasing introduced by sampling and reconstruction. Instead of having any X-ray-related artifacts, MRI has its own host of radio and magnetic artifacts. The weight and size of the patient affect the material absorption of the radiofrequency energies. MRI images may have inconsistent quantitative intensity values due to the distribution of the antenna coverage for transmission and reception and also the inhomogeneity property of the magnetic field, leading to even inaccurate geometry.

Over years, there are a lot of new advances for the MRI technique. New capabilities in fast imaging have enabled us to study perfusion and diffusion of various agents across the body, such as drug delivery in coronary arteries. These studies highlight blood flow and at the same time help create high resolution images of vascular structures. Functional MRI (fMRI) has been developed to record the physiological or biological functions of the brain tissues, which has been used a lot in studying the cerebral cortex. Moreover, the diffusion tensor imaging (DTI) reveals the direction of nerve bundles deep inside the brain white matter and ventricular muscle fibers, requiring further new research on patient-specific tractography and anatomical analysis.

3.1.4 Nuclear Medicine: SPECT and PET

In nuclear medicine imaging, the clinician injects a radioactive source into the patient and then measures its distribution using a detector array to quantify the emitted radiation. In CT and MRI, the radiation source location is external and known, while in nuclear medicine, the source is inside the body and we do not know its distribution. In addition to the geometry of the patient anatomy, nuclear medicine images provide the physiological or biological activities of the patient. In nuclear medicine imaging, we cannot use high radiation doses due to their possible consequences to the patient. Compared to CT and MRI images, nuclear medicine yields images with lower resolution and more noise in general. There is always a trade-off between the desired image quality and the suitable radiation doses.

There are two common types of 3D nuclear medicine: single photon emission computed tomography (SPECT) and positron emission tomography (PET). SPECT utilizes radiotracers which emit photons while decaying. The radioactive agents have half-lives measured in hours and can be easily transported to the clinic. Differently, PET uses radioactive isotopes which decay by positron emission with the half-lives measured in minutes. They provide physiological or biological information of the human body in addition to the anatomical structures.

3.1.5 Ultrasound

Medical ultrasonography utilizes high frequency sound waves generally in 3 to 10 megahertz that are reflected by tissue to varying degrees to produce a 2D image on a TV monitor. Ultrasound studies the function of moving structures in real time, but provides less anatomical information than CT or MRI. It is very safe to use without any ionizing radiation, and also cheap and quick to perform. Due to this reason, ultrasound is being used popularly in checking patients, especially pregnant women.

In an ultrasound instrument, a transducer is used to create the acoustic signal, and it is also used to measure the returning echo. Partial reflections can be introduced by boundaries of regions with different acoustical impedance. Ultrasound is similar to the SONAR systems used in maritime and undersea naval imaging. During image acquisition, a probe is placed and moved around the region of interest. An ultrasound facility utilizes a linear array of transducers to produce a pie-shaped slice image. Figure 3.4(a) shows the ultrasound image of a fetus in the womb, viewed at 12 weeks of pregnancy.

3.1.6 Fluoroscopy

Fluoroscopy uses a constant of input X-rays to view the internal structure of objects and produces real-time images to track the object movement. Following Wilhelm Röntgen's discovery of X-rays in 1895, the first crude flu-

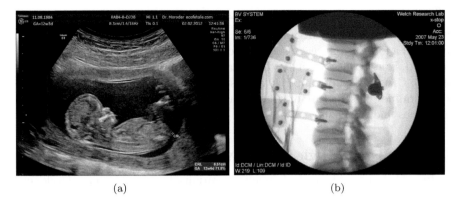

(a)   (b)

FIGURE 3.4: (a) Ultrasound image of a fetus in the womb courtesy of Dr. Wolfgang Moroder [7]; and (b) a fluoroscopy image of a lumbar spine treatment courtesy of Professor Boyle Cheng in Welch Research Lab, University of Pittsburgh.

oroscopes were created and used in experiments. In a simple fluoroscope, the patient is placed between an X-ray source and a fluorescent screen. Over the years, this scanning technique has been developed and improved. In many fluoroscopes the screen is now coupled with an X-ray image intensifier and charge-coupled device (CCD) video camera, which allows us to record the images and also play on a monitor. Due to the usage of X-rays, a form of ionizing radiation, we have to limit the radiation doses to the patient depending on the patient size and the procedure length. Again like CT, there is a trade-off between the desired image resolution and suitable X-ray doses.

Fluoroscopy has been used in many medical studies and treatments. For example, a sequence of fluoroscopy images was used in designing medical devices and surgical treatment of chronic pain in the lumbar spine [431]. As shown in Figure 3.4(b), a spinal implant device is tested at the $L2 \sim L3$ lumbar spine undergoing extension. With the aid of real-time fluoroscopy imaging, the physician can track and optimize the device design in order to release the chronic pain. Image segmentation techniques are needed to track the change of the cross-sectional foramina area.

3.1.7 Cryo-Electronic Microscopy (Cryo-EM)

Electron microscope (EM) is a type of microscope which utilizes electrons to illuminate a specimen and create an enlarged image. It is used to investigate the ultrastructure of biological and inorganic specimens such as microorganisms, cells, large molecules, metals and crystals. Cryo-electron microscopy (cryo-EM) is a form of EM with samples studied at cryogenic temperatures, generally liquid nitrogen temperatures.[2] It supports the observation of specimens in their native environment. The resolution of cryo-EM maps has been improved a lot over the years; some structures at even atomic resolution can be obtained. Due to this reason, cryo-EM has been popularly utilized in structural biology. Cryo-EM and computer reconstruction techniques are used together to determine the structure of subcellular complexes, such as rice dwarf virus, ribosomes and ion channels.

3.1.8 EBSD and High-Energy X-Rays

In addition to biomedical imaging, various techniques were developed to scan polycrystalline materials, such as electron backscatter diffraction (EBSD) and high-energy X-rays. As a microstructural-crystallographic technique, EBSD was developed to examine the crystallogrphic orientation or texture of polycrystalline materials [4]. An EBSD scanning machine generally consists of an EBSD detector with one or more phosphor screens, compact lens and low light CCD camera chip. During image acquisition, EBSD destroys the material. It scans one cross section of the material, cuts the materials to make another cross section, and then scans it. Differently, high-energy X-ray diffraction microscopy scans polycrystalline materials without physical damage to the material samples, and it can be used to study microstructure evolution or grain growth. High-energy X-rays generally have one order of magnitude higher than the traditional X-rays, and the energy level can reach 80 to 1000 keV. They can penetrate into matters deeply as a probe, permitting an in-air sample environment and operation. Figure 3.5 shows one EBSD image of the 92-grain beta titanium alloy and one high-energy X-ray image of nickel.

In summary, biomedical imaging can be used directly for clinical diagnosis. For example, CT images have been used in locating bone fracture and measuring the size via visualizing the images. Mammography has been used to screen early breast cancer by comparing a cancerous breast with a normal one, which has been estimated to reduce the breast cancer-related mortality by 20~30%. Ultrasound has no ionizing radiation, which makes it ideal and safe for checking the fetus growth in the womb for pregnant women.

As we discussed above, there are multiple imaging modalities developed with different strengths and weaknesses in imaging anatomy vs physiology and hard tissue vs soft tissue. This requires registration and alignment of different types of images. Differences in size and resolution of imaging data also need multiscale methods to couple information at various scales. In addition, many image acquisition techniques are not optimized for 3D representations. The straightforward method to improve visualization is to use better acquisition

---

[2]Under normal atmospheric pressure, nitrogen can exist as a liquid between the temperatures of 63K and 77.2K ($-346°$F and $-320.44°$F, or $-210°$C and $-195°$C). Below 63K, nitrogen freezes and becomes a solid. Above 77.2K, nitrogen boils and becomes a gas.

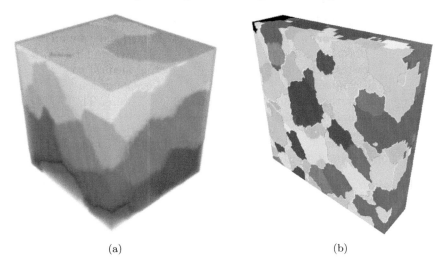

(a)   (b)

FIGURE 3.5: (a) Volume rendering of the 92-grain titanium alloy (picture from [306]), the EBSD data provided by Dr. Andrew Geltmacher from Navy Research Laboratory; and (b) the nickel model, the high-energy X-ray image data provided by Professor Robert Suter at Carnegie Mellon University [9].

techniques. During scanning, we always need to balance the image quality and acceptable doses. To have a deeper understanding of the clinical findings, we need to validate 3D visualization. In the following, we will discuss some image processing techniques to help us better understand the images using numerical computation, including basic operations, filtering, classification and segmentation, as well as registration.

3.2 Basic Operations

Image processing [68, 78, 283, 415] is the manipulation and analysis of digital image data to enhance, illuminate and extract useful information from the data stream, using numerical computation in general. In addition to medical image analysis, image processing has been applied in several other fields. In satellite surveillance analysis, it aligns multiple types of data, corresponds to a known map and highlights objects of interest. Basically, it is a screening and cartographic task. In computer vision, image processing needs to consider perspective geometry and photogrammetric distortions in analyzing camera views. In many systems, the primary tasks are autonomous navigation, target identification, threat avoidance and so on.

The main goal in biomedicine is to provide the clinician with powerful tools for image analysis and measurement. In medical image processing, the input data stream is in 3D generally. It needs both human and computers, and computers perform as assistants to us. Medical tasks include filtering, noise removal, contrast and feature enhancement; detecting medical conditions or events; as well as quantitative analysis of detected events or lesions, such as tumor volume or the size and location of a bone fracture. There are four categories of techniques in medical image processing:

1. Basic operations — operating the image using the intensity field information and its derivatives;

2. Filtering — filtering and preprocessing the data to remove noise before detection and analysis are performed;

3. Segmentation — partitioning an image into contiguous regions with cohesive properties; and

4. Registration — aligning multiple types of images to achieve a fusion of various information and better understanding of the patient.

A series of 2D slice images form a volumetric 3D image ϕ, which is a mapping from $\Re^3$ to $\Re^n$ with $n = 1$ for a scalar field. We have

$$\phi : \vec{x} \rightarrow \Re, \quad \vec{x} \subset \Re^3, \tag{3.1}$$

where $\vec{x}$ represents the domain of the volume. The image is often written as a function $\phi(x, y, z)$. Supposing F is a discrete sampling of a 2D image $\phi(x, y)$, we have

$$F_{i,j} = \phi(x_i, y_j). \tag{3.2}$$

Their partial derivatives can be obtained using the finite difference method. Supposing the grid spacing is h, we have

$$\phi_x(x_i, y_j) = \left. \frac{\partial \phi}{\partial x} \right|_{x_i, y_j} \approx \delta_x F_{i,j} \tag{3.3}$$

and

$$\phi_y(x_i, y_j) = \left. \frac{\partial \phi}{\partial y} \right|_{x_i, y_j} \approx \delta_y F_{i,j}, \tag{3.4}$$

where

$$\delta_x F_{i,j} = \frac{F_{i+1,j} - F_{i-1,j}}{x_{i+1} - x_{i-1}} = \frac{F_{i+1,j} - F_{i-1,j}}{2h} \tag{3.5}$$

and

$$\delta_y F_{i,j} = \frac{F_{i,j+1} - F_{i,j-1}}{y_{i+1} - y_{i-1}} = \frac{F_{i,j+1} - F_{i,j-1}}{2h}. \tag{3.6}$$

There are several basic operations [416] developed based on the given image domain and the derivatives. Among them, thresholding is properly the simplest point operation people use a lot on images. It ignores information about the pixel location, relying only on the pixel intensity. For example, thresholding is a binary decision made on each pixel without considering its neighbors. How to choose the optimal threshold value is the key issue here; generally people use the histogram of pixel intensity. Histogram equalization is another basic operation useful for some applications. It aims at optimizing the matching between the image and the display system, ensuring equally represented pixel intensity at each level.

By designing a stretching function or a windowing function, the brightness is adjusted and the contrast can be enhanced. Based on localized contrast manipulations, the contrast enhancement algorithm can be adaptive, multiscale, weighted localization and anisotropic. In adaptive contrast enhancement [419], a new intensity is assigned to each pixel according to an adaptive transfer function based on the local statistics such as local minimum, maximum and average. Sometimes we need to handle color images, which contain red, green and blue values at each pixel. Color maps are often used to overlay additional information on anatomical images, for example, to overlay the physiological or biological activity from SPECT on an anatomical image of the brain (MRI).

In the following sections, we will review some commonly used filtering, segmentation and registration schemes in image processing.

3.3 Filtering

Filtering is an important technique in denoising the images, including linear and nonlinear filtering. Nonlinear filtering has a better feature-preserving property than linear filtering. We will discuss both of them as follows.

3.3.1 Linear Filtering

Linear filtering [416] generally considers pixels in a linear and space-invariant manner. There are two standard techniques for filtering: convolution and Fourier transform.

Convolution. Convolution is often used to make measurements, detect features, smooth noises, and deconvolve acquisition effects like deblur the known optical artifacts from a telescope. Given the input image $\phi(x)$ and filter kernel $h(x)$, we have

$$\phi(x) \otimes h(x) = \int_{-\infty}^{\infty} \phi(x - \tau)h(\tau)d\tau. \tag{3.7}$$

The left-hand-side expression $\phi(x) \otimes h(x)$ describes an image mapping, and we have

$$\phi(x) \otimes h(x) = (\phi \otimes h)(x). \tag{3.8}$$

Similarly, the above 1D convolution can be extended to 2D and 3D. We have

$$\phi(x, y) \otimes h(x, y) = \int_{-\infty}^{\infty} \int_{-\infty}^{\infty} \phi(x - \tau, y - \nu) h(\tau, \nu) d\nu d\tau, \tag{3.9}$$

and

$$\phi(x, y, z) \otimes h(x, y, z) = \int_{-\infty}^{\infty} \int_{-\infty}^{\infty} \int_{-\infty}^{\infty} \phi(x - \tau, y - \nu, z - \omega) h(\tau, \nu, \omega) d\omega d\nu d\tau. \tag{3.10}$$

Considering the Gaussian function as a smoothing filter kernel, $g(x, \sigma) = \frac{1}{\sqrt{2\phi}\sigma} e^{-\frac{x^2}{2\sigma^2}}$, convolution can be used to denoise the images. σ is a parameter controlling the width and the height of the Gaussian kernel function. When σ becomes larger, more high frequency noise is removed, but edges and boundaries are also blurred.

The convolution operation has four important properties, including linear, commutative, associate and distributive over addition. Given three images (p, q, r) and two scalars (α, β), we have

$$(\alpha p + \beta q) \otimes r = \alpha(p \otimes r) + \beta(q \otimes r); \tag{3.11}$$

$$p \otimes q = q \otimes p; \tag{3.12}$$

$$(p \otimes q) \otimes r = p \otimes (q \otimes r); \tag{3.13}$$

and

$$p \otimes (q + r) = p \otimes q + p \otimes r. \tag{3.14}$$

In a continuous domain, differentiation can be denoted as convolution, so we have

$$\frac{\partial}{\partial x} \phi = \frac{\partial}{\partial x} \otimes \phi. \tag{3.15}$$

Differentiation enhances high frequency noise, so it is necessary to smooth the image before computing its derivative

$$h(x) \otimes \frac{\partial}{\partial x} \phi = \frac{\partial}{\partial x} h(x) \otimes \phi(x). \tag{3.16}$$

Convolution of discrete data is similar to that of the continuous form. We only need to substitute the integration with discrete summation. A discrete convolution of image P with kernel Q can be written as

$$P_{x,y} \otimes Q_{x,y} = \sum_i \sum_j P_{x-i, y-j} Q_{i,j}. \tag{3.17}$$

Convolution of discrete data is also linear, commutative, associative and distributive over addition. It can be used to smooth data and detect boundaries.

Boundaries in an image generally have a high gradient magnitude, i.e., the intensity value increases or decreases rapidly across the boundary. The gradient vector $\left[\frac{\partial \phi}{\partial x}, \frac{\partial \phi}{\partial y}\right]$ is along the steepest change direction of the image intensity. The gradient magnitude is a scalar representing a measure of the boundary strength. The second derivative is a matrix, called the Jacobian

$$
\begin{pmatrix}
\frac{\partial^2 \phi}{\partial x^2} & \frac{\partial^2 \phi}{\partial x \partial y} \\
\frac{\partial^2 \phi}{\partial y \partial x} & \frac{\partial^2 \phi}{\partial y^2}
\end{pmatrix} .
$$

From the Jacobian matrix, we define the *Laplacian* $\frac{\partial^2 \phi}{\partial x^2} + \frac{\partial^2 \phi}{\partial y^2}$. A boundary exists where the gradient is at a maximum on the boundary or the Laplacian is zero, no net change. $\frac{\partial^2 \phi}{\partial y \partial x}$ represents a decrease in the y-component of the gradient as we move along the x-direction. Similarly, $\frac{\partial^2 \phi}{\partial x \partial y}$ represents a decrease in the x-component of the gradient as we move along the y-direction.

Furthermore, the difference of Gaussians (DOG) kernel can detect edges without considering orientation. Given a 2D image $\phi(x, y)$ and two Gaussian filter kernels $g(x, y, \sigma_1)$ and $g(x, y, \sigma_2)$, where $\sigma_1 < \sigma_2$, a linear combination of these two filters yields a new image $\phi'(x, y)$ which can enhance edges. We have

$$
\phi'(x, y) = \phi(x, y) \otimes (g(x, y, \sigma_1) - g(x, y, \sigma_2)). \tag{3.18}
$$

The behavior can be modified from edge detection to contrast enhancement by using a different linear combination, for example, $2g(x, y, \sigma_1) - g(x, y, \sigma_2)$.

Fourier Transform. The Fourier transform decomposes an image into component sinusoidal spatial functions, and it maps the spatial or image domain to the frequency domain. Given an image $\phi(x)$, its Fourier transform Φ can be written as

$$
F(\phi(x)) = \int_{-\infty}^{\infty} \phi(x) e^{-i2\phi\omega x} dx = \Phi(\omega). \tag{3.19}
$$

The inverse Fourier transform is

$$
F^{-1}(\Phi(\omega)) = \int_{-\infty}^{\infty} \Phi(\omega) e^{i2\phi\omega x} d\omega = \phi(x). \tag{3.20}
$$

Fourier transform has two key properties: (1) given an image $\phi(x)$ and a scalar α, scaling in image space is equivalent to inverse scaling in frequency space, or $F(\phi(\alpha x)) = \frac{1}{\alpha}\Phi(\frac{\omega}{\alpha})$; and (2) given a kernel $h(x)$ and its Fourier transform $H(\omega)$, the convolution theory holds such that convolution in image space is equivalent to multiplication in frequency space. In other words, we have

$$
F(\phi(x) \otimes h(x)) = \Phi(\omega) H(\omega) \tag{3.21}
$$

and

$$\phi(x) \otimes h(x) = F^{-1}(F(\phi(x))F(h(x))). \tag{3.22}$$

In image processing, generally there are four important transform pairs: comb function, box filter, pyramid filter and Gaussian filter. A *comb function* is a series of pulse or delta-function, and a pulse is an infinitely narrow, infinitely high spike with finite area. Multiplying $f(x)$ by the comb function captures equally distributed samples to produce a digital image. We also know that multiplying $f(x)$ with the comb function in the image space is equivalent to convolving their Fourier transform. In the frequency space, the Fourier transform $F(f(x))$ is duplicated by the comb function at each impulse. If $F(f(x))$ is broader than the impulse spacing in the frequency space, we may have aliasing. Then we have a *Nyquist frequency*: the maximum acceptable frequency to avoid aliasing is half of the sampling frequency. A *box filter* is a reconstruction function in nearest-neighbor interpolation; it aims at resampling a discrete image to a different lattice. The *pyramid filter* is generally used for linear interpolation. It is more accurate than the box filter, but more expensive. The *Gaussian filter* has several nice features: the Fourier transform of a Gaussian is itself a Gaussian, and it has no side lobes. It is generally used to blur images.

Fourier transform can be extended to 2D and 3D. Given a 2D image $\phi(x, y)$, its Fourier transform and inverse Fourier transform can be written as

$$F(\phi(x,y)) = \int_{-\infty}^{\infty} \int_{-\infty}^{\infty} \phi(x,y)e^{-i2\phi(\omega x + \nu y)}dxdy = \Phi(\omega, \nu) \tag{3.23}$$

and

$$F^{-1}(\Phi(\omega, \nu)) = \int_{-\infty}^{\infty} \int_{-\infty}^{\infty} \Phi(\omega, \nu)e^{i2\phi(\omega x + \nu y)}d\omega d\nu = \phi(x, y). \tag{3.24}$$

The multidimensional Fourier transform is separable in each of the orthonormal basis dimensions. Eq. (3.23) can be rewritten as

$$F(\phi(x,y)) = \int_{-\infty}^{\infty} \left[\int_{-\infty}^{\infty} \phi(x,y)e^{-i2\phi\omega x}dx \right] e^{-i2\phi\nu y}dy = \Phi(\omega, \nu). \tag{3.25}$$

Following the similar manner, if we can separate the function $\phi(x, y)$ into two component functions $\phi(x, y) = \phi_1(x)\phi_2(y)$, then we can also separate the Fourier transform. We have $\Phi(\omega, \nu) = \Phi_1(\omega)\Phi_2(\nu)$, where $\Phi_1(\omega) = F(\phi_1(x))$ and $\Phi_2(\omega) = F(\phi_2(x))$. As we know that the 2D Gaussian function is separable, we have $g(x, y) = g_1(x)g_2(y)$. Its Fourier transform can be represented as $F(g(x, y)) = G(\omega, \nu) = G_1(\omega)G_2(\nu)$, where $G_1(\omega) = F(g_1(x))$ and $G_2(\omega) = F(g_2(x))$.

3.3.2 Nonlinear Filtering

Over the years, there have been many nonlinear filtering methods developed [403]. Most of them were built upon one of the following three strategies:

1. Heuristics - a set of rules are designed to operate pixels for some special purposes based on heuristic experiences;

2. Statistics - operations are designed based on statistics of the image (like the local minimum, maximum, average) and also some properties from stochastic processes; and

3. Partial differential equations (PDEs) or energy-based methods - the image is operated by solving PDEs or by minimizing a pre-defined energy function.

In the following, we will discuss a popular nonlinear filtering, Gaussian blurring. *Gaussian blurring* is a kind of low-pass transform, which performs as a filter in the frequency domain and reduces high frequency noises. Generally, the noise exists in an image with more energy in high frequencies than in the low frequency signal. An image $\phi(x, y)$ is a mapping from an image domain, usually a small square or rectangle, to a scalar field or the intensity. The Gaussian is a low-pass transform

$$g(x, y) = \frac{1}{\sqrt{2\phi\sigma^2}} e^{\frac{x^2+y^2}{2\sigma^2}}$$

and

$$G(u, v) = F(g) = e^{\frac{(x^2+y^2)\sigma^2}{2}},$$

where F represents the Fourier transform, and σ is the standard deviation or the effective width of the kernel function.

The Gaussian kernel has two nice properties: smooth and self-similar, which means that repeatedly applying Gaussian filtering can be equivalently achieved by applying a single Gaussian filter. The diffusion equation (initial value problem) can be written as

$$\frac{\partial \phi}{\partial t} = \frac{\partial^2 \phi}{\partial x^2} + \frac{\partial^2 \phi}{\partial y^2} = div(c\nabla\phi), \tag{3.26}$$

where $\phi(x, y, t)$ represents a sequence of images over time starting with $\phi(x, y, 0) = I(x, y)$ and evolving to become smoother. Here we choose $c = 1$ (isotropic diffusion). In the frequency domain, we have $\frac{\partial F}{\partial t} = -(u^2 + v^2)F$. The solution is an exponential of the form

$$F(u, v, t) = e^{-(u^2+v^2)t} F(u, v, 0) \tag{3.27}$$

and

$$\phi(x, y, t) = \frac{1}{\sqrt{4\pi t}} e^{-(x^2+y^2)/4t} \otimes f(x, y, 0). \tag{3.28}$$

We can find that its solution is equivalent to convolving the initial condition ($t=0$) with a Gaussian kernel function when $\sigma = \sqrt{2t}$.

Gaussian blurring can remove noise, but it also reduces high frequency components of the image (or features such as sharp edges) at the same time. To preserve features, we choose c as a function of $|\nabla\phi|$. $c(|\nabla\phi|)$ controls the different rates of flow along different directions, and it can be determined by the local curvature estimation. For example,

$$c(|\nabla\phi|) = \frac{1}{1 + |\nabla\phi|^2/\lambda^2},$$

where λ is an input parameter. This can give us an anisotropic filtering [38, 294]. Anisotropic filtering can also be used together with bilateral filtering [183] (a combination of domain and range filtering) to get better denoising results.

Figure 3.6 shows the filtering results of a simple foot image using a linear and nonlinear filtering. It is obvious that linear filtering removes noise and also blurs the boundary, while nonlinear filtering performs better in terms of boundary preservation. We also have a homework problem (Problem 3.4) at the end of this chapter asking students to implement simple linear and nonlinear filtering and compare them.

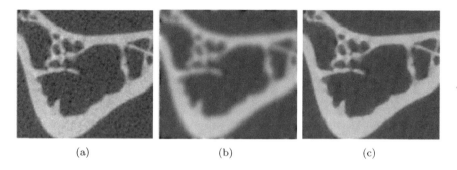

(a)   (b)   (c)

FIGURE 3.6: (a) A foot image with noise; (b) the linear filtering result; and (c) the nonlinear filtering result.

3.4   Segmentation

Segmentation [415] is an important technique in image processing to partition a digital image into multiple regions or sets of pixels based on some characteristics. Pixels in one region share similar characteristics and pixels in adjacent regions are significantly different from them. Segmentation is typically used to locate regions of interest and their boundaries in images. In particular, medical image segmentation is mainly used for locating tumors and other pathologies, measuring tissue volumes for computer-guided surgery, diagnosis and treatment planning, as well as anatomical structure study in medical school education.

In principle, there is no general solution that works for all the image segmentation problems simultaneously. We have to combine segmentation techniques with domain knowledge of a particular area like medicine in order to effectively and practically solve an image segmentation problem. The main difficult issue in segmentation is how to define and identify the region boundaries. Across the region boundaries or edges, there is often a sharp change in intensity. The gradient vector and its magnitude of the image intensity are used to detect boundaries.

3.4.1 Various Segmentation Methods

There have been many methods developed for image segmentation in the literature [80, 288]. The *watershed segmentation* [415] is a method people use a lot, which builds a series of "catchment basins" to represent the local minima of some pre-defined image metrics. The obtained watersheds are then hierarchically organized to build a graph, which connects all the individual watershed regions. The graph hierarchy can later be used to navigate and explore the entire image. In this method, the watershed transformation considers the gradient magnitude of an image as a topographic surface. Pixels with the highest gradient magnitude intensities correspond to the watershed lines or the region boundaries. Water inside a region flows downhill to a common local intensity minima, and pixels associated with a common minimum form the watershed region.

Statistical pattern recognition [415] is a technique to treat the input images as a statistical sampling of the human anatomical structure. Thresholding [27, 373] is a very common and straightforward approach used for segmentation where an image is represented as groups of pixels with values greater than or equal to the threshold and values less than the threshold. Thresholding is particularly useful for X-ray CT data where the intensity values have an intuitive mapping to the material physical density. More complex, multivalued data, such as MRI or color cryosections, require sophisticated statistical techniques. The segmentation using statistical pattern recognition can effectively partition the image into structures of interest. However, the objects are undersegmented. For example, all the bone components appear together as a single segment. Sometimes we need to take another manual editing segmentation to separate and articulate the bones individually. In contrast, *region merging* combines several small regions together considering spatial proximity, connectedness and similarity of intensity values. A *connected-thresholding or propagation approach* helps recursively aggregate all connected voxels within a user-defined intensity value range.

The *edge detection method* [262, 335] partitions an image by finding the pixels on the region boundaries or edges. However, these methods usually do not consider the spatial details and may introduce leakage of the region boundary for images with low contrast and large intensity variation. The *active surface or front evolution* method attempts to connect the margin voxels that circumscribe the regions of interest based on statistical measures of similarity. Some of them use energy-minimizing models to balance external and internal forces attempting to promote smoothness [269]. This method can easily enforce the expected geometric properties on the segmented boundaries, while some detailed features may be ignored. By using this active surface or level set method, we can apply differential geometry to the evolving front and limit the overall local curvature to produce a smooth object boundary.

The *clustering techniques* [70, 163, 230, 279] partition the image into several clusters of pixels with similar characteristics in the feature space or the maximum homogeneity. K-means [163, 195] and fuzzy c-means (FCM) [18, 84, 95] are two most popular clustering algorithms. The *k-means algorithm* partitions all pixels or voxels into K clusters, minimizing a pre-defined distance between the data points and the cluster centers. The *FCM techniques* utilize the fuzzy idea and allow an object to belong to multiple clusters simultaneously. A membership degree is also defined to measure the association between an object and its clusters. Many FCM algorithms ignore the spatial information, and they are sensitive to noise and image artifacts sometimes. Pedrycz and Waletzky [292] optimized the segmentation procedures by including the available classified information. The objective function can also be revised to make the FCM algorithm more robust to noisy images [18].

The *region-based methods* use the homogeneity of inner regions in segmentation, but they may fail in capturing the global boundary and shape properties of the object. For example, sometimes noisy boundaries and even holes can be created inside the object. *Markov random field (MRF) models* [85, 231] use the pixel correlation within a local neighborhood to decide the object region. Unlike the typical region growing methods, MRF makes use of more information of the pixel neighborhood. In [61, 134, 348], the *graph-based methods* were developed to partition the image with high efficiency especially for homogeneous regions. *Seeded region growing methods* start with a set of input seed points marking each to-be-segmented object. All pixels are allocated iteratively and the regions gradually grow until a stopping criterion is reached. Figure 3.7 shows a human brain with a tumor and the segmentation results of one slice image [420].

The *level set method* [86, 229, 290, 378, 402] is typically a partial differential equation-based variational method, which utilizes a signed function to represent the evolving contour. We can derive a similar flow for the implicit surface based on the motion equation of the contour. Level-set models are topologically flexible, but they are quite expensive in both computational time and memory. A surface in the input image ϕ is represented by an isosurface implicitly, and we have

$$S = \{\vec{x} \mid \phi(\vec{x}) = c\} \tag{3.29}$$

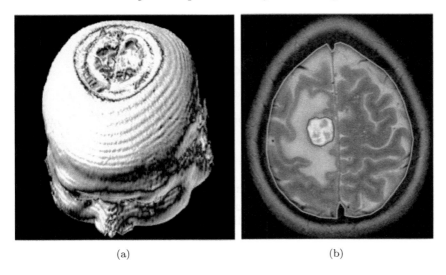

(a)   (b)

FIGURE 3.7: (a) Volume rendering of an MRI image of the human brain with a tumor (the red region); and (b) both the brain (blue) and the tumor (red) are segmented on one slice image.

where c is the isovalue. The isosurface normal can be calculated using the normalized gradient vector. For any point in the volumetric domain of the image, the normal is represented as

$$\vec{n}(\vec{x}) = \frac{\nabla \phi(\vec{x})}{|\nabla \phi(\vec{x})|}. \tag{3.30}$$

The normal is generally set to be positive if it points outward; negative otherwise. On discrete grids, we can easily compute the gradient vectors using the finite different methods like the central difference. For noisy images, we can also compute the first derivatives using convolution with a Gaussian as we discussed before. In the level set methods, we increase the dimension and represent surfaces with volumes. Like layers of an onion, a single volume generally contains a series of nested isosurfaces, but none of them can intersect with each other because a specific point in the domain can only have one unique intensity value.

The isosurface curvature can be computed using the first and second derivatives of the embedding image $\phi(\vec{x})$. The first derivatives give the normal vector $\vec{n}(\vec{x})$, which can be written in the matrix format

$$N = - \begin{bmatrix} \vec{n}_x & \vec{n}_y & \vec{n}_z \end{bmatrix}. \tag{3.31}$$

The projection of these derivatives onto the tangent plane of the isosurface defines the shape matrix β. Supposing P is the normal projection operator, we have

$$P = \vec{n} \otimes \vec{n} = \frac{1}{||\nabla\phi||^2} \begin{pmatrix} \phi_x^2 & \phi_x\phi_y & \phi_x\phi_z \\ \phi_y\phi_z & \phi_y^2 & \phi_y\phi_z \\ \phi_z\phi_z & \phi_z\phi_y & \phi_z^2 \end{pmatrix}. \tag{3.32}$$

We can then obtain the tangential projection operator $T = I - P$, and also the shape matrix

$$\beta = NT = TD^2\phi T, \tag{3.33}$$

where $D^2\phi$ is the Hessian of ϕ. From the shape matrix β, we can calculate its three real eigenvalues, $e_1 = \kappa_1$, $e_2 = \kappa_2$ and $e_3 = 0$. Their corresponding eigenvectors give the two principal directions on the tangent plane and also the normal direction. We can then obtain the mean curvature $\frac{1}{2}(\kappa_1 + \kappa_2)$, the Gaussian curvature $\kappa_1\kappa_2$ and also the total curvature or the deviation from flatness $D = \sqrt{\kappa_1^2 + \kappa_2^2}$, which is the Euclidean norm of the shape matrix β.

Deformable surfaces can be defined by an evolution differential equation on a volume. In geometric modeling, a surface is represented as a two-parameter object in the 3D space, which is a mapping

$$\vec{S}: \begin{matrix} V & \times & V \\ r & & s \end{matrix} \mapsto \begin{matrix} \Re^3. \\ x, y, z \end{matrix} \tag{3.34}$$

A deformable surface exhibits an evolution motion over time t, and we have

$$\vec{S} = \vec{S}(r, s, t). \tag{3.35}$$

The surface normal can be written as

$$\vec{N} = \vec{N}(r, s, t). \tag{3.36}$$

Then, we can obtain an evolution equation to describe the local deformation of the surface,

$$\frac{\partial\vec{S}}{\partial t} = \vec{G}(\vec{S}, \vec{S}_r, \vec{S}_s, \vec{S}_{rr}, \vec{S}_{rs}, \vec{S}_{ss}, \cdots). \tag{3.37}$$

There are many different ways to design the evolution equation. For example, the surface could move by adding a directional "forcing" function; expand or contract with a varying speed; or move along some directions determined by surface shape like curvature or force equilibrium. Surface deformations are obtained by solving the evolution partial differential equation on the 3D grids. For a point $\vec{x}$ on the deforming surface, we denote its movement as $\vec{v}$. There are two ways to represent the surface movements implicitly: static and dynamic. A single static image $\phi(\vec{x})$ contains a series of level sets, each of which corresponds to the deforming surface at one time t. We have

$$\phi(\vec{x}(t)) = k(t) \tag{3.38}$$

and then we can obtain a boundary value problem

$$\nabla\phi(\vec{x}) \cdot \vec{v} = \frac{dk(t)}{dt}. \tag{3.39}$$

Since a surface cannot pass back over time, motions in this static model must be strictly inward or outward monotonically. Differently, the dynamic approach is more flexible, which uses a one-parameter function changing over time.

$$\phi\left(\vec{x}(t), t\right) = k, \tag{3.40}$$

where k is a constant. Then we have

$$\frac{\partial\phi}{\partial t} = -\nabla\phi \cdot \vec{v}. \tag{3.41}$$

The dynamic approach is an initial value problem, which can handle forward and backward movements over time, and is computationally more expensive than the static approach. In the level set methods, the movement can be computed by solving a general form of the partial differential equation,

$$\frac{\partial\phi}{\partial t} = -\nabla \cdot \vec{v} = -\nabla\phi \cdot \vec{F}\left(\vec{x}, D\phi, D\phi^2, \cdots\right). \tag{3.42}$$

Level set methods support flexible topology change and can accommodate complex objects naturally. Over the evolution, one object can split into several smaller objects, and multiple objects can also merge together to form a bigger one. Since the second derivative information can be used to find edges, it is also used here to design a smoothing term for the evolution equation. For example,

$$\phi_t = \alpha\nabla|\nabla I| \cdot \nabla\phi + \beta|\nabla\phi|E + \gamma|\nabla\phi|\left(\epsilon - |I - T|\right), \tag{3.43}$$

where I is the input image, E is an energy function and the last term on the right is a speed term. T controls the brightness of the to-be-segmented target region, and ϵ is the threshold around T controlling expansion and contraction. Note that to ensure the second derivatives of edges can pull the surface along the desired direction, we require the initial model is close to the desired edges.

Deformable models are dynamic models. The surface evolution process can be modeled as the minimization of an energy function, which is a summation of the internal energy, the image energy and the user-defined energy. The internal energy helps keep the continuity and regularity of the surface which makes the model robust to noise; the image energy pushes the deformable model to edge features; and the user-defined energy speeds up the deformation and tries to avoid local minima. For example in the active contour models or snakes [197], an active contour is defined by a set of points with the position vector $p(s)$ and the displacement vector $d(s)$, where s is the arc-length. The energy function is in the form of

$$E = \int \left[E_{in}(d) + E_{im}(p+d) + E_u(p+d) \right] ds, \tag{3.44}$$

where E_{in}, E_{im} and E_u represent the internal, image and user-defined energy, respectively.

In the following, let us take a look at one segmentation technique based on centroid Voronoi tesselation [173].

3.4.2 Segmentation Based on Centroid Voronoi Tesselation

The centroidal Voronoi tessellation (CVT) has been applied broadly in many research fields, including image processing, data analysis, computational geometry and numerical partial differential equations [119, 121, 188]. The CVT is a special kind of Voronoi tesselation, where a generator is also the centroid or the mass center of the associated Voronoi cell. CVT-based algorithms in its simplest form are actually the k-means clustering method [195]. Du *et al.* [120] was one of the first who proposed the classic CVT algorithms and applied them on image segmentation and compression applications. The classic CVT method turns out to be sensitive to noise and may fail in segmenting noisy images. By adding an edge energy term, an edge-weighted centroidal Voronoi tessellation (EWCVT) was developed [383, 384, 385], which has been proven to be effective and efficient for image segmentation. In addition, the energy term can be replaced by a local variation energy to handle images with background inhomogeneity [384]. The EWCVT has recently been used in segmenting 3D superalloy volumes [74, 75]. However, all the above developments are still sensitive to initializations and noise, leading to inaccurate and unstable segmentation results sometimes.

In [173, 174], a new advanced model was developed by introducing the harmonic idea into the clustering energy function, namely, harmonic edge-weighted centroidal Voronoi tessellation (HEWCVT). Compared with the classic CVT and EWCVT methods, this new method has two main advantages:

- The HEWCVT-based segmentation algorithm is more stable and less sensitive to the initializations due to the soft membership of the data points; and

- The segmentation accuracy is improved by integrating the spatial information of local image features into the harmonic energy function, which compensates the effect of noise.

In the following, let us first review CVT and EWCVT schemes, and then talk about details of the HEWCVT.

Given an image $\phi(x, y)$ with the size of $M \times N$, $\Phi = \{\phi_i\}_{i=1}^{M \times N}$ denotes all the intensity values in the image. Let $C = \{c_l\}_{l=1}^{L}$ denote a set of typical colors for a color image or intensity values for a gray-scale image. The Voronoi region V_k in Φ for the level c_k $(k = 1, \ldots, L)$ can be defined as

$$V_k = \{\phi_i \in \Phi : dist\,(\phi_i, c_k) \le dist\,(\phi_i, c_l) \quad for \;\; l = 1, \cdots, L\}, \qquad (3.45)$$

where $dist\,(\phi_i, c_k)$ is a predefined distance metric between ϕ_i and c_k. Note that instead of using the physical distances between pixels, here we measure the distance in the color or intensity space. The set $V = \{V_l\}_{l=1}^{L}$ is called a *Voronoi tessellation* of the image Φ, and the set of selected color values $C = \{c_l\}_{l=1}^{L}$ is called the *Voronoi generators*.

Given a set of generators $C = \{c_l\}_{l=1}^{L}$ and a partition $U = \{U_l\}_{l=1}^{L}$ of Φ, the clustering energy $E(C; U)$ in the classic CVT [120] can be written as

$$E\,(C; U) = \sum_{l=1}^{L} \sum_{\phi_i \in U_l} |\phi_i - c_l|^2. \qquad (3.46)$$

The construction of CVTs can be viewed as a process of energy minimization. As a follow-up, EWCVT [383] adds an edge-related energy term to the CVT energy function. In the physical space, the entire image gives us an index set $D = \{P(i) : i = 1, \cdots, M \times N\}$. For the partitioned region U_l, we have the corresponding index set in the physical space, $D_l = \{P(i) : \phi_i \in U_l\}$. For each pixel $P(i)$, we denote its local neighborhood with the radius of ω as $N_\omega(P(i))$. Suppose $|N_\omega(P(i))|$ represents the number of pixels within $N_\omega(P(i))$ and $n_k(P(i))$ is the number of pixels within $D_k \cap N_\omega(P(i)) \backslash P(i)$, we then define $\tilde{n}_k(P(i)) = |N_\omega(P(i))| - n_k(P(i)) - 1$, which is the edge energy term representing the number of pixels within $N_\omega(P(i)) \backslash (D_k \cup P(i))$. Then, we have the edge-weighted CVT energy

$$E_{EW}\,(C; U) = \sum_{l=1}^{L} \sum_{\phi_i \in U_l} |\phi_i - c_l|^2 + 2\lambda \tilde{n}_l\,(P(i)), \qquad (3.47)$$

where λ is a positive weighting factor to balance the clustering energy and the edge energy.

Both the classic CVT and EWCVT are sensitive to the initializations and noise, leading to inaccurate and unstable segmentation results sometimes especially for noisy images. To resolve these issues, in [173] we introduce the harmonic ideas and define the HEWCVT energy

$$E_H(C; U) = \sum_{i=1}^{M \times N} \left(L \Big/ \sum_{l=1}^{L} dist^{-2}\,(\phi_i, c_l) \right), \qquad (3.48)$$

where

$$dist\,(\phi_i, c_k) = \sqrt{|\phi_i - c_k|^2 + 2\lambda \tilde{n}_k\,(P(i))}. \qquad (3.49)$$

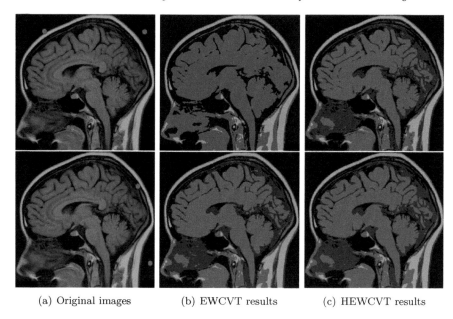

(a) Original images   (b) EWCVT results   (c) HEWCVT results

FIGURE 3.8: Segmentation results of a human brain MRI image from the BrainWeb. (a) The original images with two sets of different initializations, and each initialization contains four input generators (red dots); (b) segmentation results using EWCVT; and (c) segmentation results using HEWCVT.

HEWCVT is in fact a generalized EWCVT method, and it can be degenerated to HEWCVT and EWCVT conveniently. The construction of HEWCVT is again an energy minimization process, where generators are updated by minimizing the corresponding objective functions.

Figure 3.8 shows segmentation results of a human brain image with four clusters. For the same input image, we segment it twice with different initializations using both EWCVT and HEWCVT. From the results, we can see that CVT-based algorithms are essentially an energy minimization process. For each test, the energy function is minimized until it is converged. The EWCVT method is sensitive to the initialization. With different initializations, the results can be obviously different; see Figure 3.8(b). The energy functions also converge to different values. Compared to EWCVT, HEWCVT is more stable and insensitive to initializations. For different initializations, the HEWCVT energy functions can always converge to similar optimal results, as shown in Figure 3.8(c). In addition, HEWCVT can improve the accuracy of the segmentation by capturing more features. HEWCVT usually needs less iterations to converge to optimal results than EWCVT.

Remark. HEWCVT usually yields more accurate segmentation results than EWCVT due to the adoption of the harmonic mean. EWCVT imposes a hard membership on each data point, which only has an influence over its assigned centre. Differently in HEWCVT, all centroids partially influence the harmonic average for each data point. Such a soft membership property makes the segmentation process robust to the uncertainty or noise. Therefore compared to EWCVT, HEWCVT is less sensitive to the initializations, especially when we have a large number of clusters. In addition, the HEWCVT algorithm can be extended to generate adaptive superpixels and preserve local image features efficiently [174].

3.5   Registration

Image registration is the process of aligning or matching two images scanned at different time or using different modalities. Basically, it is to find the geometrical transform and map points from one image to another. Given a static or reference image $S(x, y)$ and a moving or target image $M(x, y)$, mathematically this can be written as

$$Determine \;\; G \;\; and \;\; L \;\; such \;\; that \;\; S(x, y) = G\left(M(L(x, y))\right), \qquad (3.50)$$

where G is the global transformation, and L represents the local transformation.

Medical image registration has many applications. For example, patients are scanned during treatment to obtain time series information that captures disease development, treatment progress and contrast bolus propagation. Aligning these time-varying images is critical in many situations. Correlating information from different image modalities is another application field for registration. For example, MRI has a better soft tissue discrimination for lesion identification, while CT provides clear bone localization useful for surgical guidance. They can be registered to combine these information together. Nuclear medicine, like PET and SPECT, provides useful physiological information for detecting abnormalities such as tumors, while CT/MRI provided anatomical structure. Therefore, PET-MRI alignment can be used to study brain tumors, and PET-CT alignment can be used to investigate radiation treatment planning.

3.5.1 Various Registration Methods

According to what criteria are used in the algorithm, there are three different kinds of image registration: landmark-based, segmentation-based and intensity-based registration. In the *landmark-based registration*, some features like points, lines, corners and crossings are marked first by the user, and then these features are used to help minimize the distance between physical points.

In *segmentation-based registration*, segmented binary structures like curves, surfaces or volumes are aligned rigidly or deformably. Differently, *intensity-based registration* minimizes an objective function which measures the similarity of two image intensities. This is an optimization method and generally it is very expensive due to its usage of full image information.

For 2D/2D or 3D/3D image registration, the spatial transformation between two images can be rigid, affine or deformable. In *rigid registration*, only rotation and translation are considered. In *affine registration*, skew and scaling are also considered in addition to rotation and translation. The most flexible non-rigid registration is also called *deformable registration*, which are free-form mappings with a regularization constraint to limit the allowable solution space. The registration process is actually a parameter optimization. We minimize a designed objective function to obtain the optimal spatial transformation parameters.

When we map a point from one image to another, the mapped point generally does not fall on the grids. An image interpolation method is needed to compute the intensity value at the mapped point. The simplest way is the linear interpolation; some people also use B-spline or other interpolation. Given a moving image $M(x)$ and a static or target image $S(x)$, a transformation $T(p)$ maps points between them for a given set of parameters p. A metric $R(p|S, M, T)$ is defined to represent the similarity between $S(x)$ and the transformed $M(x)$.

A point x in the static image $S(x)$ is mapped onto a point x' in the moving image $M(x)$. We then have

$$x' = T(x|p) \tag{3.51}$$

where

$$T : \Re^n \mapsto \Re^m \quad such \; that \quad T(S(x)) = M(x). \tag{3.52}$$

Then we can obtain a rigid transformation in 2D,

$$x' = \begin{bmatrix} x' \\ y' \end{bmatrix} = \begin{bmatrix} cos\theta & -sin\theta \\ sin\theta & cos\theta \end{bmatrix} \cdot \begin{bmatrix} x \\ y \end{bmatrix} + \begin{bmatrix} t_x \\ t_y \end{bmatrix} \tag{3.53}$$

or

$$x' = T(x|p) = T(x, y|t_x, t_y, \theta), \tag{3.54}$$

where θ is the rotation angle, and t_x and t_y are displacements along the x and y directions, respectively.

When we map a pixel from the static image onto the moving image, we need to handle three different transformations in a sequence. First, we need to map each pixel in the static image onto the static image space via a mapping T_f. Then we use the mapping T to map from the static image space to the moving image space. Finally we need to map the result from the moving image space to the discrete grid of the moving image using T_m.

Image registration aims at finding transformtion parameters p that optimize the pre-defined image similarity metric. Function derivatives are needed in many registration optimizers with respect to p. By using the chain rule, we can obtain

$$\frac{\partial R(p|S, M, T)}{\partial p_i} = \sum_j \frac{\partial R(p|S, M, T)}{\partial x'_j} \cdot \frac{\partial x'_j}{\partial p_i}, \tag{3.55}$$

where the matrix $\left[\frac{\partial x'_j}{\partial p_i}\right]$ is the *Jacobian* of the transformation. Given a transformation T mapping a point $x = \{x_1, x_2, \cdots, x_n\}$ into another point $x' = \{x'_1, x'_2, \cdots, x'_n\}$, the Jacobian matrix is defined as

$$J = \begin{bmatrix} \frac{\partial x'_1}{\partial x_1} & \frac{\partial x'_1}{\partial x_2} & \cdots & \frac{\partial x'_1}{\partial x_n} \\ \frac{\partial x'_2}{\partial x_1} & \frac{\partial x'_2}{\partial x_2} & \cdots & \frac{\partial x'_2}{\partial x_n} \\ \vdots & & & \\ \frac{\partial x'_n}{\partial x_1} & \frac{\partial x'_n}{\partial x_2} & \cdots & \frac{\partial x'_n}{\partial x_n} \end{bmatrix}. \tag{3.56}$$

The determinant $|J|$ is called the *Jacobian* of the transformation. In image registration, most times we use the following Jacobian:

$$J = \begin{bmatrix} \frac{\partial x'_1}{\partial p_1} & \frac{\partial x'_1}{\partial p_2} & \cdots & \frac{\partial x'_1}{\partial p_m} \\ \frac{\partial x'_2}{\partial p_1} & \frac{\partial x'_2}{\partial p_2} & \cdots & \frac{\partial x'_2}{\partial p_m} \\ \vdots & & & \\ \frac{\partial x'_m}{\partial p_1} & \frac{\partial x'_m}{\partial p_2} & \cdots & \frac{\partial x'_m}{\partial p_m} \end{bmatrix}, \tag{3.57}$$

where p_i are the parameters of the transformation. The transformation mapping is a function of the input point x_i and the transformation parameters p_i, so we have

$$x'_j = T(x_1, x_2, \cdots, x_n, p_1, p_2, \cdots, p_m). \tag{3.58}$$

Suppose the rotation center is C, and an affine transformation includes translation, rotation and scaling. It can be written as

$$x' = \begin{bmatrix} x' \\ y' \end{bmatrix} = D \begin{bmatrix} cos\theta & -sin\theta \\ sin\theta & cos\theta \end{bmatrix} \cdot \begin{bmatrix} x - C_x \\ y - C_y \end{bmatrix} + \begin{bmatrix} C_x \\ C_y \end{bmatrix} + \begin{bmatrix} T_x \\ T_y \end{bmatrix}. \tag{3.59}$$

Affine registration also includes the shear operation. Differential affine motion [293] and Fourier-based matching [26] are two techniques people developed to model affine registration. The former defines a quadratic error function and then minimizes the error function to obtain the optimal affine parameters. The latter is suitable for large deformation with different image contrasts.

Nonrigid or deformable registration allows more flexible deformation other than rigid and affine registration. It constructs a minimization problem with an image-matching function and a regularizing constraint. The matching function can be squared intensity difference, cross-correlation or mutual information. The regularizing constraint can be the elasticity theory, optical flow, fluid dynamics or thin-plate spline theory. For example, given a static or reference image $S(x)$ and a moving or target image $M(x)$, to find a displacement field $u(x)$ such that $S(x - u) = M(x)$, we define an L^2 energy function

$$E_{L^2}(S, M, u) = \frac{1}{2} \int_\Omega |S(x - u) - M(x)|^2 dx. \tag{3.60}$$

The energy function is then minimized to obtain

$$\frac{\partial u(x, t)}{\partial t} = -f(x, u(x, t)), \tag{3.61}$$

where the body force is

$$f(x, u(x, t)) = -[S(x - u) - M(x)]\nabla S|_{x-u}. \tag{3.62}$$

As reviewed in [65, 263, 446], there are four kinds of deformable registration techniques developed for medical image data: elastic registration, level set method, optical flow method, and diffusion-based registration. *Elastic registration* [40, 107] is actually a force equilibrium problem of elastic plates, with external forces applied to stretch the image while internal forces minimize bending and stretching via stiffness or smoothness constraints. We have

$$\mu\Delta u + (\mu + \nu)\nabla(\nabla \cdot u) = f(x, u), \tag{3.63}$$

where μ and ν are Lamé constants. This method assumes having small rotation angles and linear transformations only, therefore it is not suitable for large, nonlinear deformations. The performance on localized deformation has been improved to handle large deformations using viscous fluid models [94]. The Eulerian velocity field V is nonlinear in u, so we have

$$V = \frac{Du}{Dt} = \frac{\partial u}{\partial t} + V \cdot \nabla u. \tag{3.64}$$

The evolving reference image $T(x)$ embeds in a viscous fluid governed by the Navier-Stokes equation. Based on viscous fluid registration, a new energy function was designed with Jacobian maps introduced to achieve a much more stable registration [414]. The *level set method* is a numerical technique that can easily track shapes with topology change [284] and also combine segmentation together with registration [118, 273]. In particular, multiscale, multigrid and multilevel methods [98] as well as edge-matching techniques [275] were developed in matching 2D and 3D images. For example, various multiscale solvers were developed to minimize a pre-defined energy function [182, 221]. As a suitable method for deformation in temporal sequences of images [41, 167], the *optical flow method* assumes that the corresponding intensity values in two images are the same and estimates the deformation as an image velocity or displacement. The *diffusion-based registration* considers the contours and other features in an image as membranes, and another image as deformable grids with geometrical constraints [25, 371].

There are several open source software packages available online. Insight Segmentation and Registration Toolkit (ITK) [13] is an open-source and cross-platform system for image analysis, provided by National Library of Medicine. Automated Image Registration (AIR) [2] supports automatic registration of 2D and 3D images; its source codes in C are available on its website. FLIRT (FMRIB's Linear Image Registration Tool) [5] is an automatic, robust and accurate image registration tool supporting linear or affine brain registration.

In the following, we will first take the optical flow registration method as an example to describe in detail how we use registration in two applications: cardiac modeling [434] and dynamic lung modeling with tumor tracking [433]. Then, we will also talk about multiscale methods.

3.5.2 Optical Flow Method with Applications

The optical flow method is also a fast mono-modality nonrigid registration technique. Given an image $I(x)$ with the intensity values preserved over time, we have

$$I(x, t) = I(x + v\delta t, t + \delta t), \qquad (3.65)$$

where v is the velocity vector or optical flow, and δt is a small perturbation over time t. By expanding the right-hand side of the equation in Taylor series and ignoring the second-order and higher terms, we can obtain the optical flow constraint,

$$C_{op} = I_x \cdot v + I_t = 0. \qquad (3.66)$$

It means that for the corresponding points of two images, we assume they retain the same original intensity value. The registration process minimizes a global variational energy,

$$E(v) = \int_\Omega \left(C_{op}^2 + |v_x|^2 \right) d\Omega = \int_\Omega F(x, v, v_x) d\Omega, \qquad (3.67)$$

where $v_x = \nabla v$ and $C_{op} = I_x \cdot v + I_t$. Substituting v with $v' = v + \epsilon w$, we can obtain

$$\frac{dE(v')}{d\epsilon} = \int_\Omega \frac{d}{d\epsilon} F(x, v', v_x') d\Omega. \qquad (3.68)$$

By using the chain rule, we can derive

$$
\begin{aligned}
\frac{dF(x, v', v'_x)}{d\epsilon} &= \frac{\partial F}{\partial v'}\frac{dv'}{d\epsilon} + \frac{\partial F}{\partial v'_x}\frac{dv'_x}{d\epsilon} = \frac{\partial F}{\partial v}w + \frac{\partial F}{\partial v_x}\frac{d}{d\epsilon}\left(\frac{dv'}{dx}\right) \\
&= \frac{\partial F}{\partial v}w + \frac{\partial F}{\partial v_x}\frac{d}{dx}\left(\frac{dv'}{d\epsilon}\right) = \frac{\partial F}{\partial v}w + \frac{\partial F}{\partial v_x}\frac{dw}{dx}. \quad (3.69)
\end{aligned}
$$

The second term on the right in the result can be derived further by checking

$$
\frac{d}{dx}\left(\frac{\partial F}{\partial v_x}w\right) = \frac{d}{dx}\left(\frac{\partial F}{\partial v_x}\right)w + \frac{\partial F}{\partial v_x}\frac{dw}{dx}, \quad (3.70)
$$

and we obtain

$$
\frac{\partial F}{\partial v_x}\frac{dw}{dx} = \frac{d}{dx}\left(\frac{\partial F}{\partial v_x}w\right) - \frac{d}{dx}\left(\frac{\partial F}{\partial v_x}\right)w. \quad (3.71)
$$

After applying the boundary condition $\frac{\partial F}{\partial v_x}\big|_{\partial\Omega} = 0$, we have

$$
\frac{dE(v)}{d\epsilon} = \int_\Omega \left(\frac{\partial F}{\partial v} - \frac{d}{dx}\frac{\partial F}{\partial v_x}\right)wd\Omega = 0, \quad (3.72)
$$

where $F(x, v, v_x) = C_{op}^2 + |v_x|^2$ and $C_{op} = I_x \cdot v + I_t$. Plugging them into the above equation, we obtain the governing equation

$$
I_x(I_x \cdot v + I_t) - v_{xx} = 0, \quad (3.73)
$$

which can be solved using various numerical methods like the finite difference method and finite element method.

Cardiac Image Registration. An edge-enhanced optical flow algorithm [434] has been developed based on Thirion's diffusing model, also known as the "demons" algorithm [312, 371, 381], for cardiac image registration applications. Given a static image $S(x)$ and a moving image $M(x)$, we define a passive force $\vec{f}_p$ and an active force $\vec{f}_a$. For any grid point in the image, the "demons" force $\vec{f}_d$ can be computed as a summation of these two forces:

$$
\vec{f}_d = \vec{f}_p + \vec{f}_a = (M - S) \times \left(\frac{\vec{\nabla}S}{|\vec{\nabla}S|^2 + (S - M)^2} + \frac{\vec{\nabla}M}{|\vec{\nabla}M|^2 + (S - M)^2}\right). \quad (3.74)
$$

To accurately capture the boundary edge of an isocontour, we use the intensity information at the contour boundary to define a demon function $k(M)$ for each contour point P in S, as shown in Figure 3.9. We have

$$
k(M) = \begin{cases} -1 & M < S_{in} \\ \frac{(M - S_{in}) + (M - S_{out})}{S_{out} - S_{in}} & S_{in} \leq M \leq S_{out} \\ 1 & M > S_{out} \end{cases} \quad (3.75)
$$

where $S_{in} = S(P - \vec{n}_S)$ and $S_{out} = S(P + \vec{n}_S)$. Based on this demon function, we define an edge-enhanced "demons" force $\vec{f}_e$ to improve the alignment along the boundaries

$$\vec{f}_e = k(M)(\vec{n}_S + \vec{n}_M), \qquad (3.76)$$

where $\vec{n}_S$ and $\vec{n}_M$ are the normal vectors at the contour point P in S and M, respectively. The "demons" force $\vec{f}_d$ and the edge-enhanced "demons" force $\vec{f}_e$ are coupled together to align images in our deformable registration algorithm.

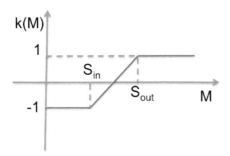

FIGURE 3.9: The demon function $k(M)$ to enhance the boundary edge of an isocontour. S_{in} and S_{out} are defined using the intensity information at the boundary.

We have tested our registration algorithm on two 2D cardiac images from different patients; see Figure 3.10. For these two images, the intensity range and the object position and orientation are different. To address these problems, we introduced a global transformation and a scaling factor before applying the deformable registration. The edge-enhanced demons force helps a lot in matching isocontour boundaries, while there are still some boundaries that are hard to align due to their blur definition and large deformation. In addition, we also applied smoothing operations after each registration. In [434], we built a cubic Hermite model for the whole heart structure as at atlas, then we applied the deformable registration to deform the atlas and match with the new patient images. In addition, deformable image registration can also be applied to diffusion tensor reorientation [21, 296]. Given the static image $S(x)$, the moving image $M(x)$, and the diffusion tensor field for $S(x)$, we can obtain the reoriented diffusion tensor field for $M(x)$ using our optical flow approach [434].

Dynamic Lung Modeling and Tumor Tracking. According to the statistics [352], lung cancer is one of the leading causes of cancer death in the United States - about one in every three cancer deaths. Generally there are three kinds of treatment: radiation therapy, laser therapy and cryosurgery. Radiation therapy exposes cancer cells to high-energy X-rays and damages their DNA structures and genetic functions. In this way, these cancer cells lose

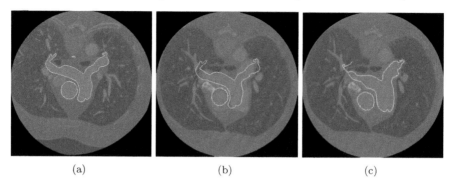

(a)   (b)   (c)

FIGURE 3.10: Image registration between two 2D images. (a) Isocontour (pink) extracted from Image 1; (b) the original isocontour (blue) overlaid with Image 2 before registration; (c) the deformed isocontour (green) overlaid with Image 2 after registration.

the ability to continue growing or dividing. Laser therapy uses high-intensity light to increase the local temperature and kills the cancer cells. Cryotherapy decreases the local temperature to destroy abnormal or diseased tissue. Due to the organ motion during respiration, it is critical to track the tumor motion, which can help us decrease the treatment margin and minimize the normal tissue exposure. In this way, we can kill the cancer cells as much as possible while minimizing the damage to the surrounding healthy tissues. Therefore, dynamic lung modeling and tumor tracking are very critical for the treatment planning. Deformable image registration techniques are needed in tracking dose deposited in the target tumor volume over time.

We have developed a systematic computational framework for dynamic lung modeling and tumor tracking using an optical flow registration coupled with geometric smoothing [433]; see the pipeline in Figure 3.11. Given a set of 4D CT images with 10 phases, we segment the lung and tumor slice by slice from one stable exhale phase like Phase 5, and then reconstruct the surface triangular model for them. Noise generally exists in the constructed 3D surface models, therefore we utilize geometric flows (or geometric partial differential equations) [429, 441] to smooth and improve the surface mesh. Figure 3.12 shows one constructed lung model with tumor. After that, we embed the surface model within volumetric grids using the signed distance function technique, which calculates the shortest distance from each grid point to the surface and also assigns positive or negative signs to grids inside and outside the boundary. From the volumetric grid data, we use our octree-based mesh generation methods [427, 428] to generate surface and volumetric meshes for the lung and the tumor. Both triangular and tetrahedral meshes are then used in the following deformable registration for dynamic lung modeling and tumor tracking. Finally, the tracking results can be used to optimize the probe or beam angle and dose distribution in the treatment.

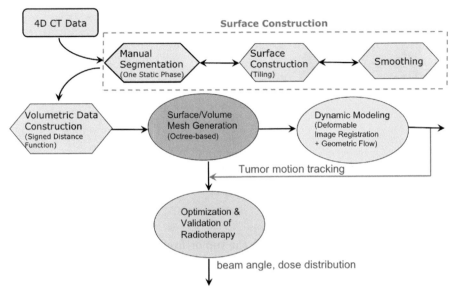

FIGURE 3.11: Pipeline of dynamic lung modeling and tumor tracking for treatment plan optimization.

In CT images, the intensity values directly reflect the tissue density. Here we choose an intensity-based algorithm for radiotherapy application [381]. Our registration method works for both triangular and tetrahedral meshes; we deform the Phase 5 image and match it with images at other phases. Based on an optical flow method, it is also known as the "demons" algorithm [371]. To obtain smooth geometry during image registration, we include a geometric smoothing procedure into our registration besides a Gaussian filter on the image domain. Here, we minimize a pre-defined energy function [433],

$$\mathcal{E}(\mathbf{x}) = \int_{\Omega} \left(g(\mathbf{x}(u,v,w)) - 1 \right)^2 du dv dw, \qquad (3.77)$$

where $\Omega = [0,1]^3$, $\mathbf{x}(u,v,w)$ is the position vector of any grid point in the given CT image. The $g(x)$ term includes the local geometry information,

$$g(\mathbf{x}) = g_{uu}g_{vv}g_{ww} + 2g_{uv}g_{vw}g_{uw} - (g_{uw}^2 g_{vv} + g_{vw}^2 g_{uu} + g_{uv}^2 g_{ww}) \qquad (3.78)$$

where

$$g_{uu} = \mathbf{x}_u^T \mathbf{x}_u, \qquad g_{uv} = \mathbf{x}_u^T \mathbf{x}_v, \qquad g_{uw} = \mathbf{x}_u^T \mathbf{x}_w,$$
$$g_{vv} = \mathbf{x}_v^T \mathbf{x}_v, \qquad g_{vw} = \mathbf{x}_v^T \mathbf{x}_w, \qquad g_{ww} = \mathbf{x}_w^T \mathbf{x}_w.$$

An L^2-gradient flow is then constructed by minimizing the energy function $\mathcal{E}(x)$. This geometric smoothing technique is coupled with the optical flow registration.

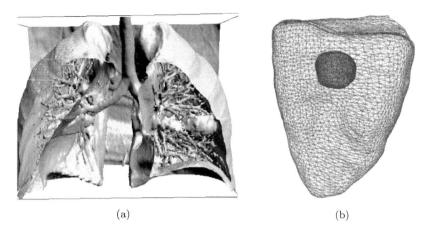

(a)   (b)

FIGURE 3.12: (a) The lung and the tumor from the CT image (Phase 5) with many other unnecessary objects; and (b) the constructed tetrahedral mesh of the lung with the tumor (only surface triangular mesh is visualized).

We have applied our algorithm to 2D lung images [433]. For example, we took the same coronal cross section at images of two phases. In addition, we applied our algorithm to the tetrahedral lung mesh and calculated the volume at each time phase. We generated the tetrahedral mesh directly from Phase 5 image data. The registration results yield a good match with medical data, with the volume change reaching the maximum at Phase 0 through the 10 phases of one cycle. Furthermore, we applied our algorithm to the triangle surface mesh and calculated the maximum displacement of the left lung, the right lung and the tumor during respiration. The tumor motion is significant during one breath cycle, which can even reach 1 to 2 centimeters. The moving trajectory of the tumor will be used to control the movement of the radiation beams and dose distribution, and thus to optimize the radiation therapy for image-guided lung cancer treatment planning.

3.5.3 Multiscale Methods

Due to the multiresolution nature of the background grids, features in the given images can be aligned at multiple scales [135, 182, 221, 343]. Given the static image $S(x)$ and the moving image $M(x)$, cubic B-spline basis functions are generally used to build the spatial transformation $f(x)$ between them. An energy function is defined to minimize the difference between $S(x)$ and $M(x)$, and so we have

$$
\begin{aligned}
E(x) \;=\; & \int_{\Omega} \left(S(x) - M(f(x)) \right)^2 d\Omega \\
+ \; & \lambda_1 \int_{\Omega} \left(\|f_{,u}(x)\|^2 + \|f_{,v}(x)\|^2 \right) d\Omega \\
+ \; & \lambda_2 \int_{\Omega} \left(\|f_{,u}(x)\|^2 \|f_{,v}(x)\|^2 - (< f_{,u}(x), f_{,v}(x) >)^2 \right) d\Omega, \quad (3.79)
\end{aligned}
$$

where λ_1 and λ_2 are two weighting factors. $f_{,u}(x)$ and $f_{,v}(x)$ denote the derivatives with respect to two parametric directions u and v, and $< \cdot, \cdot >$ represents the inner product operator. The right-hand side of this equation is basically a summation of the fidelity term and two regularization terms.

In finite element-based methods [135, 221, 343], the energy function is minimized to derive an L^2-gradient flow. The spatial domain is then discretized using the finite element method, and the temporal domain is discretized using an explicit Euler scheme. Alternatively, the energy function can be handled directly to derive its differential formula, with a dynamic multilevel scheme developed to improve the algorithm efficiency [182]. In this method, the gradient of the deforming image is used to design a new coefficient factor for the fidelity term to accelerate the evolution process.

Figure 3.13 shows the registration results for a pair of brain images. The picture in (c) shows the initial difference between the reference image and the target image. The matching effect can be observed from the evolving results in (d-h), and also from the difference between the deformed result and the given target image as shown in (i).

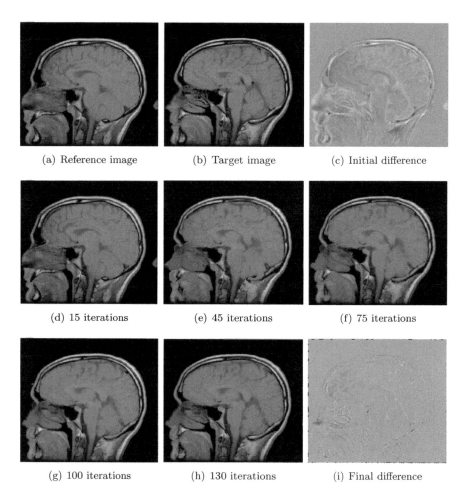

FIGURE 3.13: Brain image registration results. Registering the target image (b) from the reference image (a). (d-h) are the registration results after 15, 45, 75, 100 and 130 iterations, respectively. (c) shows the initial difference between (a) and (b), and (i) shows the difference between (h) and (b).

Homework

Problem 3.1. CT and MRI are two important scanning techniques in medical imaging. Please summarize the pros and cons of these two different scanning techniques in terms of

1. Scanning mechanism;

2. Homogeneity vs inhomogeneity;

3. Slice orientation (the z-direction);

4. Artifacts; and

5. Facility price.

Problem 3.2. Answer the following questions briefly:

1. In CT bone is much brighter than soft tissue, while in MRI skin and fat are brighter than bone. Explain why.

2. Given a cadaver (very dry), you are asked to scan one knee joint structure. Which method would you like to choose? Explain why MRI cannot be used?

Problem 3.3. Answer the following questions briefly:

1. Contrast enhancement, filtering, segmentation and registration are four important image processing techniques. For each of them, please describe what it is used for.

2. In image processing, boundary detection is the most critical step. Given a 2D image $\phi(x, y)$, how do you detect boundaries of different regions?

Problem 3.4. Implement the following linear and nonlinear filters using the finite different method to smooth the image $\phi(x, y)$, and apply your code to the given 2D image. Please output your results in .pgm format and visualize them using IrfanView or other software.

1. Implement the following linear filtering:

$$\partial_t \phi - \nabla^2 \phi = 0,$$

and generate two results using different iterations.

2. Implement the following nonlinear filtering

$$\partial_t \phi - div(g(|\nabla \phi|) \nabla \phi) = 0,$$

where

$$g(|\nabla\phi|) = \frac{1}{1 + |\nabla\phi|^2/\lambda^2},$$

λ is an input parameter. Generate two results using different iterations and λ values (for example, 1 and 10), compare and discuss your results.

Please summarize your results and show all the .pgm files in images (including the input foot.pgm for comparison).

Problem 3.5. Conduct a research survey on image segmentation and registration. Pick one method from the literature, implement and test it on some 2D images.

Appendix: pgm format description

P2	– pgm file, gray-scale image
# Created by IrfanView	– comment
149 136	– dimensions, dimX, dimY
255	– the max intensity value, the range is [0, 255]
41 28 41 66 85 66 57 66 66	– the intensity value at each grid point,
...	(dimX*dimY) values

Chapter 4

Fundamentals to Geometric Modeling and Meshing

In this chapter, fundamentals to geometric modeling and mesh generation are first introduced, followed by geometric objects and transformations, curves and surfaces, as well as spline-based modeling. Since isocontouring is important for image visualization and modeling, we introduce two important isocontouring methods: marching cubes and dual contouring. These two methods will also be used in the next two chapters for mesh generation.

4.1   Fundamentals

Let us first review the fundamentals to the representation and manipulation of sets of basic geometric elements such as points, line segments, polygons and polyhedra.

4.1.1 Three Spaces

There are three commonly used spaces: vector space, affine space and Euclidean space. The linear *vector space* consists of only two objects, scalars and vectors. The *affine space* includes an additional element, the point, to the vector space. The *Euclidean space* further adds distance to the affine space. In the following, let us review the definition and properties of scalars and these three spaces in detail.

Scalars. A scalar field consists of ordinary real numbers and their operations. Let S represent a set of elements or scalars, like α, β and γ. We have $S = \{\alpha, \beta, \gamma, \cdots\}$. Two fundamental operations are defined to operate these scalars: *addition* and *multiplication*. $\forall \alpha, \beta \in S$, we have

$$\textit{Addition} \quad : \quad \alpha + \beta \in S, \tag{4.1}$$

$$\textit{Multiplication} \quad : \quad \alpha \cdot \beta \in S. \tag{4.2}$$

These operations have three important properties: *commutative, associative* and *distributive* properties. $\forall \alpha, \beta, \gamma \in S$, we have

$$\textit{Commutative}: \quad \alpha + \beta \quad = \quad \beta + \alpha, \tag{4.3}$$

$$\alpha \cdot \beta \quad = \quad \beta \cdot \alpha, \tag{4.4}$$

$$\textit{Associative}: \quad \alpha + (\beta + \gamma) \quad = \quad (\alpha + \beta) + \gamma, \tag{4.5}$$

$$\alpha \cdot (\beta \cdot \gamma) \quad = \quad (\alpha \cdot \beta) \cdot \gamma, \tag{4.6}$$

$$\textit{Distributive}: \quad \alpha \cdot (\beta + \gamma) \quad = \quad (\alpha \cdot \beta) + (\alpha \cdot \gamma). \tag{4.7}$$

Generally with no ambiguity, we use $\alpha\beta \Leftrightarrow \alpha \cdot \beta$. There are two special scalars: *the additive inverse* (0) and *the multiplicative inverse* (1). $\forall \alpha \in S$, we have

$$\alpha + 0 \quad = \quad 0 + \alpha = \alpha, \tag{4.8}$$

$$\alpha \cdot 1 \quad = \quad 1 \cdot \alpha = \alpha. \tag{4.9}$$

It is obvious that each element has an additive inverse $(-\alpha)$ and a multiplicative inverse (α^{-1}). We have

$$\alpha + (-\alpha) \quad = \quad 0, \tag{4.10}$$

$$\alpha \cdot \alpha^{-1} \quad = \quad 1. \tag{4.11}$$

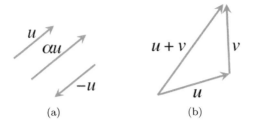

(a)   (b)

FIGURE 4.1: (a) Examples of vectors; and (b) vector-vector addition.

The real numbers together with their ordinary addition and multiplication operations form a scalar field.

Vector Spaces. In addition to scalars, a *vector space* contains a second type of entity called vectors. *Vectors* have two basic operations, vector-vector addition and scalar-vector multiplication. *Vector-vector addition* has three properties: *closed*, *commutative* and *associate*. Let V represent a vector space. $\forall u, v, w \in V$, we have

$$Closed \quad : \quad u + v \in V, \tag{4.12}$$
$$Commutative \quad : \quad u + v = v + u, \tag{4.13}$$
$$Associative \quad : \quad u + (v + w) = (u + v) + w. \tag{4.14}$$

Figure 4.1 shows some examples of vectors and vector-vector addition. 0 is a special vector defined such that $\forall u \in V$, $u + 0 = u$. Each vector u has an additive inverse $-u$ such that $u + (-u) = 0$. *Scalar-vector multiplication* is another operation defined such that, for any scalar α and any vector u, αu is a vector in V. The scalar-vector operation has the *distributive* property:

$$distribution: \quad \alpha(u + v) \quad = \quad \alpha u + \alpha v, \tag{4.15}$$
$$(\alpha + \beta)u \quad = \quad \alpha u + \beta u. \tag{4.16}$$

A n-tuple of scalars (or real or complex numbers) can be written as $v = (v_1, v_2, \cdots, v_n)$. Vector-vector addition and scalar-vector multiplication are defined as

$$u + v \quad = \quad (u_1, u_2, \cdots, u_n) + (v_1, v_2, \cdots, v_n)$$
$$= \quad (u_1 + v_1, u_2 + v_2, \cdots, u_n + v_n), \tag{4.17}$$
$$\alpha v \quad = \quad (\alpha v_1, \alpha v_2, \cdots, \alpha v_n). \tag{4.18}$$

This n-dimensional vector space is denoted as $\Re^n$. We can use matrix algebra to manipulate vector operations.

Given n vectors $u_1, u_2, \cdots, u_n$, we can write a linear combination of them as

$$u = \alpha_1 u_1 + \alpha_2 u_2 + \cdots + \alpha_n u_n, \tag{4.19}$$

where $\alpha_1, \alpha_2, \cdots, \alpha_n$ are all scalars. If the only set of scalars satisfying

$$\alpha_1 u_1 + \alpha_2 u_2 + \cdots + \alpha_n u_n = 0$$

is $\alpha_1 = \alpha_2 = \cdots = \alpha_n = 0$, then the vectors are *linearly independent* and they form a basis of a space, where n is the dimension of the space.

If $\{v_1, v_2, \cdots, v_n\}$ is a basis for V, any vector v in this space can be expressed uniquely using these basis vectors. We have

$$v = \beta_1 v_1 + \beta_2 v_2 + \cdots + \beta_n v_n, \tag{4.20}$$

where $\beta_1, \beta_2, \cdots, \beta_n$ are scalars. If $\{v_1', v_2', \cdots, v_n'\}$ is another basis for V, there is a representation of v in terms of the basis vectors

$$v = \beta_1' v_1' + \beta_2' v_2' + \cdots + \beta_n' v_n'. \tag{4.21}$$

There exists an $n \times n$ transformation matrix M such that

$$\begin{bmatrix} \beta_1' \\ \beta_2' \\ \vdots \\ \beta_n' \end{bmatrix} = M \begin{bmatrix} \beta_1 \\ \beta_2 \\ \vdots \\ \beta_n \end{bmatrix}. \tag{4.22}$$

Note that this matrix changes representations from one set of basis vectors to another through a simple linear transformation with only scalar operations.

Affine Spaces. A vector space lacks geometric concepts such as *location* and *distance*; see Figure 4.2. Vectors have only magnitude and direction; they do not have a position. A vector can be represented using a set of basis vectors that defines a coordinate system. An *affine space* adds the third type of entity

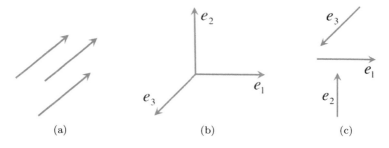

(a)   (b)   (c)

FIGURE 4.2: Affine spaces. (a) Identical vectors; (b) basis vectors located at the origin; and (c) arbitrary placement of basis vectors.

to a vector space, *points* like P, Q, R, $\cdots$. *Point-point subtraction* gives a vector in V; we have $v = P - Q$. For each pair of v and Q, we can find a P such that $P = v + Q$, which defines a *vector-point addition* operation. We can then obtain a consequence of the *head-to-tail axiom*: given three points P, Q and R, we have $(P - Q) + (Q - R) = P - R$ as shown in Figure 4.3(a).

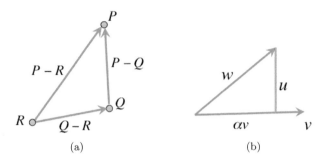

(a)   (b)

FIGURE 4.3: (a) Head-to-tail axiom of vectors; and (b) projection between two vectors.

A frame contains a point P_0 and a set of vectors $\{v_1, v_2, \cdots, v_n\}$, which defines a basis for the vector space. Given a frame, an arbitrary vector in this space can be written uniquely as

$$v = \alpha_1 v_1 + \alpha_2 v_2 + \cdots + \alpha_n v_n, \tag{4.23}$$

and an arbitrary point can be represented as

$$P = P_0 + \beta_1 v_1 + \beta_2 v_2 + \cdots + \beta_n v_n. \tag{4.24}$$

The two sets of scalars, $\{\alpha_1, \cdots, \alpha_n\}$ and $\{\beta_1, \cdots, \beta_n\}$, provide the representations of the vector and the point. P_0 can be considered as the origin of the frame which performs as a reference, and all points are defined from this origin point.

Euclidean Spaces. Given scalars $\{\alpha, \beta, \gamma, \cdots\}$ and vectors $\{u, v, w, \cdots\}$, the *Euclidean space* introduces a new operation, the *inner or dot product*, which produces a real number by combining two vectors. We have

$$\text{Commutative} \quad : \quad u \cdot v = v \cdot u, \tag{4.25}$$
$$\text{Associative} \quad : \quad (\alpha u + \beta v) \cdot w = \alpha u \cdot w + \beta v \cdot w. \tag{4.26}$$

We also have $v \cdot v > 0$ if $v \neq 0$, and otherwise $v \cdot v = 0$. Two different vectors u and v are defined as *orthogonal* if $u \cdot v = 0$. The magnitude or the length of a vector v is usually computed as $|v| = \sqrt{v \cdot v}$. For any two points P and Q, we can construct a vector $P - Q$ and we have

$$|P - Q| = \sqrt{(P - Q) \cdot (P - Q)}. \tag{4.27}$$

The angle between two vectors u and v is defined as

$$cos\theta = \frac{u \cdot v}{|u||v|} = \begin{cases} 0 & u \perp v, \\ 1 & u \parallel v. \end{cases} \tag{4.28}$$

The concept of *projection* is introduced by finding the shortest distance from a point to a line or a plane. Given two vectors v and w, we can divide one vector w into two components, one parallel and one orthogonal to the other vector v. As shown in Figure 4.3(b), we have

$$\begin{aligned} w &= \alpha v + u, \\ u \cdot v &= 0, \\ w \cdot v &= (\alpha v + u) \cdot v = \alpha v \cdot v + u \cdot v = \alpha v \cdot v. \end{aligned}$$

Therefore, we can obtain

$$\alpha = \frac{w \cdot v}{v \cdot v}.$$

The vector αv is the projection of w onto v, and we can also obtain the orthogonal component

$$u = w - \frac{w \cdot v}{v \cdot v} v. \tag{4.29}$$

Given a set of basis vectors $\{a_1, a_2, \cdots, a_n\}$ in an n-dimensional space, we can use the *Gram–Schmidt orthogonalization* scheme to build another set of *orthonormal* basis $\{b_1, b_2, \cdots, b_n\}$, in which each vector has a unit length and is orthogonal to each other. We have

$$b_i \cdot b_j = \begin{cases} 1 & if \ i = j, \\ 0 & otherwise. \end{cases} \tag{4.30}$$

The Gram–Schmidt orthogonalization scheme is processed iteratively. We aim at finding a vector $b_2 = a_2 + \alpha b_1$ which is orthogonal to b_1 when choosing a proper α value. We have $b_2 \cdot b_1 = 0 = a_2 \cdot b_1 + \alpha b_1 \cdot b_1$, therefore

$$\alpha = -\frac{a_2 \cdot b_1}{b_1 \cdot b_1} = -a_2 \cdot b_1$$

and

$$b_2 = a_2 - \frac{a_2 \cdot b_1}{b_1 \cdot b_1} b_1 = a_2 - (a_2 \cdot b_1) b_1.$$

We have constructed the orthogonal vector by removing the parallel component, which is also the projection of a_2 onto b_1. The general iterative step can be written as

$$b_k = a_k + \sum_{i=1}^{k-1} a_i b_i,$$

which is orthogonal to $\{b_1, b_2, \cdots, b_{k-1}\}$. There are a total of $k-1$ orthogonal conditions

$$\alpha_i = -\frac{a_k \cdot b_i}{b_i \cdot b_i}.$$

During iterations, each vector should be normalized by replacing b_i by $b_i/|b_i|$.

The *cross-product* is another important operator for vectors. Given two nonparallel vectors u and v in a 3D space, the cross-product gives a new vector w which is orthogonal to both u and v, and we have $w \cdot u = w \cdot v = 0$. In a coordinate system, if u has components $\{\alpha_1, \alpha_2, \alpha_3\}$ and v has components $\{\beta_1, \beta_2, \beta_3\}$, then the cross-product is represented as

$$w = u \times v = \begin{bmatrix} \alpha_2\beta_3 - \alpha_3\beta_2 \\ \alpha_3\beta_1 - \alpha_1\beta_3 \\ \alpha_1\beta_2 - \alpha_2\beta_1 \end{bmatrix}. \tag{4.31}$$

In the right-hand coordinate system, the cross-product of x-axis and y-axis gives the third axis, the z-axis.

4.1.2 Matrices

A *matrix* is an $n \times m$ array of scalars, representing n rows and m columns. n and m are referred as the number of rows and the number of columns. If $m = n$, then we obtain a square matrix of dimension n. The matrix A can be written using its elements. We have $A = [a_{ij}]$ where $i = 1, \cdots, n$ and $j = 1, \cdots, m$. The transpose of an $n \times m$ matrix A is a $m \times n$ matrix which can be obtained by interchanging the rows and columns of the matrix. We denote the transpose matrix as $A^T = [a_{ji}]$. A column matrix is a matrix with only one column ($m = 1$), and a row matrix is a matrix with only one row ($n = 1$). For example, $b = [b_i]$ is a column matrix and b^T is a row matrix. There are three basic matrix operations: scalar-matrix multiplication, matrix-matrix addition and matrix-matrix multiplication. For an any-size matrix A, the *scalar-matrix multiplication* is to simply multiply each element in the matrix by a scalar α, and we have $\alpha A = [\alpha a_{ij}]$. The *matrix-matrix addition* adds the corresponding elements of two matrices with the same dimensions. We have $C = A + B = [a_{ij} + b_{ij}]$. The *matrix-matrix multiplication* is to multiply an $n \times l$ matrix A with an $l \times m$ matrix B to obtain an $n \times m$ matrix. We have $C = AB = [c_{ij}]$, where $c_{ij} = \sum_{k=1}^{l} a_{ik}b_{kj}$. This operation requires the number of columns in A equal to the number of rows in B.

The scalar-matrix multiplication operation follows a number of rules that hold for any matrix A and scalars α and β. We have

$$\alpha(\beta A) = (\alpha\beta)A, \tag{4.32}$$
$$\alpha\beta A = \beta\alpha A. \tag{4.33}$$

The matrix-matrix addition operation satisfies the *commutative* and *associative* properties. For any $n \times m$ matrices A, B and C, we have

$$
\begin{aligned}
Commutative \quad &: \quad A + B = B + A, && (4.34) \\
Associative \quad &: \quad A + (B + C) = (A + B) + C. && (4.35)
\end{aligned}
$$

Differently, the matrix-matrix multiplication operation is *associative, but not commutative*. We have

$$
\begin{aligned}
A(BC) \quad &= \quad (AB)C, && (4.36) \\
AB \quad &\neq \quad BA. && (4.37)
\end{aligned}
$$

In computer graphics and visualization, matrices represent transformations of objects like translation and rotation. The order of transformation is not commutative, and different orders yield different results. For example, a rotation followed by a translation is different from a translation followed by a rotation.

The identity matrix I is a special square matrix with $1s$ on the diagonal and $0s$ elsewhere. We have $I = [a_{ij}]$, where

$$
a_{ij} = \begin{cases} 1 & if \ \ i = j, \\ 0 & otherwise. \end{cases}
$$

We also have $AI = A$ and $IB = B$. Row and column matrices are two other special matrices. A vector or a point in 3D can be represented as the column matrix

$$
p = \begin{bmatrix} x \\ y \\ z \end{bmatrix}. \tag{4.38}
$$

The transpose of p is the row matrix, $p^T = [x \ y \ z]$. The product of an n-dimensional square matrix and a column matrix with the same dimension yields a new n-dimensional column matrix. A square matrix is often used to represent a transformation of the point or vector, and we have $p' = Ap$. $p' = ABCp$ represents a sequence of transformations. Note that the order of transformations is not commutative. Since $(AB)^T = B^T A^T$, we can obtain $p'^T = p^T C^T B^T A^T$.

For a matrix A, if another matrix B exists satisfying $BA = I$, then B is called the inverse of A and A is nonsingular. We have $B = A^{-1}$. A non-invertible matrix is *singular*. The inverse of a square matrix exists if and only if it has a nonzero determinant. The time complexity of the determinant calculation is $O(n^3)$ for an n-dimensional matrix whose rank is n. For a non-square matrix, its row (or column) rank is the maximum number of linearly independent rows (or columns).

4.1.3 Change of Representations

Suppose we have an n-dimensional vector space, and $\{u_1, u_2, \cdots, u_n\}$ and $\{v_1, v_2, \cdots, v_n\}$ are two bases for the vector space. A given vector v can be represented as

$$v = \alpha_1 u_1 + \alpha_2 u_2 + \cdots + \alpha_n u_n \qquad (4.39)$$

or

$$v = \beta_1 v_1 + \beta_2 v_2 + \cdots + \beta_n v_n. \qquad (4.40)$$

The representations of v can also be written as $v = [\alpha_1 \ \alpha_2 \ \cdots \ \alpha_n]^T$ or $v' = [\beta_1 \ \beta_2 \ \cdots \ \beta_n]^T$ depending on which basis is used. In the following, let us take a look at how to convert v to v'. The basis vectors $\{u_1, u_2, \cdots, u_n\}$ can be represented using the basis $\{v_1, v_2, \cdots, v_n\}$. There exists a set of scalars γ_{ij} satisfying

$$u_i = \gamma_{i1} v_1 + \gamma_{i2} v_2 + \cdots + \gamma_{in} v_n, \qquad (4.41)$$

where $i = 1, \cdots, n$. This expression can be rewritten in the matrix form, and we obtain

$$\begin{bmatrix} u_1 \\ u_2 \\ \vdots \\ u_n \end{bmatrix} = A \begin{bmatrix} v_1 \\ v_2 \\ \vdots \\ v_n \end{bmatrix}, \qquad (4.42)$$

where A is the $n \times n$ matrix, or $A = [\gamma_{ij}]$. We can use column matrices to represent v and v', and we have

$$v = a^T \begin{bmatrix} u_1 \\ u_2 \\ \vdots \\ u_n \end{bmatrix} \qquad and \qquad v' = b^T \begin{bmatrix} v_1 \\ v_2 \\ \vdots \\ v_n \end{bmatrix}, \qquad (4.43)$$

where $a = [\alpha_i]$ and $b = [\beta_i]$. We have $b^T = a^T A$ and A is the transformation matrix between these two bases.

4.1.4 Eigenvalues and Eigenvectors

Supposing λ is the eigenvalue and u is the corresponding eigenvector of matrix M, we have $Mu = \lambda u$ and

$$Mu - \lambda u = Mu - \lambda IU = (M - \lambda I)u = 0. \qquad (4.44)$$

The above equation has a nontrivial solution if and only if the determinant is zero or $|M - \lambda I| = 0$. For an $n \times n$ matrix M, the determinant leads to a degree-n polynomial, which has n roots. For each eigenvalue, we can find an eigenvector corresponding to it. With all distinct eigenvalues, their corresponding eigenvectors form a basis of an n-dimensional vector space.

Supposing T is a nonsingular matrix, we consider $Q = T^{-1}MT$. Then we have $Qv = T^{-1}MTv = \lambda v$ and $MTv = \lambda Tv$. Therefore, Q and M share the same eigenvalues, and the eigenvectors of Q are the transformations of the eigenvectors of M. We call M and Q a pair of *similar matrices*.

Eigenvalues and their corresponding eigenvectors have a geometric meaning. Considering an ellipsoid centered at the origin, whose axes align with the three coordinate axes. We have

$$\lambda_1 x^2 + \lambda_2 y^2 + \lambda_3 z^2 = 1, \tag{4.45}$$

where λ_1, λ_2 and λ_3 are all positive. We can also rewrite it in the matrix form and obtain

$$\begin{bmatrix} x & y & z \end{bmatrix} \begin{bmatrix} \lambda_1 & 0 & 0 \\ 0 & \lambda_2 & 0 \\ 0 & 0 & \lambda_3 \end{bmatrix} \begin{bmatrix} x \\ y \\ z \end{bmatrix} = 1. \tag{4.46}$$

It is obvious that λ_1, λ_2 and λ_3 are the three eigenvalues of the diagonal matrix. They also represent the length inverse of the major and minor axes of the ellipsoid. When we rotate the ellipsoid (a rigid body motion), we obtain a new ellipsoid with the length of axes unchanged.

4.2   Geometric Objects and Transformations

After reviewing the fundamentals to geometric modeling, let us talk about some basic geometric objects like points, lines, planes, coordinate systems and transformations.

4.2.1 Scalars, Points, and Vectors

General geometric objects include lines, polygons and polyhedra. Their relationships can be described using three fundamental types: scalars, points and vectors. As reviewed in the previous section, a *point* is a location in the space, and a mathematical point has no size or shape. A *scalar* is a real or complex number to specify quantities such as distance. Addition and multiplication are two basic operations defined for scalars. They satisfy the commutative and associative properties. A *vector* is a directed line segment or a quantity with direction and magnitude, such as velocity and force, but no fixed position in the space.

Vectors have two basic operations, the addition of two vectors (head-to-tail) and the multiplication of a scalar with a vector. A zero vector has the magnitude of zero with undefined orientation. Given a vector u, its inverse vector $-u$ has the same length but with the opposite direction. From the viewpoint of geometry, point-vector addition moves a directed line segment from one point to another, and point-point subtraction yields a line segment or vector. From the viewpoint of mathematics, the vector space has two distinct entities, vectors and scalars. The Euclidean space adds the distance or a measure of size to it, while the affine space introduces the concept of points.

Scalar, point and vector are three geometric abstract data types. Generally, we use Greek letters α, β, γ, $\cdots$ to denote scalars, uppercase letters P, Q, R, $\cdots$ to define points, and lowercase letters u, v, w, $\cdots$ to denote vectors. According to vector-scalar multiplication, the direction of αv is the same as v if α is positive and the opposite if α is negative. We have $|\alpha v| = |\alpha||v|$. Point-point subtraction forms a vector, we have $v = P - Q$ and $P = v + Q$. Vector-vector addition is basically the head-to-tail rule, and we have $u + v = (P - Q) + (Q - R) = P - R$.

The *inner or dot product* of u and v is written as $u \cdot v$. When $u \cdot v = 0$, u and v are orthogonal to each other. In a Euclidean space, the magnitude square of a vector can be computed by the dot product, $|u|^2 = u \cdot u$. The angle between two vectors is computed by $cos\theta = (u \cdot v)/(|u||v|)$. In addition, $|u|cos\theta = u \cdot v/|v|$ is the projection of u onto v. The dot product describes the shortest distance from the endpoint of u to the line segment v.

In the vector space, vectors are *linearly independent* if we cannot represent one in terms of the others using the scalar-vector addition operation. The maximum number of linearly independent vectors in the space is defined as the dimension of the vector space. Given two nonparallel vectors u and v, their *cross-product* determines a vector w which is orthogonal to both u and v, and we have $w = u \times v$. Then we call u, v and w mutually orthogonal. The angle between u and v in the plane containing them is given by $|sin\theta| = |u \times v|/(|u||v|)$.

4.2.2 Lines and Planes

In an affine space, the sum of a point P_0 and a vector v yields a *line*, which can be represented as

$$P(\alpha) = P_0 + \alpha v, \tag{4.47}$$

where α is a scalar value. This is the parametric representation of a line by varying the parameter α to generate points on the line; see Figure 4.4(a). The vector can also be represented by two points such that $v = R - Q$, and then we have

$$P = P_0 + \alpha(P_1 - P_0) = \alpha P_1 + (1 - \alpha)P_0 \tag{4.48}$$

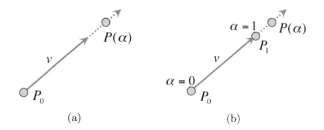

FIGURE 4.4: (a) A line represented by the sum of a point and a vector; and (b) a line represented by an affine sum of two points.

or

$$P = \alpha_1 P_0 + \alpha_2 P_1, \tag{4.49}$$

where $\alpha_1 + \alpha_2 = 1$. This gives an *affine sum*; see Figure 4.4(b).

In geometric modeling, *convexity* is an important property for geometric objects. Given a set of points, we can check each line segment connecting any two points in the object. If any point on the segment also lies in the object, we say this set of points forms a *convex* object. Otherwise, they form a *concave* object. We can get a better understanding by using the affine sums. For a scalar $0 \leq \alpha \leq 1$, the affine sum describes the line segment connecting R and Q as shown in Figure 4.5(a), and it is obvious that this line segment is a convex object. We can extend the affine sum definition to n points, and obtain

$$P = \alpha_1 P_1 + \alpha_2 P_2 + \cdots + \alpha_n P_n. \tag{4.50}$$

By induction, we can show that this sum is affine if and only if

$$\alpha_1 + \alpha_2 + \cdots + \alpha_n = 1. \tag{4.51}$$

The set of points formed by the affine sum of n points, with the additional restriction $\alpha_i \geq 0$ where $i = 1, 2, \cdots, n$, is called the *convex hull* of the set of points. It is also the smallest convex object with all the given points included; see Figure 4.5(b).

A 2D *plane* in the affine space can be considered as a direct extension of the 1D parametric line. It is well known that three noncolinear points define a unique plane. As shown in Figure 4.6, P, Q and R are three such points, and the line segment connecting P and Q can be written as

$$S(\alpha) = \alpha P + (1 - \alpha)Q, \tag{4.52}$$

where $0 \leq \alpha \leq 1$. We choose an arbitrary point on this line segment and connect it with R to form another line segment. Using a second parameter β, we can represent the line segment SR as

$$T(\beta) = \beta S + (1 - \beta)R, \tag{4.53}$$

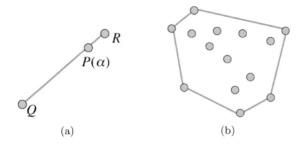

(a)   (b)

FIGURE 4.5: Convexity. (a) A line; and (b) a convex hull.

where $0 \leq \beta \leq 1$. Combining the above two equations, we then obtain the parametric representation of a 2D plane,

$$
\begin{aligned}
T(\alpha, \beta) &= \beta \left(\alpha P + (1 - \alpha) Q \right) + (1 - \beta) R \\
&= P + \beta (1 - \alpha)(Q - P) + (1 - \beta)(R - P). \quad (4.54)
\end{aligned}
$$

Note that $Q - P$ and $R - P$ are two arbitrary vectors. It is obvious that a plane can also be represented in terms of a point and two nonparallel vectors, and we have

$$
T(\alpha, \beta) = P_0 + \alpha u + \beta v, \quad (4.55)
$$

where $0 \leq \alpha, \beta \leq 1$. It is easy to observe that all the points $T(\alpha, \beta)$ lie in the triangle formed by P, Q and R. If a point P' lies in the plane, then we have $P' - P_0 = \alpha u + \beta v$. The normal of the plane can be computed using the cross-product, so we have $n = u \times v$ and $n \cdot (P - P_0) = 0$.

4.2.3 Coordinate Systems and Frames

A 3D vector space consists of three linearly independent basis vectors $\{v_1, v_2, v_3\}$, which can be used to represent any vector w uniquely. We have

$$
w = \alpha_1 v_1 + \alpha_2 v_2 + \alpha_3 v_3, \quad (4.56)
$$

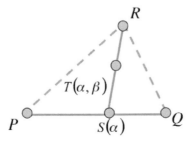

FIGURE 4.6: Parametric representation of a plane in an affine space.

where $\{\alpha_1, \alpha_2, \alpha_3\}$ are three scalars, representing the components of w with respect to the basis $\{v_1, v_2, v_3\}$, respectively. The representation can be written in the matrix form,

$$w = a^T \begin{bmatrix} v_1 \\ v_2 \\ v_3 \end{bmatrix}, \tag{4.57}$$

where $a = [\alpha_1, \alpha_2, \alpha_3]^T$. These three basis vectors $\{v_1, v_2, v_3\}$ define a *coordinate system*. Together with the origin, the basis vectors define a *frame*. In a given frame, every vector can be represented as

$$w = \alpha_1 v_1 + \alpha_2 v_2 + \alpha_3 v_3 \tag{4.58}$$

and similarly, every point can be represented as

$$P = P_0 + \beta_1 v_1 + \beta_2 v_2 + \beta_3 v_3. \tag{4.59}$$

Given three basis vectors $\{e_1, e_2, e_3\}$, a vector v can be represented by $(\alpha_1, \alpha_2, \alpha_3)$ and we have $v = \alpha_1 e_1 + \alpha_2 e_2 + \alpha_3 e_3$, where $e_1 = (1, 0, 0)^T$, $e_2 = (0, 1, 0)^T$ and $(0, 0, 1)^T$. We can represent any vector v as a column matrix a or a 3-tuple $(\alpha_1, \alpha_2, \alpha_3)$. These 3-tuples form a Euclidean space $\Re^3$.

Given two sets of bases $\{v_1, v_2, v_3\}$ and $\{u_1, u_2, u_3\}$, each basis vector in one set can be represented in terms of the other set. We can find nine scalar components $\{\gamma_{ij}\}$ satisfying

$$
\begin{aligned}
u_1 &= \gamma_{11} v_1 + \gamma_{12} v_2 + \gamma_{13} v_3, \\
u_2 &= \gamma_{21} v_1 + \gamma_{22} v_2 + \gamma_{23} v_3, \\
u_3 &= \gamma_{31} v_1 + \gamma_{32} v_2 + \gamma_{33} v_3.
\end{aligned}
$$

Then we have

$$\begin{bmatrix} u_1 \\ u_2 \\ u_3 \end{bmatrix} = M \begin{bmatrix} v_1 \\ v_2 \\ v_3 \end{bmatrix} \quad where \quad M = \begin{bmatrix} \gamma_{11} & \gamma_{12} & \gamma_{13} \\ \gamma_{21} & \gamma_{22} & \gamma_{23} \\ \gamma_{31} & \gamma_{32} & \gamma_{33} \end{bmatrix}. \tag{4.60}$$

We assume a vector w is represented in terms of these two bases, and then we have $w = \alpha_1 v_1 + \alpha_2 v_2 + \alpha_3 v_3$ and $w = \beta_1 u_1 + \beta_2 u_2 + \beta_3 u_3$. We can write them as

$$w = a^T \begin{bmatrix} v_1 \\ v_2 \\ v_3 \end{bmatrix} \quad where \quad a = \begin{bmatrix} \alpha_1 \\ \alpha_2 \\ \alpha_3 \end{bmatrix}, \tag{4.61}$$

and

$$w = b^T \begin{bmatrix} u_1 \\ u_2 \\ u_3 \end{bmatrix} \quad where \quad b = \begin{bmatrix} \beta_1 \\ \beta_2 \\ \beta_3 \end{bmatrix}. \tag{4.62}$$

Then we have

$$w = b^T M \begin{bmatrix} v_1 \\ v_2 \\ v_3 \end{bmatrix} = a^T \begin{bmatrix} v_1 \\ v_2 \\ v_3 \end{bmatrix}, \tag{4.63}$$

or $a = M^T b$ and $b = (M^T)^{-1} a$.

A point P with the position vector (x, y, z) is generally represented in a 3D frame defined by P_0, v_1, v_2 and v_3. We have

$$P = \begin{bmatrix} x \\ y \\ z \end{bmatrix} \quad or \quad P = P_0 + x v_1 + y v_2 + z v_3. \tag{4.64}$$

We can also use the *homogeneous coordinates* or a 4D column matrix to represent points and vectors in 3D. In the frame $\{v_1, v_2, v_3, P_0\}$, any point P can be represented as

$$P = \alpha_1 v_1 + \alpha_2 v_2 + \alpha_3 v_3 + P_0 \tag{4.65}$$

or

$$P = \begin{bmatrix} \alpha_1 & \alpha_2 & \alpha_3 & 1 \end{bmatrix} \begin{bmatrix} v_1 \\ v_2 \\ v_3 \\ P_0 \end{bmatrix}. \tag{4.66}$$

P is represented by the column matrix $P = [\alpha_1 \ \alpha_2 \ \alpha_3 \ 1]^T$. In the same frame, any vector w can be described as

$$w = \delta_1 v_1 + \delta_2 v_2 + \delta_3 v_3 = \begin{bmatrix} \delta_1 & \delta_2 & \delta_3 & 0 \end{bmatrix} \begin{bmatrix} v_1 \\ v_2 \\ v_3 \\ P_0 \end{bmatrix}. \tag{4.67}$$

Given two frames $\{v_1, v_2, v_3, P_0\}$ and $\{u_1, u_2, u_3, Q_0\}$, we can represent the basis vectors and the reference point of one frame in terms of the other. We have

$$
\begin{aligned}
u_1 &= \gamma_{11} v_1 + \gamma_{12} v_2 + \gamma_{13} v_3, \\
u_2 &= \gamma_{21} v_1 + \gamma_{22} v_2 + \gamma_{23} v_3, \\
u_3 &= \gamma_{31} v_1 + \gamma_{32} v_2 + \gamma_{33} v_3, \\
Q_0 &= \gamma_{41} v_1 + \gamma_{42} v_2 + \gamma_{43} v_3 + P_0.
\end{aligned}
$$

Then, we have

$$\begin{bmatrix} u_1 \\ u_2 \\ u_3 \\ Q_0 \end{bmatrix} = M \begin{bmatrix} v_1 \\ v_2 \\ v_3 \\ P_0 \end{bmatrix} \quad where \quad M = \begin{bmatrix} \gamma_{11} & \gamma_{12} & \gamma_{13} & 0 \\ \gamma_{21} & \gamma_{22} & \gamma_{23} & 0 \\ \gamma_{31} & \gamma_{32} & \gamma_{33} & 0 \\ \gamma_{41} & \gamma_{42} & \gamma_{43} & 1 \end{bmatrix} \tag{4.68}$$

and

$$
b^T
\begin{bmatrix}
u_1 \\
u_2 \\
u_3 \\
Q_0
\end{bmatrix}
= b^T M
\begin{bmatrix}
v_1 \\
v_2 \\
v_3 \\
P_0
\end{bmatrix}
= a^T
\begin{bmatrix}
v_1 \\
v_2 \\
v_3 \\
P_0
\end{bmatrix}.
\tag{4.69}
$$

Therefore, we obtain $a = M^T b$, where M is called the transformation matrix between these two frames.

4.2.4 Transformations

Translation is an operation to relocate a point P by a fixed distance following a given direction to a new location P', which only needs to specify a displacement vector d. We have $P' = P + d$. This definition does not need a reference to a frame or representation.

For *rotation*, we need to specify a particular point as the origin, and the rotation operation will be performed about it. For example, a 2D point at (x, y) is rotated about the origin by an angle θ to the new position (x', y'). We can describe this rotation in the polar form as

$$
x = \rho cos\phi, \qquad x' = \rho cos(\theta + \phi), \tag{4.70}
$$
$$
y = \rho sin\phi, \qquad y' = \rho sin(\theta + \phi). \tag{4.71}
$$

Then, we can rewrite them in the matrix form to obtain

$$
\begin{bmatrix} x' \\ y' \end{bmatrix} =
\begin{bmatrix} cos\theta & -sin\theta \\ sin\theta & cos\theta \end{bmatrix}
\begin{bmatrix} x \\ y \end{bmatrix}.
\tag{4.72}
$$

The rotation operation has three features: (1) during rotation only the origin is unchanged, which is called the fixed point; (2) generally rotation is in 3D and the counterclockwise rotation direction is defined as positive; and (3) a 2D rotation in the plane is equivalent to a 3D rotation about the z-axis.

Both rotation and translation are *rigid-body transformations*. They can change the location and orientation of an object, but they cannot change the shape of the object. *Scaling* is an affine and nonrigid-body transformation; it can be uniform or nonuniform. A sequence of *scaling, translation, rotation* and *shear* forms an arbitrary affine transformation. Scaling consists of a fixed point and a scalar factor a. If $a > 1$, the object gets longer in the specific direction; if $0 \le a < 1$, the object gets shrinked in that direction; and if $a < 0$, the object gets reflected about the fixed point.

We can handle transformations in the homogeneous coordinates, $q = p + \alpha v$. In a frame, each affine transformation can be represented using a 4×4

matrix,

$$M = \begin{bmatrix} \alpha_{11} & \alpha_{12} & \alpha_{13} & \alpha_{14} \\ \alpha_{21} & \alpha_{22} & \alpha_{23} & \alpha_{24} \\ \alpha_{31} & \alpha_{32} & \alpha_{33} & \alpha_{34} \\ 0 & 0 & 0 & 1 \end{bmatrix}. \tag{4.73}$$

Translation relocates points to new positions using a displacement vector. When we move the point p to p', we have $p' = p + d$. In the homogeneous coordinates, we have

$$p = \begin{bmatrix} x \\ y \\ z \\ 1 \end{bmatrix}, \quad p' = \begin{bmatrix} x' \\ y' \\ z' \\ 1 \end{bmatrix} \quad and \quad d = \begin{bmatrix} \alpha_x \\ \alpha_y \\ \alpha_z \\ 0 \end{bmatrix}. \tag{4.74}$$

These equations can be rewritten as

$$x' = x + \alpha_x,$$
$$y' = y + \alpha_y,$$
$$z' = z + \alpha_z.$$

Using matrix multiplication, we have

$$p' = Tp \quad where \quad T = \begin{bmatrix} 1 & 0 & 0 & \alpha_x \\ 0 & 1 & 0 & \alpha_y \\ 0 & 0 & 1 & \alpha_z \\ 0 & 0 & 0 & 1 \end{bmatrix}. \tag{4.75}$$

T is called the *translation matrix*. If we relocate a point by the vector d and then relocate it again by the vector $-d$, then we will return to the original location. Therefore, we have

$$T^{-1}(\alpha_x, \alpha_y, \alpha_z) = T(-\alpha_x, -\alpha_y, -\alpha_z) = \begin{bmatrix} 1 & 0 & 0 & -\alpha_x \\ 0 & 1 & 0 & -\alpha_y \\ 0 & 0 & 1 & -\alpha_z \\ 0 & 0 & 0 & 1 \end{bmatrix}. \tag{4.76}$$

Both scaling and rotation have a fixed point during the transformation. We set the fixed point as the origin. A *scaling matrix* with a fixed point at the origin allows for an independent scaling along the coordinate axes,

$$x' = \beta_x x, \quad y' = \beta_y y \quad and \quad z' = \beta_z z.$$

Then, we have $p' = Sp$ where

$$S = S(\beta_x, \beta_y, \beta_z) = \begin{bmatrix} \beta_x & 0 & 0 & 0 \\ 0 & \beta_y & 0 & 0 \\ 0 & 0 & \beta_z & 0 \\ 0 & 0 & 0 & 1 \end{bmatrix}. \tag{4.77}$$

It is obvious that we also have

$$S^{-1}(\beta_x, \beta_y, \beta_z) = S(\frac{1}{\beta_x}, \frac{1}{\beta_y}, \frac{1}{\beta_z}). \tag{4.78}$$

Rotation is a bit more complicated. Let us first take a look at rotation with a fixed point at the origin. There are three degrees of freedom, each of them corresponding to the rotation about one axis. We can also use matrix computation to handle the rotations, but note that matrix multiplication is not commutative. For a rotation about the z-axis by an angle θ, we have

$$
\begin{aligned}
x' &= xcos\theta - ysin\theta, \\
y' &= xsin\theta + ysin\theta, \\
z' &= z.
\end{aligned}
$$

Rewrite them into the matrix form to obtain

$$p' = R_z p, \quad where \quad R_z = R_z(\theta) = \begin{bmatrix} cos\theta & -sin\theta & 0 & 0 \\ sin\theta & cos\theta & 0 & 0 \\ 0 & 0 & 1 & 0 \\ 0 & 0 & 0 & 1 \end{bmatrix}. \tag{4.79}$$

Similarly, we have

$$R_x = R_x(\theta) = \begin{bmatrix} 1 & 0 & 0 & 0 \\ 0 & cos\theta & -sin\theta & 0 \\ 0 & sin\theta & cos\theta & 0 \\ 0 & 0 & 0 & 1 \end{bmatrix} \tag{4.80}$$

and

$$R_y = R_y(\theta) = \begin{bmatrix} cos\theta & 0 & sin\theta & 0 \\ 0 & 1 & 0 & 0 \\ -sin\theta & 0 & cos\theta & 0 \\ 0 & 0 & 0 & 1 \end{bmatrix}. \tag{4.81}$$

For a sequence of rotations about the three axes, we have $R = R_z R_y R_x$. We also have $R^{-1}(\theta) = R(-\theta)$ and $R^{-1}(\theta) = R^T(\theta)$, therefore we can obtain the orthogonal matrix $R^{-1} = R^T$.

Each *shear* operation is characterized by a single angle θ, and we have $x' = x + ycot\theta$, $y' = y$ and $z' = z$, which can be written in a shear matrix

$$H_x(\theta) = \begin{bmatrix} 1 & cot\theta & 0 & 0 \\ 0 & 1 & 0 & 0 \\ 0 & 0 & 1 & 0 \\ 0 & 0 & 0 & 1 \end{bmatrix}. \tag{4.82}$$

We also have $H_x^{-1}(\theta) = H_x(-\theta)$.

An affine transformation can be obtained by multiplying or concatenating a sequence of basic transformations together. For example, we perform three successive transformations on a point p and obtain a new point q. We have $q = CBAp = Mp$, where $M = CBA$. Concatenation of transformations is very useful. A rotation about a fixed point but not at the origin can be expressed by three transformations: translation to the origin, rotation about the origin, and then translation back. The transformation matrix can be written as

$$
\begin{aligned}
M &= T(p_f)R_Z(\theta)T(-p_f) \\
&= \begin{bmatrix}
\cos\theta & -\sin\theta & 0 & x_f - x_f\cos\theta + y_f\sin\theta \\
\sin\theta & \cos\theta & 0 & y_f - x_f\sin\theta - y_f\cos\theta \\
0 & 0 & 1 & 0 \\
0 & 0 & 0 & 1
\end{bmatrix},
\end{aligned}
\qquad (4.83)
$$

where $p_f = [x_f \ y_f]^T$ is the translation vector and θ is the rotation angle. A general rotation is much more complex. An arbitrary rotation about the origin can be decomposed into three successive rotations about the three axes, $R = R_x R_y R_z$. We can first perform a rotation about the z-axis, then perform a rotation about the y-axis, and finally perform a rotation about the x-axis. The transformation matrix can be written as $R = R_x R_y R_z$. For a rotation about an arbitrary axis, suppose the vector about which we rotate is defined by two points in the space, $u = p_2 - p_1$. We first normalize it to obtain

$$
v = \frac{u}{|u|} = \begin{bmatrix} \alpha_x \\ \alpha_y \\ \alpha_z \end{bmatrix},
\qquad (4.84)
$$

where $\alpha_x^2 + \alpha_y^2 + \alpha_z^2 = 1$. We can perform two rotations to align the rotation axis v with the z-axis, and then rotate about the $z-$axis. Finally we undo the two rotations for aligning. The entire procedure can be represented by a sequence of five rotations, and we have

$$
R = R_x(-\theta_x)R_y(-\theta_y)R_z(\theta)R_y(\theta_y)R_x(\theta_x).
\qquad (4.85)
$$

4.3 Curves and Surfaces

There are three main types of representations for objects like curves and surfaces. We have (1) the explicit representation: $y = f(x)$; (2) the implicit representation: $f(x, y) = 0$; and (3) the parametric representation: $x = x(u)$, $y = y(u)$, $z = z(u)$. We will discuss each of them as follows.

4.3.1 Explicit Representation

In 2D, the explicit form of a curve expresses one variable (the dependent variable) in terms of the other (the independent variable). In the x, y space, we write the representation in the form of

$$y = f(x). \tag{4.86}$$

Similarly, we can also express x as a function of y,

$$x = g(y). \tag{4.87}$$

We can represent a line explicitly,

$$y = mx + h, \tag{4.88}$$

where m is the slope and h is the y-intercept. Note that this equation does not hold for vertical lines. A circle of radius r centered at the origin is another example. Using an explicit representation, we can only write one equation for half of the circle

$$y = \sqrt{r^2 - x^2}, \tag{4.89}$$

where $0 \leq |x| \leq r$. The other half can be written in the similar way,

$$y = -\sqrt{r^2 - x^2}. \tag{4.90}$$

In 3D, the explicit representation of a curve needs two equations. Suppose x is the independent variable, we have two dependent variables y and z, which can be expressed by x. We have

$$y = f(x) \quad and \quad z = g(x). \tag{4.91}$$

A surface requires two independent variables in its explicit representation, $z = f(x, y)$. The explicit representation has limitations, for example, we may not be able to represent a curve or surface explicitly. The equations $y = ax + b$ and $z = cx + d$ describe a line in 3D, but they cannot represent a line in a plane with constant x. Similarly, a surface represented by an equation of $z = f(x, y)$ cannot represent a sphere. This is because a given pair of x and y can correspond to zero, one or two points on the sphere.

4.3.2 Implicit Representations

In general, most curves and surfaces have implicit representations. In 2D, an implicit curve can be represented by

$$f(x, y) = 0. \tag{4.92}$$

In 3D, a line can be represented by

$$ax + by + c = 0 \tag{4.93}$$

and in 2D, a circle centered at the origin can be written as

$$x^2 + y^2 - r^2 = 0. \tag{4.94}$$

The function f can be considered as a membership function that distinguishes points belonging to the curve from others in the space. Unlike the explicit form, the implicit form is less coordinate system-dependent and it can represent all lines and circles.

In 3D, the implicit equation $f(x, y, z) = 0$ represents a surface. For example, a plane can be written as

$$ax + by + cz + d = 0, \tag{4.95}$$

where a, b, c and d are constants. A sphere with radius r centered at the origin can be represented as

$$x^2 + y^2 + z^2 - r^2 = 0. \tag{4.96}$$

In 3D, we can represent a curve as the intersection of two surfaces,

$$f(x, y, z) = 0 \quad and \quad g(x, y, z) = 0. \tag{4.97}$$

Sometimes, we use intersection of algebraic surfaces to represent lines or curves. Algebraic surfaces are surfaces represented by a polynomial function $f(x, y, z)$. For example, each term of a quadric surface can have degree up to 2. They can represent objects like spheres, disks and cones. These objects intersect each other with lines or curves.

4.3.3 Parametric Representation

The parametric form of a curve uses an independent variable or the parameter u to represent the curve. In 3D, there are three explicit functions for a curve in terms of u:

$$x = x(u), \quad y = y(u) \quad and \quad z = z(u). \tag{4.98}$$

Compared to other representations, the parametric form stays the same in 2D and 3D. We can visualize how the point $p(u) = [x(u)\ y(u)\ z(u)]^T$ moves as u varies. As shown in Figure 4.7(a), the derivative can be considered as the tracing velocity, pointing along the tangential direction of the curve,

$$\frac{dp(u)}{du} = \begin{bmatrix} \frac{dx(u)}{du} \\ \frac{dy(u)}{du} \\ \frac{dz(u)}{du} \end{bmatrix}. \tag{4.99}$$

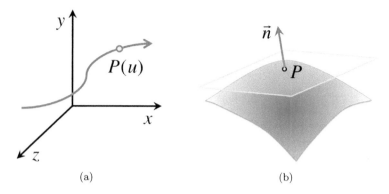

(a)   (b)

FIGURE 4.7: (a) A parametric curve $p(u)$; and (b) a parametric surface with its tangent plane and normal $\vec{n}$ at point p.

In the parametric forms, surfaces require two parameters u and v. We can describe a surface using three equations in terms of u and v,

$$x = x(u, v), \quad y = y(u, v) \quad and \quad z = z(u, v); \tag{4.100}$$

or we can write in the column matrix

$$p(u, v) = \begin{bmatrix} x(u, v) \\ y(u, v) \\ z(u, v) \end{bmatrix}. \tag{4.101}$$

As u and v vary over some interval, we obtain all the points $p(u, v)$ moving on the surface.

We can use the derivative vectors to compute the tangent plane at each point on the surface. We have

$$\frac{\partial p}{\partial u} = \begin{bmatrix} \frac{\partial x(u,v)}{\partial u} \\ \frac{\partial y(u,v)}{\partial u} \\ \frac{\partial z(u,v)}{\partial u} \end{bmatrix} \quad and \quad \frac{\partial p}{\partial v} = \begin{bmatrix} \frac{\partial x(u,v)}{\partial v} \\ \frac{\partial y(u,v)}{\partial v} \\ \frac{\partial z(u,v)}{\partial v} \end{bmatrix}. \tag{4.102}$$

If these two vectors are not parallel to each other, we can use their cross-product to compute the normal at each point as shown in Figure 4.7(b). We have

$$n = \frac{\partial p}{\partial u} \times \frac{\partial p}{\partial v}. \tag{4.103}$$

The parametric form of curves and surfaces is the most flexible and robust in geometric modeling. It has been used in computer-aided design intensively.

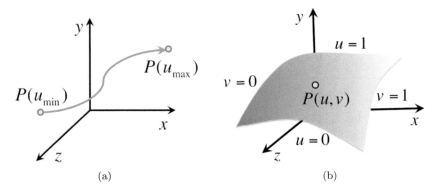

FIGURE 4.8: (a) A curve segment $p(u)$ for $u_{min} \leq u \leq u_{max}$; and (b) a surface patch $p(u,v)$ for $0 \leq u \leq 1$ and $0 \leq v \leq 1$.

Note that we do not fully remove all the dependencies on a particular coordinate system or frame, but we can develop a local system that changes as a point moves on the curve or surface. Parametric forms are not unique in general, as a given curve or surface can be represented in different ways. We usually use polynomials in u to represent curves and polynomials in u and v to represent surfaces.

Consider a curve in the form $p(u) = [x(u) \ y(u) \ z(u)]^T$. A polynomial parametric curve of degree n can be written as

$$p(u) = \sum_{k=0}^{n} u^k c_k,$$
(4.104)

where each c_k has independent x, y and z components. We have $c_k = [c_{xk} \ c_{yk} \ c_{zk}]^T$. The $n+1$ column matrices $\{c_k\}$ are the coefficients of p, yielding $3(n+1)$ degrees of freedom. The curves can be defined for any range interval of u such as $u_{min} \leq u \leq u_{max}$. Without loss of generality, we assume that $0 \leq u \leq 1$; see Figure 4.8(a). Similarly, we can define a parametric polynomial surface as

$$p(u,v) = \begin{bmatrix} x(u,v) \\ y(u,v) \\ z(u,v) \end{bmatrix} = \sum_{i=1}^{n} \sum_{j=0}^{m} c_{ij} u^i v^j.$$
(4.105)

We need to specify $3(n+1)(m+1)$ coefficients to determine a particular surface $p(u,v)$. Generally we take $n = m$ to define a surface patch as shown in Figure 4.8(b), letting u and v vary over the rectangle range $0 \leq u, v \leq 1$. Any surface patch can be considered as a collection of curves generated by holding either u or v constant and varying the other. An important strategy in computer-aided design is to first define parametric polynomial curves and then use them to generate surfaces with similar geometric characteristics.

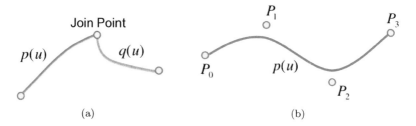

FIGURE 4.9: (a) A join point between $p(v)$ and $q(v)$; and (b) a curve seg-ment $p(u)$ with its control points P_0, P_1, P_2 and P_3.

4.3.4 Design Criteria

Parametric forms have been used popularly in computer-aided geometric design due to its five important properties:

1. Local control of object shape;

2. Surface smoothness and continuity;

3. Being able to evaluate derivatives;

4. Stability; and

5. Ease of rendering.

Suppose we want to build an airplane model using flexible wood strips. We can first build a set of cross sections, and then connect them to form the body of the airplane. For each cross section, we need to sketch a desired curve consisting of several curve segments. Each segment should be smooth itself. We also want a certain smoothness at the joint point shared by two neighboring segments; see Figure 4.9(a). The smoothness is usually defined using the derivatives along the curve. For a polynomial curve

$$p(u) = \sum_{k=0}^{3} c_k u^k, \qquad (4.106)$$

all derivatives exist and can be computed analytically. Therefore, the only places we need to handle are at the joint points.

Instead of designing all the segments using a single global representation, we usually design each segment individually and then connect all the segments together. In this way, a change can only affect a surrounding local region and such a local control leads to a stable design. Generally we handle a small number of control points in designing our shapes. In addition, the designed curves pass through or interpolate some of the control points and come close to others, as shown in Figure 4.9(b). In computer-aided design, we tend to obtain smooth curves instead of interpolating all the control points.

4.4 Spline-Based Modeling

For parametric polynomial curves, we need to choose the degree of the curve. A high degree curve contains many parameters and evaluation of such a curve is expensive. A too-low-degree curve does not have enough parameters to work with. Most designers choose cubic polynomial curves. A cubic parametric polynomial can be represented as

$$p(u) = c_0 + c_1 u + c_2 u^2 + c_3 u^3 = \sum_{k=0}^{3} c_k u^k = u^T c, \qquad (4.107)$$

where

$$c = \begin{bmatrix} c_0 \\ c_1 \\ c_2 \\ c_3 \end{bmatrix}, \quad u = \begin{bmatrix} 1 \\ u \\ u^2 \\ u^3 \end{bmatrix} \quad and \quad c_k = \begin{bmatrix} c_{kx} \\ c_{ky} \\ c_{kz} \end{bmatrix}. \qquad (4.108)$$

c is a column matrix consisting of the polynomial coefficients. We will need to determine them using the control points. We seek to solve 12 equations in 12 unknowns. Since x, y and z are independent, we group these 12 equations into three independent sets, and each set contains our equations in four unknowns. The design of a particular cubic polynomial curve is based on the given control points. The requirements of the design include: (1) interpolating conditions or passing some given points; (2) interpolating some derivatives at certain parameter values; and (3) satisfying smoothness conditions at the joint points. Sometimes, we also need to satisfy the requirement that the curve passes close to several control points, yielding more than one curve solution.

In the following, let us talk about five basic types of cubic polynomial curves and surfaces: interpolating polynomial, Hermite polynomial, Bézier polynomial, B-splines and subdivisions.

4.4.1 Interpolating Curves and Surfaces

Suppose we have four control points p_0, p_1, p_2 and p_3 in 3D. We have $p_k = [x_k \ y_k \ z_k]^T$, where $k = 0, \cdots, 3$. We aim at finding the coefficients c such that the polynomial $p(u) = u^T c$ interpolates the four given control points. We have four 3D interpolating points, so we can obtain 12 equations in 12 unknowns. Generally, we choose the equally spaced u values such as

$u = \{0, \frac{1}{3}, \frac{2}{3}, 1\}$. The four equations are

$$p_0 = p(0) = c_0, \tag{4.109}$$

$$p_1 = p(\tfrac{1}{3}) = c_0 + \frac{1}{3}c_1 + (\tfrac{1}{3})^2 c_2 + (\tfrac{1}{3})^3 c_3, \tag{4.110}$$

$$p_2 = p(\tfrac{2}{3}) = c_0 + \frac{2}{3}c_1 + (\tfrac{2}{3})^2 c_2 + (\tfrac{2}{3})^3 c_3, \tag{4.111}$$

$$p_3 = p(1) = c_0 + c_1 + c_2 + c_3. \tag{4.112}$$

We can write them in the matrix form

$$p = Ac, \tag{4.113}$$

where

$$p = \begin{bmatrix} p_0 \\ p_1 \\ p_2 \\ p_3 \end{bmatrix} \quad and \quad A = \begin{bmatrix} 1 & 0 & 0 & 0 \\ 1 & \frac{1}{3} & (\frac{1}{3})^2 & (\frac{1}{3})^3 \\ 1 & \frac{2}{3} & (\frac{2}{3})^2 & (\frac{2}{3})^3 \\ 1 & 1 & 1 & 1 \end{bmatrix}. \tag{4.114}$$

A is a nonsingular matrix, so we invert it to obtain the *interpolating geometry matrix* M_I and we have

$$c = M_I p, \tag{4.115}$$

where

$$M_I = A^{-1} = \begin{bmatrix} 1 & 0 & 0 & 0 \\ -5.5 & 9 & -4.5 & 1 \\ 9 & -22.5 & 18 & -4.5 \\ -4.5 & 13.5 & -13.5 & 4.5 \end{bmatrix}. \tag{4.116}$$

Given a sequence of control points, we choose not to define a single interpolating curve with a high degree to avoid oscillation. Instead, we define a set of cubic interpolating curves. Each curve is defined using a group of four control points; see Figure 4.10. At the join points, the derivatives may not be continuous. Therefore, we need to handle the continuity at the join points specifically.

We substitute the interpolating coefficients into the polynomial and obtain

$$p(u) = u^T c = u^T M_I p, \tag{4.117}$$

which can be rewritten as

$$p(u) = b(u)^T p, \tag{4.118}$$

where

$$b(u) = M_I^T u. \tag{4.119}$$

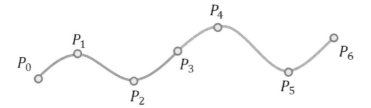

FIGURE 4.10: Joining of two interpolating segments. Each segment has four control points.

$b(u)$ is a column matrix consisting of the four *blending polynomials*

$$b(u) = \begin{bmatrix} b_0(u) \\ b_1(u) \\ b_2(u) \\ b_3(u) \end{bmatrix}.$$

(4.120)

Each blending polynomial is a cubic polynomial. We can express $p(u)$ in terms of these blending polynomials and we have

$$\begin{aligned} p(u) &= b_0(u)p_0 + b_1(u)p_1 + b_2(u)p_2 + b_3(u)p_3 \\ &= \sum_{i=1}^{3} b_i(u)p_i. \end{aligned}$$

(4.121)

The polynomials blend the individual contributions of each control point together, showing the effect of a given control point on the curve. However, the blending functions lack the derivative continuity at the join points shared by two adjacent segments. As shown in Figure 4.11, we have

$$b_0(u) = -\frac{9}{2}\left(u - \frac{1}{3}\right)\left(u - \frac{2}{3}\right)(u - 1),$$

(4.122)

$$b_1(u) = \frac{27}{2}u\left(u - \frac{2}{3}\right)(u - 1),$$

(4.123)

$$b_2(u) = -\frac{27}{2}u\left(u - \frac{1}{3}\right)(u - 1),$$

(4.124)

$$b_0(u) = \frac{9}{2}u\left(u - \frac{1}{3}\right)\left(u - \frac{2}{3}\right).$$

(4.125)

The interpolating curve can be easily extended to an interpolating surface patch. A bicubic surface patch can be represented as

$$p(u, v) = \sum_{i=0}^{3}\sum_{j=0}^{3} u^i v^j c_{ij},$$

(4.126)

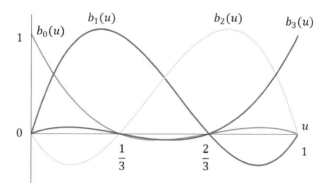

FIGURE 4.11: Four blending functions of an interpolating curve.

where c_{ij} is a three-element column matrix of the x, y and z coefficients. The surface patch can also be written as

$$p(u, v) = u^T C v, \tag{4.127}$$

where

$$C = [c_{ij}] \quad and \quad v = \begin{bmatrix} 1 \\ v \\ v^2 \\ v^3 \end{bmatrix}. \tag{4.128}$$

A bicubic polynomial patch is then defined by the 48 elements of C, or 16 three-element vectors.

Given 16 3D control points p_{ij} where $i = 0, \cdots, 3$ and $j = 0, \cdots, 3$, we can use them to define an interpolating surface patch as shown in Figure 4.12. Suppose u and v are equally spaced in the parametric space, like $\{0, \frac{1}{3}, \frac{2}{3}, 1\}$. We obtain three sets of 16 equations in 16 unknowns. For example, the three independent equations for $u = v = 0$ are

$$p_{00} = \begin{bmatrix} 1 & 0 & 0 & 0 \end{bmatrix} C \begin{bmatrix} 1 \\ 0 \\ 0 \\ 0 \end{bmatrix} = c_{00}. \tag{4.129}$$

Instead of writing down all the equations and solving them, here we process in a more direct way. Considering $v = 0$, we can get a curve in u that interpolates p_{00}, p_{10}, p_{20} and p_{30}. Using the results of interpolating curves, we can write this curve as

$$p(u, 0) = u^T M_I \begin{bmatrix} p_{00} \\ p_{10} \\ p_{20} \\ p_{30} \end{bmatrix} = u^T C \begin{bmatrix} 1 \\ 0 \\ 0 \\ 0 \end{bmatrix}. \tag{4.130}$$

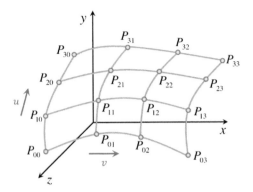

FIGURE 4.12: Cubic interpolating surface patch with 16 control points.

Similarly, $v = \frac{1}{3}, \frac{2}{3}, 1$ define three other interpolating curves, and each curve can be handled in the same manner. Putting all these four curves together, we write all 16 equations as

$$u^T M_I P = u^T C A^T, \tag{4.131}$$

where $A = M_I^{-1}$. We then have $C = M_I P M_I^T$ and the surface can be represented as

$$p(u, v) = u^T M_I P M_I^T v. \tag{4.132}$$

Blending polynomials can also be extended from curves to surfaces. Note that $M_I^T u$ describes the interpolating blending functions. A surface patch can be represented as

$$p(u, v) = \sum_{i=0}^{3} \sum_{j=0}^{3} b_i(u) b_j(v) p_{ij}, \tag{4.133}$$

where $b_i(u) b_j(v)$ represents a blending patch. This is basically a *tensor-product* technique. Variables in the tensor-product surface $p(u, v)$ are separable, and we have $p(u, v) = f(u)g(v)$, where $f(u)$ and $g(v)$ are row and column matrices, respectively.

4.4.2 Hermite Curves and Surfaces

Different from interpolating curves, *Hermite curves* start with only two control points (p_0 and p_3) and their derivatives as shown in Figure 4.13(a). We aim at building a curve interpolating these two points at $u = 0$ and $u = 1$, respectively. We have the following two conditions:

$$p(0) = p_0 = c_0, \tag{4.134}$$
$$p(1) = p_3 = c_0 + c_1 + c_2 + c_3. \tag{4.135}$$

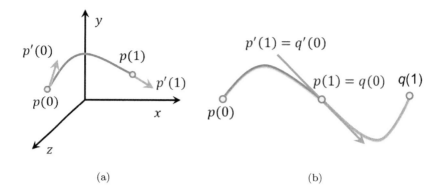

FIGURE 4.13: (a) Definition of the Hermite curve $p(u)$; and (b) two Hermite curve segments share the same derivative at a join point.

We can get two other conditions using the derivatives of the function at $u = 0$ and $u = 1$. We can easily derive the derivative of the polynomial,

$$p'(u) = \begin{bmatrix} \frac{dx}{du} \\ \frac{dy}{du} \\ \frac{dz}{du} \end{bmatrix} = c_1 + 2uc_2 + 3u^2 c_3. \tag{4.136}$$

If the two derivatives are denoted as p'_0 and p'_3, then the two other conditions are

$$p'_0 = p'(0) = c_1, \tag{4.137}$$
$$p'_3 = p'(1) = c_1 + 2c_2 + 3c_3. \tag{4.138}$$

We can write all these four equations in the matrix form

$$q = \begin{bmatrix} p_0 \\ p_3 \\ p'_0 \\ p'_3 \end{bmatrix} = \begin{bmatrix} 1 & 0 & 0 & 0 \\ 1 & 1 & 1 & 1 \\ 0 & 1 & 0 & 0 \\ 0 & 1 & 2 & 3 \end{bmatrix} c. \tag{4.139}$$

Then, we can obtain

$$c = M_H q, \tag{4.140}$$

where M_H is the *Hermite geometry matrix*

$$M_H = \begin{bmatrix} 1 & 0 & 0 & 0 \\ 0 & 0 & 1 & 0 \\ -3 & 3 & -2 & -1 \\ 2 & -2 & 1 & 1 \end{bmatrix}. \tag{4.141}$$

The resulting Hermite polynomial is then obtained by

$$p(u) = u^T M_H q. \tag{4.142}$$

According to the definition of Hermite polynomials, two neighboring curve segments share both the same function value and the derivative at the join point; see Figure 4.13(b). Therefore, both the resulting function and the first derivative are continuous over all the segments.

We then can obtain the increased smoothness of the Hermite representation by writing the polynomial as

$$p(u) = b(u)^T q, \tag{4.143}$$

where the new blending functions $b(u)$ are written as

$$b(u) = M_H^T u = \begin{bmatrix} 2u^3 - 3u^2 + 1 \\ -2u^3 + 3u^2 \\ u^3 - 2u^2 + u \\ u^3 - u^2 \end{bmatrix}. \tag{4.144}$$

None of these four polynomials is zero inside the interval $(0, 1)$, and they are much smoother than the interpolating polynomial blending functions.

Following the same manner, we can define a bicubic Hermite surface patch via a tensor product,

$$p(u, v) = \sum_{i=0}^{3} \sum_{j=0}^{3} b_i(u) b_j(v) q_{ij}, \tag{4.145}$$

where $Q = [q_{ij}]$ is the extension of q to surface. Four elements in Q are chosen to interpolate the patch corners, while the others match certain derivatives at the patch corners.

Supposing the left polynomial in Figure 4.14(a) is $p(u)$ and the right one is $q(u)$, we obtain various continuity conditions by matching the polynomials and their derivatives at $u = 1$ for $p(u)$, and at $u = 0$ for $q(u)$. To have continuous functions, we must satisfy

$$p(1) = \begin{bmatrix} p_x(1) \\ p_y(1) \\ p_z(1) \end{bmatrix} = q(0) = \begin{bmatrix} q_x(0) \\ q_y(0) \\ q_z(0) \end{bmatrix}. \tag{4.146}$$

All the parametric components must be the same at the join point. Such property is called C^0 parametric continuity; see Figure 4.14(a).

When we consider derivatives, we require

$$p'(1) = \begin{bmatrix} p'_x(1) \\ p'_y(1) \\ p'_z(1) \end{bmatrix} = q'(0) = \begin{bmatrix} q'_x(0) \\ q'_y(0) \\ q'_z(0) \end{bmatrix}. \tag{4.147}$$

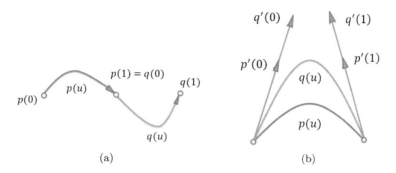

FIGURE 4.14: Geometry continuity between two parametric curve segments. (a) C^0 parametric continuity; and (b) G^1 parametric continuity.

When we match all the three parametric equations and the first derivatives, we then obtain C^1 parametric continuity. For a curve in the 3D space, the derivative at a point defines the tangent line at that point. If we only require the derivatives to be proportional or

$$p'(1) \propto q'(0), \tag{4.148}$$

then they point along the same direction, but they may have different magnitudes. Such continuity is called G^1 geometric continuity. The same idea can be extended to higher derivatives such as C^n and G^n geometric continuity. As shown in Figure 4.14(b), two different curves $p(u)$ and $q(u)$ share the same endpoints, and their tangents at the endpoints follow the same direction.

4.4.3 Bézier Curves and Surfaces

The Bézier curve is another type of spline. Given four control points as shown in Figure 4.15(a), p_0, p_1, p_2 and p_3. Suppose we interpolate the two endpoints with a cubic polynomial $p(u)$, then we have

$$p_0 = p(0), \tag{4.149}$$
$$p_3 = p(1). \tag{4.150}$$

Rather than interpolating the two middle control points (p_1 and p_2), we use them to compute the tangents at $u = 0$ and $u = 1$. As shown in Figure 4.15(a), in the parametric space we choose the linear approximations and obtain two equations

$$p'(0) = \frac{p_1 - p_0}{1/3} = 3(p_1 - p_0), \tag{4.151}$$
$$p'(1) = \frac{p_3 - p_2}{1/3} = 3(p_3 - p_2). \tag{4.152}$$

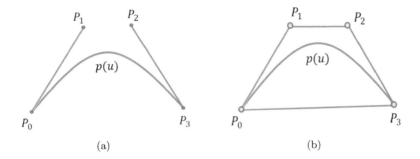

(a) (b)

FIGURE 4.15: (a) Definition of a Bézier curve with approximating tangents at the two endpoints; and (b) a Bézier polynomial with its convex hull.

We apply them to the derivatives of the parametric polynomial $p(u) = u^T c$ at the two endpoints, and then obtain the following two conditions:

$$3p_1 - 2p_0 = c_1, \qquad (4.153)$$
$$3p_3 - 3p_2 = c_1 + 2c_2 + 3c_3. \qquad (4.154)$$

From the two interpolation conditions, we have

$$p_0 = c_0, \qquad (4.155)$$
$$p_3 = c_0 + c_1 + c_2 + c_3. \qquad (4.156)$$

Again, we have three sets of four equations in four unknowns, which can be written in the matrix form

$$c = M_B p, \qquad (4.157)$$

where M_B is the Bézier geometry matrix

$$M_B = \begin{bmatrix} 1 & 0 & 0 & 0 \\ -3 & 3 & 0 & 0 \\ 3 & -6 & 3 & 0 \\ -1 & 3 & -3 & 1 \end{bmatrix}. \qquad (4.158)$$

The cubic Bézier polynomial is thus represented as

$$p(u) = u^T M_B p. \qquad (4.159)$$

Given a sequence of control points, $p_0, \cdots, p_n$, we use p_0, p_1, p_2 and p_3 to build the first curve; p_3, p_4, p_5 and p_6 for the second curve; and so on. It is obvious that we have C^0-continuity. If we use the same tangents on the left and right of a join point, we can automatically obtain C^1-continuity.

From the four blending functions in Figure 4.16, we can observe important advantages to the Bézier curve. The curve can be written as

$$p(u) = b(u)^T p, \qquad (4.160)$$

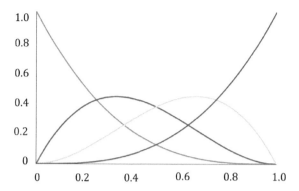

FIGURE 4.16: Blending polynomials of a cubic Bézier curve.

where

$$b(u) = M_B^T u = \begin{bmatrix} (1-u)^3 \\ 3u(1-u)^2 \\ 3u^2(1-u) \\ u^3 \end{bmatrix}. \tag{4.161}$$

These four polynomials are actually Bernstein polynomials

$$b_{kd}(u) = \frac{d!}{k!(d-k)!} u^k (1-u)^{d-k}. \tag{4.162}$$

Bernstein polynomials have two remarkable properties. First, the polynomials can only be zero at the two endpoints, $u = 0$ and $u = 1$. Consequently each blending polynomial must be smooth and positive for $0 < u < 1$. Second, they satisfy the partition of unity property $\sum_{i=0}^{d} b_{id}(u) = 1$ in the interval $b_{id} \leq 1$. Therefore, $p(u)$ lies in the convex hull defined by the four control points. We have $p(u) = \sum_{i=0}^{3} b_i(u) p_i$ as shown in Figure 4.15(b).

Using a tensor product of blending functions, we generate the Bézier surface patches. If P is a 4×4 array of control points, $P = [p_{ij}]$, then the corresponding Bézier patch is defined as

$$\begin{aligned} p(u,v) &= \sum_{i=0}^{3} \sum_{j=0}^{3} b_i(u) b_j(v) p_{ij} \\ &= u^T M_B P M_B^T v. \end{aligned} \tag{4.163}$$

As shown in Figure 4.17(a), the patch is fully contained in the convex hull defined by the control points and interpolates four corners p_{00}, p_{03}, p_{30} and p_{33}. The other conditions can be obtained using various derivatives at the patch corners.

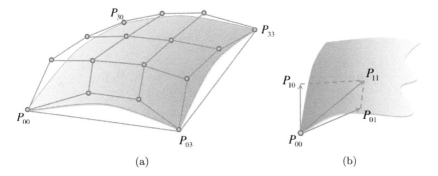

FIGURE 4.17: (a) A Bézier surface patch fully contained in the convex hull of the control points; and (b) the twist at a corner of the Bézier surface patch.

Considering one corner like $u = v = 0$, we can evaluate $p(u, v)$ and the first partial derivatives to get

$$p(0,0) = p_{00}, \tag{4.164}$$

$$\frac{\partial p}{\partial u}(0,0) = 3(p_{10} - p_{00}), \tag{4.165}$$

$$\frac{\partial p}{\partial v}(0,0) = 3(p_{01} - p_{00}), \tag{4.166}$$

$$\frac{\partial^2 p}{\partial u \partial v}(0,0) = 9(p_{00} - p_{01} + p_{10} - p_{11}). \tag{4.167}$$

We can obtain the first three conditions from extensions of the Bézier curve. The last one represents a measure of the tendency of the patch to divert from being flat or to twist at the corner; see Figure 4.17(b).

4.4.4 Cubic B-Splines

Cubic Bézier curves and surface patches have only C^0-continuity at the join points or patch edges for surfaces. There are several possible ways to resolve this limitation. We can choose high degree polynomials, short intervals and more polynomial segments, or use the same control points but do not require the polynomials to interpolate any control points.

Consider four control points in a sequence of control points $\{p_{i-2}, p_{i-1}, p_i, p_{i+1}\}$. The previous approaches use them to define a cubic curve, which interpolates p_{i-2} and p_{i+1} and spans from p_{i-2} to p_{i+1} with $u \in [0, 1]$. Suppose for $u \in [0, 1]$, we only span the distance between the two middle control points or from p_{i-1} to p_i; see Figure 4.18(a). Similarly, we use $\{p_{i-3}, p_{i-3}, p_{i-1}, p_i\}$ to span between p_{i-2} and p_{i-1}, and $\{p_{i-1}, p_i, p_{i+1}, p_{i+2}\}$ to span between p_i and p_{i+1}; see the five control points in Figure 4.18(b). Suppose $p(u)$ is the curve to span between p_i and p_{i+1}, and $q(u)$ is the curve to its left between p_{i-2} and p_{i-1}. We then can match conditions at the join point using $p(0)$

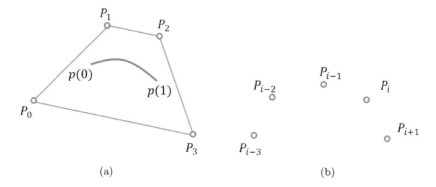

(a)   (b)

FIGURE 4.18: (a) Four control points define a cubic B-spline curve within the convex hull and between the two middle points; and (b) five control points.

and $q(1)$. We aim at constructing a matrix M, with which the desired cubic polynomial is written as

$$p(u) = u^T M p, \tag{4.168}$$

where p is the matrix of control points,

$$p = \begin{bmatrix} p_{i-2} \\ p_{i-1} \\ p_i \\ p_{i+1} \end{bmatrix}. \tag{4.169}$$

$q(u)$ can be represented using the same matrix M, and we have

$$q(u) = u^T M q, \tag{4.170}$$

where

$$q = \begin{bmatrix} p_{i-3} \\ p_{i-2} \\ p_{i-1} \\ p_i \end{bmatrix}. \tag{4.171}$$

Generally, we can set up a set of conditions at $p(0)$ matching with conditions at $q(1)$, and we can also set up conditions matching various derivatives at $p(1)$. In the following, let us derive the most popular matrix using symmetric approximations at the join point. Therefore, any evaluation of conditions at $q(1)$ cannot use p_{i-3} because it does not exist in $p(u)$. Similarly, we cannot use p_{i+1} in any condition at $p(0)$. We can obtain two conditions satisfying this symmetry condition,

$$p(0) = q(1) = \frac{1}{6}(p_{i-2} + 4p_{i-1} + p_i), \tag{4.172}$$

$$p'(0) = q'(1) = \frac{1}{2}(p_i - p_{i-2}). \tag{4.173}$$

$p(u)$ can be represented in terms of the coefficient array c,

$$p(u) = u^T c, \qquad (4.174)$$

and these conditions are

$$c_0 = \frac{1}{6}(p_{i-2} + 4p_{i-1} + p_i), \qquad (4.175)$$

$$c_1 = \frac{1}{2}(p_i - p_{i-2}). \qquad (4.176)$$

We can apply the symmetric conditions at $p(1)$ and obtain

$$p(1) = c_0 + c_1 + c_2 + c_3 = \frac{1}{6}(p_{i-1} + 4p_i + p_{i+1}), \qquad (4.177)$$

$$p'(1) = c_1 + 2c_2 + 3c_3 = \frac{1}{2}(p_{i+1} - p_{i-1}). \qquad (4.178)$$

We now obtain four equations for the coefficients c, which can be used to compute the *B-spline geometry matrix* M_S, and we have

$$M_S = \begin{bmatrix} 1 & 4 & 1 & 0 \\ -3 & 0 & 3 & 0 \\ 3 & -6 & 3 & 0 \\ -1 & 3 & -3 & 1 \end{bmatrix}. \qquad (4.179)$$

This B-spline geometry matrix produces a polynomial with several important properties, which can be observed by examining the blending polynomials as shown in Figure 4.19(a),

$$b(u) = M_S^T u = \frac{1}{6} \begin{bmatrix} (1-u)^3 \\ 4 - 6u^2 + 3u^3 \\ 1 + 3u + 3u^2 - 3u^3 \\ u^3 \end{bmatrix}. \qquad (4.180)$$

We can show that the blending functions satisfy partition of unity, or we have

$$\sum_{i=0}^{3} b_i(u) = 1, \qquad (4.181)$$

where $0 \le b_i(u) \le 1$ in the interval $0 \le u \le 1$. The curve lies in the convex hull of the control points. Note that the curve only spans part of the range of the convex hull; see Figure 4.18(a). The defined curve is C^1-continuous; actually it is C^2-continuous. This can be verified by computing the second derivative $p''(u)$ at $u = 0$ and $u = 1$. Compared to Bézier cubic, computing cubic B-splines is much more time-consuming.

Each control point contributes to the resulting cubic B-spline in its four

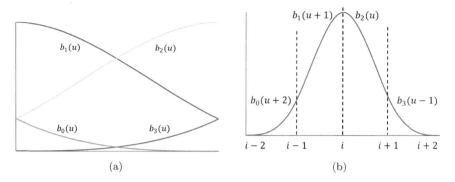

(a)   (b)

FIGURE 4.19: (a) Four blending functions of a cubic B-spline curve; and
(b) cubic B-spline basis functions.

adjacent intervals. This property provides a guarantee on the locality of the
spline, which means that if we change a single control point, we only affect
the resulting curve in its four adjacent intervals. The total contribution of
a control point can be represented as $B_i(u)p_i$, where $B_i(u)$ is a piecewise
polynomial function shown in Figure 4.19(b),

$$B_i(u) = \begin{cases} 0 & u < i - 2, \\ b_0(u+2) & i-2 \le u < i-1, \\ b_1(u+1) & i-1 \le u < i, \\ b_2(u) & i \le u < i+1, \\ b_3(u-1) & i+1 \le u < i+2, \\ 0 & u \ge i+2. \end{cases} \tag{4.182}$$

Given control points $P_0, \cdots, P_m$, we can represent the entire spline curve
as

$$p(u) = \sum_{i=1}^{m-1} B_i(u)p_i. \tag{4.183}$$

For the set of functions $B(u - i)$, each member is obtained by shifting a
single function as shown in Figure 4.20. They form a set of basis functions
for all the cubic B-spline curves. Given a set of control points, we use a linear
combination of these basis functions to construct a piecewise polynomial curve
$p(u)$ spanning the entire interval.

Similarly, B-spline surfaces can be defined via a tensor product. Given the
B-spline blending functions, the surface patch is represented as

$$p(u, v) = \sum_{i=0}^{3} \sum_{j=0}^{3} b_i(u)b_j(v)p_{ij}. \tag{4.184}$$

As shown in Figure 4.21, the B-spline patch is defined only over the central

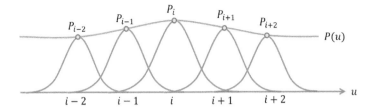

FIGURE 4.20: Approximation function over interval as a linear combination of basis functions.

area. We need to perform the work nine times of what we do for the Bézier patch. Due to the inheritance of the convex hull property and the additional continuity of the B-spline curves, the B-spline surface patch is smoother than a Bézier surface patch.

Given a set of control points $P_0, \cdots, P_m$, we tend to build a general B-spline function $p(u) = [x(u) \ y(u) \ z(u)]^T$ over an interval $u_{min} \leq u \leq u_{max}$, which should be smooth and also close to the control points. Suppose we have a set of knots $\{u_k\}$ such that

$$u_{min} = u_0 \leq u_1 \leq \cdots \leq u_n = u_{max}. \tag{4.185}$$

The sequence $u_0, u_1, \cdots, u_n$ is called the knot array. In splines, the function $p(u)$ is a degree-d polynomial in the form

$$p(u) = \sum_{j=0}^{d} c_{jk} u^j, \tag{4.186}$$

where $u_k < u < u_{k+1}$. To define a degree-d spline, we need to define $n(d+1)$ 2D coefficients c_{jk}. We apply various continuity requirements at the knots and also interpolation requirements at control points to obtain the required conditions.

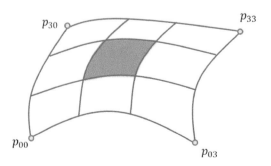

FIGURE 4.21: A cubic B-spline surface patch.

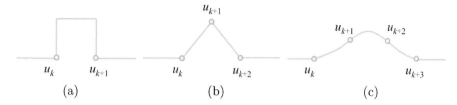

FIGURE 4.22: (a) A degree-0 basis function; (b) a degree-1 basis function; and (c) a degree-2 basis function.

B-splines can be defined recursively using a set of basis or blending functions. Each basis function is nonzero only over the local region spanned by a few knots. We have

$$p(u) = \sum_{i=0}^{m} B_{id}(u) p_i,$$ (4.187)

where each basis function $B_{id}(u)$ is a degree-d polynomial and equals to zero outside the interval (u^i_{min}, u^i_{max}).

The terminology *B-splines* comes from *basis splines*, where the functions $\{B_{id}(u)\}$ form a set of basis for the given knot sequence and degree. As shown in Figure 4.22, B-splines are defined recursively by the Cox–deBoor recursion, and we have

$$B_{k0} = \begin{cases} 1 & u_k \leq u \leq u_{k+1} \\ 0 & otherwise \end{cases}$$ (4.188)

and

$$B_{kd} = \frac{u - u_k}{u_{k+d} - u_k} B_{k,d-1}(u) + \frac{u_{k+d} - u}{u_{k+d+1} - u_{k+1}} B_{k+1,d-1}(u).$$ (4.189)

B-splines are C^{d-1}-continuous at the knots. The convex hull property holds because the basis functions satisfy partition of unity and are nonnegative. We have

$$\sum_{i=0}^{m} B_{id}(u) = 1 \quad and \quad 0 \leq B_{id}(u) \leq 1.$$ (4.190)

According to the locality of the splines, each $B_{id}(u)$ is nonzero in only $d + 1$ intervals, and each control point affects only $d + 1$ intervals. In addition, each point on the resulting curve is within the convex hull defined by these $d + 1$ control points. Note that each recursion is a linear interpolation of basis functions produced in the previous step, and the linear interpolation of degree-k polynomials yields a degree-$(k + 1)$ polynomial.

As shown in Figure 4.23(a), uniform knot interval in the knot vector yields uniform splines. Given the uniform knot sequence $\{0, 1, 2, \cdots, n\}$, the cubic B-spline can be obtained from the Cox–deBoor formula with equally spaced knots. For closed spline curves in Figure 4.23(b), we use the periodic nature of the control points to define the spline over the entire knot sequence. Again, each spline basis function is obtained by shifting a single function.

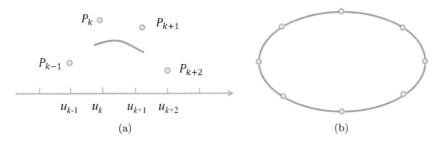

(a) (b)

FIGURE 4.23: (a) Uniform B-spline; and (b) periodic uniform B-spline.

In addition, we also have non-uniform B-splines with non-uniform knot intervals. For example, repeated knots happen in the knot vector sometimes, which can pull the spline closer to the control point associated with the knot. If a knot at the end has a multiplicity of $d+1$, the degree-d B-spline interpolates or passes through that point. Repeating knots at the ends can force the splines to interpolate the endpoints; everywhere else we can still keep uniform knots. Such kind of spline is an *open spline*. The knot vector $\{0, 0, 0, 0, 1, 2, \cdots, n-1, n, n, n, n\}$ is often used to define a cubic *open spline*. In $\{0, 0, 0, 0, 1, 1, 1, 1\}$, the cubic B-spline becomes the cubic Bézier curve. Internal knots can also be repeated.

The definition of non-uniform B-splines can be easily extended to 2D and 3D. We can also represent the B-splines using four dimensions. Consider a control point in 3D, $p_i = [x_i \ y_i \ z_i]$. The weighted homogeneous-coordinate representation of this point is

$$q_i = w_i \begin{bmatrix} x_i \\ y_i \\ z_i \\ 1 \end{bmatrix}. \tag{4.191}$$

Here, we use the weights w_i to increase or decrease the contribution of a particular control point. We can use these weighted points to build a four-dimensional B-spline. We obtain the first three components of the resulting spline using the B-spline representation of the weighted points,

$$q(u) = \begin{bmatrix} x(u) \\ y(u) \\ z(u) \end{bmatrix} = \sum_{i=0}^{n} B_{id}(u) w_i p_i. \tag{4.192}$$

The w component is the scalar B-spline polynomial which can be derived from the set of weights,

$$w(u) = \sum_{i=0}^{n} B_{id}(u)w_i. \tag{4.193}$$

In homogeneous coordinates, this representation contains a w that may not be 1, and we can rationalize it to obtain

$$p(u) = \frac{1}{w(u)}q(u) = \frac{\sum_{i=0}^{n} B_{id}(u)w_i p_i}{\sum_{i=0}^{n} B_{id}(u)w_i}. \tag{4.194}$$

Now each component of $p(u)$ becomes a rational function in terms of u. This yields a *non-uniform rational B-spline* (NURBS) curve. NURBS curves have several important properties. They retain all the properties of 3D B-splines, such as the convex hull and continuity properties. NURBS curves can be handled exactly in perspective views. This property is very useful for computer graphics and visualization. It has been shown that quadrics is a special case of quadratic NURBS curve, therefore NURBS curves can be used to represent most widely used curves and also be extended to surfaces.

Moreover, in recent years a new technique named T-spline [334] has been developed based on NURBS, supporting local refinement with T-junctions. It has been used in computer-aided design and isogeometric analysis [179]. We will discuss them later in Chapters 7 and 8.

4.4.5 Subdivision

First, let us take a look at recursive subdivision of a cubic Bézier polynomial. As we all know, the curve must lie inside the convex hull of the control points; see Figure 4.24(a). We can split the curve into two separate polynomials, the left one $l(u)$ and the right one $r(u)$, which are both valid over half of the original interval and also cubic. Now we need to rescale the parameter u for l and r. The convex hull for l and r must lie inside the convex hull for p, which comes from the *variation-diminishing property* of the Bézier curve.

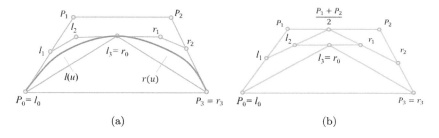

FIGURE 4.24: (a) Convex hulls and control points; and (b) recursive subdivision of Bézier polynomials.

Let us first find the convex hull for $l(u)$; the computation for $r(u)$ is similar. We start with

$$p(u) = u^T M_B \begin{bmatrix} p_0 \\ p_1 \\ p_2 \\ p_3 \end{bmatrix}, \tag{4.195}$$

where

$$M_B = \begin{bmatrix} 1 & 0 & 0 & 0 \\ -3 & 3 & 0 & 0 \\ 3 & -6 & 3 & 0 \\ -1 & 3 & -3 & -1 \end{bmatrix}. \tag{4.196}$$

The polynomial $l(u)$ must interpolate the two endpoints $p(0)$ and $p\left(\frac{1}{2}\right)$. Therefore, we have

$$l(0) \quad = \quad l_0 = p_0, \tag{4.197}$$
$$l(1) \quad = \quad l_3 = p\left(\frac{1}{2}\right) = \frac{1}{8}(p_0 + 3p_1 + 3p_2 + p3). \tag{4.198}$$

In particular, at $u = 0$ the slope of l must match with the slope of p. Note that the parameter for p only spans the range $\left(0, \frac{1}{2}\right)$ while u varies over $(0, 1)$. Implicitly, we have the substitution $\bar{u} = 2u$ and $d\bar{u} = 2du$. Then we can obtain

$$l'(0) = 3(l_1 - l_0) = p'(0) = \frac{3}{2}(p_1 - p_0). \tag{4.199}$$

Similarly, at the midpoint we obtain

$$l'(1) = (l_3 - l_2) = p'\left(\frac{1}{2}\right) = \frac{3}{8}(-p_0 - p_1 + p_2 + p_3). \tag{4.200}$$

These four equations can be solved analytically and also be expressed geometrically. In the following, we construct both the left and right sets of control points as shown in Figure 4.24(b). Note that the interpolation condition requires that

$$l_0 = p_0, \tag{4.201}$$
$$r_3 = p_3. \tag{4.202}$$

We can verify that the slopes on the left and right lead to

$$l_1 = \frac{1}{2}(p_0 + p_1), \tag{4.203}$$
$$r_2 = \frac{1}{2}(p_2 + p_3). \tag{4.204}$$

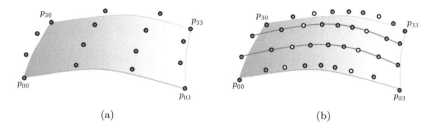

(a)   (b)

FIGURE 4.25: Subdivision of Bézier surfaces. (a) Before subdivision with 16 control points (blue dots); and (b) after subdivision with some control points retained (8 blue dots), some control points discarded (8 white dots) and some new control points created (20 orange dots).

Then we can also obtain the interior points

$$l_2 = \frac{1}{2}\left(l_1 + \frac{1}{2}(p_1 + p_2)\right), \tag{4.205}$$

$$r_1 = \frac{1}{2}\left(r_2 + \frac{1}{2}(p_1 + p_2)\right). \tag{4.206}$$

Finally we obtain the shared middle point

$$l_3 = r_0 = \frac{1}{2}(l_2 + r_1). \tag{4.207}$$

We can observe that subdivision approaches support adaptive refinement and we can subdivide only one side if required. Recursive subdivision is used a lot in geometric modeling and rendering. The subdivision algorithm can be easily extended to Bézier surfaces. Given a cubic surface patch with 16 control points as shown in Figure 4.25(a), four points in a row or column determine a Bézier curve that can be subdivided. In 2D, we should subdivide the patch into four smaller patches and calculate new control points; see Figure 4.25(b). Some control points are discarded and some are retained after the subdivision.

In curve subdivision as shown in Figure 4.26, we start with four control points, p_0, p_1, p_2 and p_3, and obtain seven new control points, s_0, $\cdots$, s_6. We can consider each set of points as defining a piecewise-linear curve. The second curve is a refinement of the first. We continue the subdivision process iteratively and converge to the B-spline in the limit. We can obtain

$$\begin{aligned}
s_0 &= p_0, & s_4 &= \tfrac{1}{4}(p_1 + 2p_2 + p_3),\\
s_1 &= \tfrac{1}{2}(p_0 + p_1), & s_5 &= \tfrac{1}{2}(p_2 + p_3),\\
s_2 &= \tfrac{1}{4}(p_0 + 2p_1 + p_2), & s_6 &= p_3.\\
s_3 &= \tfrac{1}{8}(p_0 + 3p_1 + 3p_2 + p_3),
\end{aligned}$$

The subdivision technique and its benefits are not only limited to B-splines.

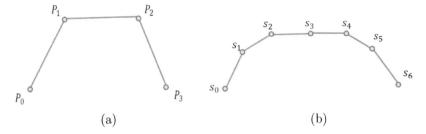

(a)   (b)

FIGURE 4.26: Piecewise-linear curve. (a) Four control points; and (b) after one subdivision step.

Over the years, a variety of methods have been developed to generate subdivision curves and the refined curves converge to a smooth curve. Subdivision can also be extended to surfaces represented by triangles, rectangles and star-shaped polygons.

A subdivision surface represents a smooth surface via the specification of a coarser piecewise linear polygonal mesh. The smooth *limit surface* can be computed from the coarse control mesh as the limit of iteratively subdividing each polygonal face into smaller ones. It is recursive in nature. Starting from a given polygonal control mesh, a refinement scheme is applied to the mesh with new vertices and faces produced.

The limit surface is defined by the initial mesh and that after subdivision, the newly generated control points may not lie on the limit surfaces. There are several subdivision schemes developed. Catmull-Clark subdivision is based on bi-cubic uniform B-splines and works for arbitrary quadrilateral meshes. As shown in Figure 4.27(a), it subdivides a quadrilateral element into four smaller identical ones using the edge middle points and the face center. For arbitrary initial control meshes, this scheme generates limit surfaces that are C^2-continuous everywhere except at extraordinary nodes where the continuity is C^1.

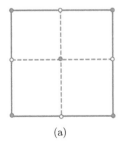 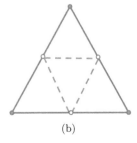

(a)   (b)

FIGURE 4.27: (a) Catmull-Clark subdivision for quadrilaterals; and (b) Loop's subdivision for triangles.

Doo-Sabin subdivision uses the analytical expression of bi-quadratic uniform B-spline surface to generate C^1-continuous limit surfaces with arbitrary topology for arbitrary initial control meshes. Loop's subdivision only works for triangles, and it uses a quartic box-spline of six direction vectors to provide a rule to generate C^2-continuous limit surfaces everywhere except at extraordinary nodes where the continuity is C^1. As shown in Figure 4.27(b), a triangle is subdivided into four smaller identical ones using middle edge points. Such mid-edge subdivision scheme generates C^1-continuous limit surfaces on initial meshes with arbitrary topology.

4.5   Isocontouring and Visualization

In recent years, the scanning techniques have been developed rapidly. For each spatial grid point in the image, it may have a scalar value (like X-ray intensity), a vector (like velocity in a fluid) or even a tensor (like stresses or strains) attached to it. Their scientific and medical applications require fast and efficient visualization of large scalar, vector and tensor fields, as well as high-fidelity geometric modeling from volumetric images. For example, we need to handle large size datasets like the visible-women data, which are as large as $1734 \times 512 \times 512$. Volumetric data are not geometric models and they need to be translated to geometries. In addition to static images, the scanned images may represent dynamic or time-varying information, which need even more sophisticated visualization techniques. Visualization of high-dimensional data like vector or tensor fields is challenging.

Geometric models exist in images implicitly, and image visualization is basically a display of implicit functions or isocontouring. Consider the implicit function $f(x, y) = 0$. Given an x, in general we cannot directly find the corresponding y. However, given a pair of x and y, we can easily check if they lie on an isocontour or not. In many applications, we have the height function in the form of $z = f(x, y)$. To find all the points with a given height c, we need to solve the implicit equation $g(x, y) = f(x, y) - c = 0$. As we all know, sometimes we may have the analytical solution of an implicit function, but there is no general solution to convert it to the explicit representation or represent one variable using the others. For a given c, the function $g(x, y)$ describes a curve corresponding to a constant $z = c$ in the equation of $f(x, y)$, which is the contour curve or isocontour.

In image visualization, many times we need to extract some isocontours or isosurfaces. In the following we will discuss two main isocontouring methods, Marching Cubes and Dual Contouring.

4.5.1 Marching Cubes

Given discretized sampling data or scanned images, we need a technique to extract the contour from them. The Marching Squares technique is such an isocontouring technique for 2D images, which is also a simplified version of the Marching Cubes technique [256] in 3D. The Marching Squares technique displaces isocurves or contours for function $f(x, y) = c$. For a continuous function $f(x, y)$, we first sample it on a regular grid with $\{f_{i,j}(x, y)\}$, where i and j are indices of the grids along the x and y directions, respectively. $\{f_{i,j}(x, y)\}$ can also be obtained directly from scanned images. For each pixel cell as shown in Figure 4.28(a), it has four corners and each corner has a function value. We have $f_{i,j}(x, y)$, $f_{i+1,j}(x, y)$, $f_{i+1,j+1}(x, y)$ and $f_{i,j+1}(x, y)$. These function values are compared with the isovalue c and we color the corner green if its function value is greater than or equal to c; otherwise we color it white. By using this way, we convert the scalar field into binaries.

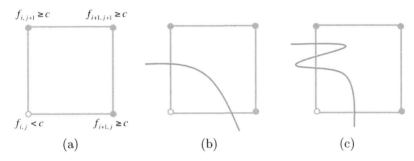

(a)   (b)   (c)

FIGURE 4.28: (a) Coloring the corners for a binary representation by comparing the function values with the isovalue; (b-c) Principle of Occum's Razor.

In a pixel, if an edge has two ending points with different colors, we call such an edge *sign change edge*. An isocontour must intersect a sign change edge an odd number of times. As shown in Figure 4.28(b-c), we always pick the simplest interpretation that we assume only one crossing happens, following the "*Principle of Occum's Razor.*" We then check all the possibilities of a pixel configuration. Since each pixel has four corners and each corner has only two possibilities, therefore we have $2^4 = 16$ possibilities. Considering the symmetry and complementary, they can be reduced to only four unique cases, as shown in Figure 4.29. All the other configurations can be mapped into these four unique cases. It is obvious that the first three unique cases yield a unique and simple interpretation. However, the last unique case has two diagonal corners with the same color, yielding two possible interpretations as shown in Figure 4.30(a-b). For such an ambiguity case, the isocontour topology can have many different local possibilities; see Figure 4.30(c-e).

This is actually a sampling problem because we are limited by the sampling resolution and there is no enough local information to get the details. We do not know the mathematical solution either without extra information. Due

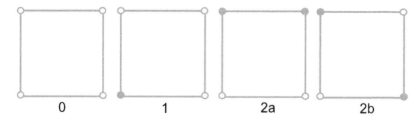

FIGURE 4.29: Four unique cases of Marching Squares.

to this reason, "wrong" interpretations may be adopted yielding holes in a surface when we extend Marching Squares to Marching Cubes for volumetric data. Super-sampling can be a way to provide more resolution information to certain local regions. Another way to resolve this topology ambiguity problem is to check a larger local area and use the neighborhood information to help us decide the correct topology.

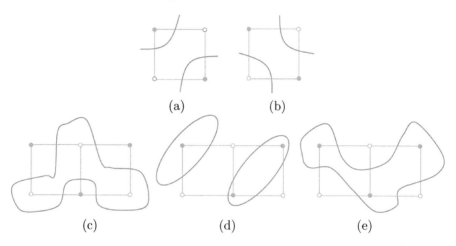

FIGURE 4.30: Various ambiguous cases in Marching Squares.

Once we decide the correct interpretation for each pixel cell, next we need to compute the intersection point between the sign change edge and the iso-contour. Here we choose a linear interpolation using the function values at the two ending points and also the isovalue c. For example, for an edge with two corners, we have $f(x_i, y_j) = a$ and $f(x_{i+1}, y_j) = b$, where $a > c$ and $b < c$. Suppose the x spacing is Δx, then we can obtain the intersection point using a line segment,

$$x = x_i + \frac{(a - c)\Delta x}{a - b}. \tag{4.208}$$

The Marching Squares technology has several advancements. To extract isocontour efficiently, we can start with a cell intersected by the isocontour, and then follow this contour to adjacent cells as necessary to complete the contour. In this way, we only need to visit cells along the isocontour, ignoring all the cells away from the boundary. In Marching Squares, all the cells can be dealt with independently, therefore it is straightforward to extend it to 3D for volumetric data, Marching Cubes.

As shown in Figure 4.31(a), a volumetric dataset is discretized sampling data from a set of measurements or scanning. It can also be obtained by evaluating the function at a set of points. Suppose our samples are taken at equally spaced points in x, y and z, we have

$$x_i = x_0 + i\Delta x,$$
$$y_i = y_0 + j\Delta y,$$
$$z_i = z_0 + k\Delta z,$$

and

$$f_{ijk} = f(x_i, y_j, z_k).$$

Each f_{ijk} can be considered as the average value of the scalar field inside this parallelepiped volume element, or voxel.

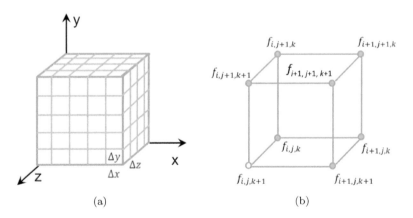

(a) (b)

FIGURE 4.31: (a) A volumetric dataset; and (b) a voxel cell.

Computer visualization and engineering calculation, like finite element analysis, require high-fidelity visualization. There are two main approaches to visualize a volumetric dataset: direct volume rendering and isocontouring. Direct volume rendering displaces each voxel with assigned color and opacity in producing an image and all voxels contribute to the final image. Differently, isocontouring extracts the isosurface $f(x, y, z) = c$ using only a subset of the voxels, where c is the isovalue. Similar to Marching Squares, for a given isovalue c, we color each cell corner and convert to binaries by comparing its

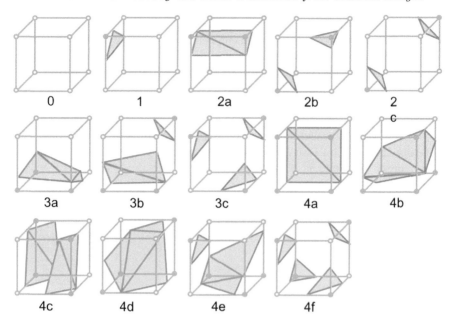

FIGURE 4.32: 14 unique cases of Marching Cubes.

function value with the isovalue; see the example in Figure 4.31(b). A cell has a total of 8 corners and each corner has two possibilities, therefore we have a total of $2^8 = 256$ possible corner coloring configurations. Considering symmetry and complementary, these configurations can reduce to 14 unique cases, as shown in Figure 4.32. For each case, we first compute all the intersection points between the isosurface and the cell edges using linear interpolation, and then use the triangular polygons to tessellate these intersections. We obtain a local triangular mesh passing through the cell. In most cases, an isosurface is extracted in the form of piecewise linear approximation for the modeling and rendering purpose. The Marching Cubes algorithm (MC) [256] visits each cell in a volume and performs local triangulation based on the sign configuration of the eight vertices. To avoid visiting unnecessary cells, accelerated algorithms [35, 407] were developed to minimize the time to search for contributing cells.

When two diagonal corners have the same configuration as shown in Figure 4.33, we have ambiguities as in 2D. The same configuration has two possible interpretations, leading to completely different shapes and topologies. The wrong selection of an interpretation may leave holes to a smooth surface. We need more local information to help us decide the correct topology in that local region. The isosurfaces of a function defined by the trilinear interpolation inside a cubic cell may have a complicated shape and topology which cannot be reconstructed correctly using MC. The function values of face and body saddles in a cell can be used to decide the correct topology and consistent triangulation of an isosurface in the cell [276]. Lopes and Brodlie [255] provided a more accurate triangulation.

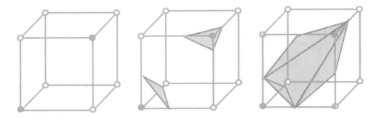

FIGURE 4.33: Ambiguitiy examples in 3D.

The MC technique has several drawbacks. It generates uniform and denser triangles than are really needed, therefore we have to simplify or decimate the obtained mesh. In addition, many badly shaped triangles are usually generated. Edge contraction is a common method for triangle decimation. Sometimes we use local surface smoothness and the aspect ratio of triangles as the criteria for simplification. Similar to edge contraction, the grid snapping method [274] merges mesh vertices with grids, improving the aspect ratio of the elements. In [49], a planar points cloud was taken as input to generate triangular meshes with bounded aspect ratios. In addition, various methods like quality mesh generation (QMG) [272] were proposed to mesh a d-dimensional region with a bounded aspect ratio. We can also resample the surface to create a new set of points cloud and then use Delaunay triangulation [109] to remesh it (details of Delaunay triangulation will be given in the next chapter).

Moveover, the MC technique cannot preserve sharp features in the data. An adaptive isosurface can be generated by triangulating cells with different levels. When the adjacent cubes have different resolution levels, the cracking problem happens. To keep the face gap-free, the mass center of the coarser triangle is inserted, which helps build a set of triangles to represent the local isosurface seamlessly [401]. The chain-gang algorithm [216] was described to render isosurface of super adaptive resolution, which can resolve the data discontinuities. In [291], progressive multiresolution representation and recursive subdivision were coupled together, and the edge bisection method was applied to construct and smooth isosurfaces. In addition, a surface wave-front propagation technique [408] was developed to produce multiresolution meshes with good aspect ratios.

4.5.2 Dual Contouring

Unlike Marching Cubes, the enhanced distance field representation and the extended MC algorithm [207] can detect and reconstruct sharp features on the isosurface. By combining SurfaceNets [155] and the extended marching cubes algorithm [207], the octree-based dual contouring method [189] was developed to generate adaptive isosurfaces with good aspect ratio and also preserve sharp features. First, we use the isosurface accuracy as a metric to build adaptive octree structure. For regions with more detailed features, higher level octree cells are used. As shown in Figure 4.34, the dual contouring method defines a quadratic error function in each leaf cell

$$QEF(x) = \sum_i (n_i \cdot (x - p_i))^2 , \qquad (4.209)$$

where p_1 and p_2 are two intersection points of the isocontour (the orange curve) with the sign change edges (the red edges), and n_1 and n_2 are two unit normal vectors at p_1 and p_2, respectively. A *sign change edge* is a cell edge whose two endpoints lie on different sides of the isocontour. A minimizer point (the red dot) is then computed by minimizing this error function, which is also the intersection of two tangent lines in 2D. Note that the two tangent lines may not intersect within the current cell, leading to possible self-intersection or bad isocontouring result. To resolve this issue, we choose the cell center or the average of the intersection points as the minimizer.

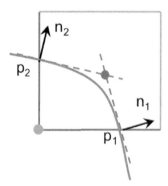

FIGURE 4.34: A minimizer point is computed in a cell by minimizing the pre-defined quadratic error function in Eq. (4.209). The orange curve is the isocontour, the two red edges are sign change edges, and the red dot is the minimizer point.

As we all know that isocontour passes across all the sign change edges, therefore we analyze each sign change edge for isocontouring. As shown in Figure 4.35, in 2D each sign change edge is always shared by two leaf cells no matter if it is uniform or adaptive; while in 3D it is shared by four cells in the uniform case and three or four cells in the adaptive case. We connect the minimizer points associated with these leaf cells to approximate the isocontour piecewise. The resulting isosurface consists of all-quadrilateral elements for the uniform case and a hybrid of triangles and quadrilaterals for the adaptive case. Compared to Marching Cubes, the octree-based dual contouring method

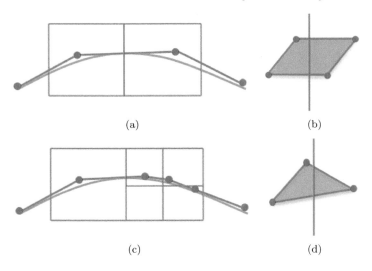

(a)   (b)

(c)   (d)

FIGURE 4.35: Analyzing sign change edges to extract isocontours in 2D (a, c) and isosurfaces in 3D (b, d). The orange curve is the isocontour, the red edges are sign change edges, and the red dots are minimizer points.

yields adaptive meshes with better aspect ratio, which makes it suitable for finite element mesh generation.

Based on these isocontouring algorithms, in the following two chapters we will discuss how to build triangular and tetrahedral, as well as quadrilateral and hexahedral meshes from volumetric images for finite element applications.

Homework

Problem 4.1. Given a triple of noncollinear points P_0, P_1 and P_2 in the plane, any other point Q in the plane can be expressed uniquely as an affine combination of these three:

$$Q = \omega_0 P_0 + \omega_1 P_1 + \omega_2 P_2,$$

where

$$\omega_0 + \omega_1 + \omega_2 = 1.$$

The triple $(\omega_0, \omega_1, \omega_2)$ of scalars is called the barycentric coordinates of Q relative to P_i ($i = 0$, 1 and 2). For example, in Figure 4.36, the Q coordinates are $(0.5, 0, 0.5)$.

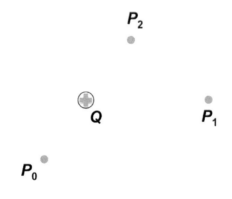

FIGURE 4.36: Problem 1.

1. Give a drawing that illustrates the regions of the plane that are associated with each of the possible sign classes ($+$ or $-$) of the three barycentric coordinates. That is, label the region in which all three coordinates are positive with $(+, +, +)$, label the region in which only the first coordinate is negative with $(-, +, +)$, etc.

2. Which sign class(es) are missing from the diagram? Explain why?

Problem 4.2. Answer the following two questions regarding the parametric continuity (see Figure 4.37).

1. Explain the differences between C^1 continuity and G^1 continuity for a parametric curve.

2. Suppose that we join two Bézier curves of degree 2 end-to-end, using the control points sequence (P_0, P_1, P_2) and (P_2, P_3, P_4), respectively.

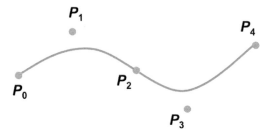

FIGURE 4.37: Problem 4.2.

Exactly what conditions must be satisfied by these five points for the combined curve to have C^1 parametric continuity at the point at which they are joined. How about G^1 parametric continuity? Prove your answer carefully by showing the continuity of the derivatives at this point. [Hint: The Bézier curve of degree 2 determined by (P_0, P_1, P_2) is $(1 - t)^2 P_0 + 2(1 - t)t P_1 + t^2 P_2$.]

Chapter 5

Image-Based Triangular and Tetrahedral Meshing

The rapid development of finite element simulations in computational medicine, biology, material sciences and engineering has increased the need for quality mesh generation. For medical images, the boundary of regions of interest is often defined implicitly by isocontours. Therefore, in this chapter we will first review unstructured triangular and tetrahedral mesh generation. After that, we introduce octree-based mesh generation techniques from scanned images, their extension to multiple-material domains, how to resolve topology ambiguities and improve the mesh quality, as well as dual contouring-based mesh generation with guaranteed angle range.

5.1   A Review of Unstructured Triangular and Tetrahedral Mesh Generation

In mesh generation, the terms *grid* and *mesh* are used interchangeably in general [372]. Numerical solution of differential equations needs to discretize the analysis domain into discrete points and elements such as triangles, quadrilaterals, tetrahedra and hexahedra. Finite difference, finite element and finite volume methods are three popular techniques people use a lot in numerical computation. There are two kinds of grids or meshes: structured and unstructured meshes. A *structured mesh* has the same valence number for all the interior nodes (e.g., 6 for triangles and 4 for quads), which is the number of elements sharing this point. On the other hand, *unstructured meshes* relax this node valence requirement and allow various numbers of elements to share a single vertex.

There are three main methods for triangular and tetrahedral mesh generation: octree-based methods, Delaunay triangulation and advancing front methods. In the *octree-based methods* [339, 428], the quadtree and octree data structure are used to generate triangular and tetrahedral meshes in 2D and 3D, respectively. *Delaunay triangulation* [109] starts from a given points cloud and connects them to form triangular and tetrahedral elements using an "empty circle" criterion in 2D and an "empty sphere" criterion in 3D. *Advancing front methods* start from a given boundary triangular mesh and generate tetrahedral meshes by inserting points inward [145, 254, 336]. The boundary surface mesh is preserved in the final mesh. In the following, we will review these three methods in detail.

5.1.1 Octree-Based Method

The octree technique was first developed by Mark Shepherd's group at Rensselaer Polytechnic Institute in the 1980s. It recursively subdivides cubes containing the geometry until a desired resolution is achieved [339]. There are many ways to define the stopping criterion. In a quadtree cell as shown in Figure 5.1, suppose $f_1(x)$ is the real geometry and $f_2(x)$ is a linear approximation to the geometry by connecting two intersection points on the cell edges. We can compute the maximal distance $d = max|f_1(x) - f_2(x)|$ or the area $A = \int_{x_1}^{x_2} |f_1(x) - f_2(x)| \cdot \Delta x dx$ as the criterion.

As shown in Figure 5.2(a, b), irregular cells are created along the boundary, and we need to compute a significant number of surface intersections. Triangular and tetrahedral elements are generated from both the irregular boundary cells and the regular interior cells. The octree technique does not preserve a pre-defined surface mesh. The resulting mesh follows the orientation of the cubes in the octree structure, and a different octree orientation yields a different mesh.

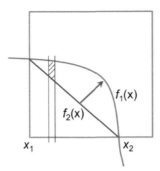

FIGURE 5.1: Surface error calculation in a quadtree cell. The red curve $f_1(x)$ is the real isocontour and the blue line $f_2(x)$ is a linear approximation by connecting two intersection points on the cell edges.

To ensure the element size does not change too dramatically, we limit the maximum level difference of neighboring quadtree or octree cells to be at most one, and obtain strongly-balanced quadtree or octree. Figure 5.2(c) shows a strongly-balanced adaptive quadtree, leading to adaptive triangular meshes as shown in Figure 5.2(d). The Marching Cubes technique [256] is used to connect all the intersection points along the boundary. Both the irregular boundary cells and the regular interior cells are triangulated. The resulting meshes may have many small angles. Smoothing and cleanup operations can be employed to improve the element aspect ratio.

As reviewed in Chapter 4.5, Dual Contouring [189] is another isocontouring method that can be used to generate finite element meshes. It produces meshes with better aspect ratios than Marching Cubes because the mesh vertices are free to move inside the cube, while in Marching Cubes they are restricted to grid edges. For each cell, Dual Contouring computes a minimizer point by minimizing a pre-defined quadratic energy function; see Eq. (4.209). There are two methods to generate triangular meshes for the interior domain of the isocontour in 2D [428, 441]. In Method 1 as shown in Figure 5.3(a), we analyze sign change edges, interior edges in the boundary cell and also interior cells. For each sign change edge, it is shared by two cells and we obtain two minimizers. Together with the interior endpoint of this sign change edge, they construct a yellow triangle. For each interior edge in the boundary cell, the minimizer point in the boundary cell and the two endpoints of this edge form a pink triangle. For interior cells, we just split them into triangles (blue triangles). In Method 2 as shown in Figure 5.3(b), we do not distinguish boundary and interior cells. Instead, we analyze all the sign change edges and interior edges. We use the same way to handle each sign change edge: two minimizer points and the interior endpoint of this edge form a yellow triangle. For an interior cell, we set the center point as the minimizer. When we analyze each interior edge, we first identify the two cells sharing this edge, and then use the minimizer

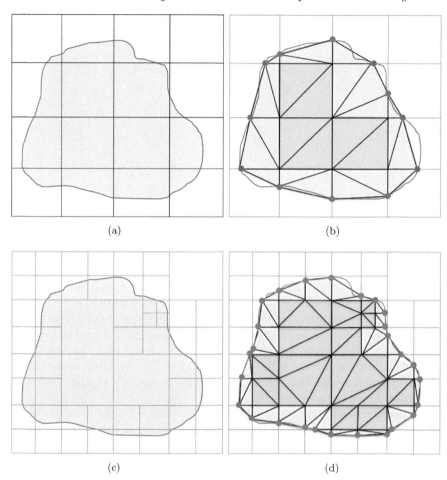

(a)   (b)

(c)   (d)

FIGURE 5.2: (a) A uniform quadtree structure; (b) the resulting triangular mesh based on (a) using Marching Cubes; (c) an adaptive quadtree structure; and (d) the resulting triangular mesh based on (c) using Marching Cubes. In (b) and (d), the red dots are intersection points between the boundary with the cell edges. The interior cell region is marked in blue, and the boundary cell region is marked in yellow.

of each cell and the two endpoints of this edge to construct a blue triangle. Compared to Method 1, we can observe that Method 2 produces a bit more vertices and elements in the interior cell region, but with better aspect ratios.

There are several software packages developed based on the octree technique. For example, Scientific Computation Research Center (SCOREC) in Rensselaer Polytechnic Institute [15] developed several software like PHASTA. QMG is a software package developed by Professor Steve Vavasis at University

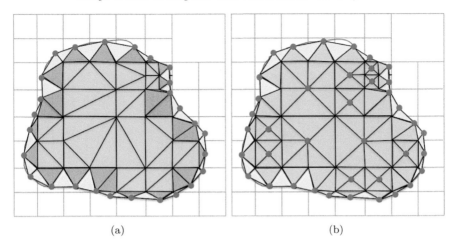

(a)   (b)

FIGURE 5.3: Two Dual Contouring Methods to generate triangular meshes. (a) Method 1. The yellow triangles are associated to sign change edges, the pink triangles are associated to interior edges in boundary cells, and the blue triangles are associated to other interior edges; and (b) Method 2. The yellow triangles are associated to sign change edges, and the blue triangles are associated to all interior edges.

of Waterloo [14]. In our Computational Biomodeling Laboratory at Carnegie Mellon University [6], we have developed a series of meshing software packages including the LBIE-Mesher (Level-Set Boundary Interior and Exterior Mesh Generator).

5.1.2 Delaunay Triangulation and Voronoi Diagram

Delaunay Triangulation. Invented by Boris Delaunay in 1934, Delaunay triangulation [109] of a point set P in a plane is a triangulation $DT(P)$ such that no point in P lies inside the circumcircle of any triangle in $DT(P)$. In other words, Delaunay triangulations maximize the minimum angle of the triangulation, tending to avoid "sliver" triangles. Figure 5.4 shows a Delaunay triangulation in a plane with circumcircles.

According to the Delaunay definition, if the circumcircle of a triangle formed by three points in P does not contain any other vertices, then we say this circumcircle is *empty*. Other vertices are allowed to be on the perimeter of the circumcircle, but not inside. If all the circumcircles of an input points cloud are empty, then we call their formed triangulation *Delaunay triangulation*. This Delaunay condition is called "*empty circle*." There are some special cases. For a set of colinear points, there is no Delaunay triangulation according to the empty circle condition. For four points on the same circle like vertices of a rectangle, their Delaunay triangulation is not unique. We can have two

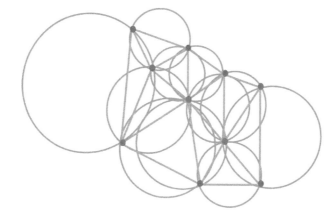

FIGURE 5.4: A Delaunay triangulation in a plane with circumcircles. The red dots are the input points cloud.

possible triangulations by choosing different diagonals to split the rectangle into two triangles, and both of them satisfy the Delaunay condition.

We can also generalize the metrics other than Euclidean, but we cannot guarantee a Delaunay triangulation exists or is unique. The Delaunay triangulation can be extended to an n-dimensional Euclidean space, where $n \geq 2$. Given a d-dimensional points cloud P, a Delaunay triangulation is a triangulation $DT(P)$ such that no point in P lies inside the circumhypersphere of any simplex in $DT(P)$. In 2D, if no three points in P are colinear and no four points lie on the same circle, then we can obtain a unique Delaunay triangulation for P. For an n-dimensional set of points, a unique Delaunay triangulation requires that no $n+1$ points lie on the same hyperplane and no $n+2$ points lie on the same hypersphere.

Given n points, its d-dimensional Delaunay triangulation has the following properties. The Delaunay triangulation contains at most $O(n^{d/2})$ simplices, and the union of all simplices forms the convex hull of the given points. Suppose there are m vertices on the convex hull in the 2D plane, then any triangulation of these vertices has at most $2n - 2 - m$ triangles with one exterior face. According to the Delaunay criterion, Delaunay triangulation maximizes the minimum angle, and any circumcircle does not contain any other input points in its interior (they can be on the perimeter). If a circle passing through two input points is empty, then the line segment of these two points is an edge of a Delaunay triangulation. The Delaunay triangulation in d-dimensional spaces is the projection of the convex hull of point projections onto a $(d+1)$-dimensional paraboloid. For a point p inside the convex hull of a Delaunay triangulation, the closest vertex to p does not need to be the corners of the triangle containing p.

From the above properties, we can observe an important feature for Delaunay triangulation. For example, Figure 5.5(a) shows two triangles ABD and

BCD with a common edge BD. If $\alpha + \beta \leq 180°$, then the triangles satisfy the Delaunay condition; otherwise, they are not Delaunay triangles. We can use the *flipping technique* to make them meet the Delaunay condition by just switching the common edge from BD to AC; see Figure 5.5(b, c).

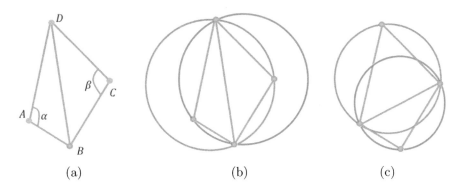

(a)   (b)   (c)

FIGURE 5.5: The flipping technique of Delaunay triangulation. (a) Two triangles with one common edge; and (b, c) the flipping technique is used to make the triangulation meet the Delaunay condition.

In computing Delaunay triangulation, it is important to have fast operations to detect when a point lies inside a triangle's circumcircle, and an efficient data structure is required to store triangles and edges. In 2D, we can detect if point D lies within the circumcircle of A, B and C (ordered in counter clockwise) by evaluating the determinant,

$$\begin{vmatrix} A_x & A_y & A_x^2 + A_y^2 & 1 \\ B_x & B_y & B_x^2 + B_y^2 & 1 \\ C_x & C_y & C_x^2 + C_y^2 & 1 \\ D_x & D_y & D_x^2 + D_y^2 & 1 \end{vmatrix} = \begin{vmatrix} A_x - D_x & A_y - D_y & (A_x - D_x)^2 + (A_y - D_y)^2 \\ B_x - D_x & B_y - D_y & (B_x - D_x)^2 + (B_y - D_y)^2 \\ C_x - D_x & C_y - D_y & (C_x - D_x)^2 + (C_y - D_y)^2 \end{vmatrix}, \quad (5.1)$$

which is positive if and only if D lies inside the circumcircle.

The most straightforward way to implement Delaunay triangulation is to repeatedly insert one vertex at a time and retriangulate the affected elements surrounding that vertex. The time complexity is $O(n^2)$. This procedure can be speeded up using a sweepline method, in which we sort vertices by one coordinate and add them in that order. The time complexity is improved to $O(n^{3/2})$ although the worst case is still $O(n^2)$. Another efficient incremental algorithm stores the entire history of triangulation in a tree structure. In the divide-and-conquer algorithm, we recursively split vertices into two sets. The Delaunay triangulation is computed for each set, and then we merge them together. The merging operation can be done in $O(n)$ and the total running time is improved to $O(nlogn)$. Figure 5.6(a) shows an example of Delaunay triangulation results.

Delaunay triangulation has many applications. In modeling terrain and

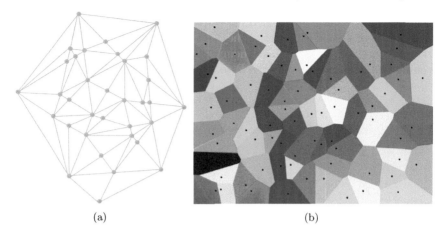

(a)   (b)

FIGURE 5.6: (a) A Delaunay triangulation; and (b) a Voronoi diagram.

many other objects starting from a points cloud, Delaunay triangulation produces good quality triangles with the minimum angle maximized. Due to
its angle guarantee and efficient mesh generation procedure, Delaunay-based
methods have been used broadly to generate finite element meshes for many
finite element simulations. There are a lot of advances based on the basic Delaunay triangulation, aiming at generating quality triangular and tetrahedral
meshes. Delaunay refinement aims to refine the triangles or tetrahedra locally
by inserting new nodes in order to maintain the Delaunay criterion. Different approaches were studied to introduce and define new nodes [87, 92, 344].
Sliver exudation [88] was developed to eliminate those slivers. A deterministic algorithm [87] was developed to generate a weighted Delaunay mesh with
no poorly shaped tetrahedra including slivers. Shewchuk [345] used the constrained Delaunay triangulation (CDT) to resolve the problem of enforcing
boundary conformity. Delaunay refinement [344], edge removal and multiface
removal optimization algorithms [346] were utilized to improve the tetrahedral
quality. Shewchuk [347] drew some valuable conclusions on quality measures
for finite element analysis. In recent years, Chrisochoides and Chernikov performed many studies on parallel 2D and 3D Delaunay meshing with guaranteed quality [90, 91, 142] and also applied their techniques to medical image
applications.

Voronoi Diagram. The Delaunay triangulation of a points cloud P has a
dual graph of the Voronoi tessellation, the Voronoi diagram [379]. A Voronoi
diagram was named after a Russian mathematician George F. Voronoi who
defined the n-dimensional case in 1908. It is also called a Voronoi tessellation, a Voronoi decomposition, or a Dirichlet tessellation (named after Leje-
une Dirichlet [114]). It is a special decomposition of a metric space determined
by distances to a specified points cloud. The informal use of Voroni diagrams

started from Descartes in 1644. Later in 1850, Dirichlet used 2D and 3D Voronoi diagrams when he conducted a study of quadratic forms. In 1854, a British physician named John Snow used Voronoi diagrams to investigate how the Soho Cholera epidemic was spread out through the infected broad street pump.

Given a set of points S, its *Voroni diagram* is the partition of the plane, where a local region $V(p)$ is created for each point in S such that all points in $V(p)$ are closer to p than to any other point in S. In the Euclidean space, we find one point in S that is closest to x. Sometimes, we can find two or more points that are equally close to x. If there are only two points a and b in S, then the point set equidistant from them forms a hyperplane, which is the perpendicular bisector of the line segment connecting a and b. In 2D, it splits the plane into two half-planes; one contains all the points closer to a than to b, and the other one contains all the points closer to b than to a.

Suppose c is a point in S, then all the points closer to c than to any other point in S form a convex polytope called the Dirichlet domain or *Voronoi cell* for c. Such polytopes tessellate the entire space and are called the *Voronoi tessellation* or Voronoi diagrams. In 2D, it is easy to plot them; see Figure 5.6(b) for a Voronoi diagram result generated from a real-time interactive Voronoi and Delaunay diagram generation software [16].

Voronoi diagrams have several properties. For a given set of points S, the dual graph of a Voronoi diagram is the Delaunay triangulation. Each point associates with one cell in the Voronoi diagram, and the closest pair of points corresponds to two neighboring cells. Only when the Voronoi cells of two points share an infinitely long edge, they are adjacent on the convex hull. For regular lattices of 2D and 3D points, we have many familiar Voronoi tessellations. For example, a 2D lattice yields an irregular honeycomb tessellation with equal hexagons and point symmetry. A regular triangular lattice gives regular triangular meshes; a rectangular lattice reduces hexagons to rectangles in rows and columns; and a square lattice leads to a regular tessellation of squares.

Like Delaunay triangulation, Voronoi diagrams can be extended to metrics other than Euclidean although the Voronoi tessellation is not guaranteed to exist. They can also be defined by measuring distances to objects instead of points. Such Voronoi diagrams are called the *medial axis*, which have been used in image segmentation, geometric modeling and mesh generation. In material sciences, polycrystalline microstructures in metallic alloys are usually represented in Voronoi tessellations. Given n points in a d-dimensional space, the Voronoi diagram requires $O(n^{d/2})$ storage space. Generally this is not feasible for $d > 2$.

Voronoi diagrams have been used in many application fields. We can build a point location data structure based on the Voronoi diagram, then we can efficiently find the object closest to a given query point. Nearest-neighbor queries have many applications. For example, we can easily find the nearest hospital or the most similar object in a database, and also derive the capacity of a wireless network. In polymer physics, we can use it to represent free volume of the polymer. They were used in geophysics and meteorology to study spatially distributed data like rainfall measurements, which are called Thiessen polygons after an American meteorologist Alfred H. Thiessen. In condensed matter physics, Voronoi tessellations are also called Wigner-Seitz unit cells. In general metric spaces, the cells are called metric fundamental polygons.

There are many available links and source codes available online. Here we list some of them as follows.

- Real-time interactive Voronoi and Delaunay diagrams with source code. http://www.cs.cornell.edu/Info/People/chew/Delaunay.html

- Delaunay triangulation and Voronoi diagram in CGAL, the Computational Geometry Algorithms Library

- Applet for calculation and visualization of convex hull, Delaunay triangulations and Voronoi diagrams in space

- 2D C Delaunay code and Java applet. http://goanna.cs.rmit.edu.au/~gl/research/comp_ geom/delaunay /delaunay.html

- C++ 2D Delaunay code. http://www.compgeom.com/~piyush/scripts/triangle/

- Mathworld on Delaunay triangulation. http://mathworld.wolfram.com/DelaunayTriangulation.html

- Qhull for computing Delaunay triangulations in 2-d, 3-d, etc. http://www.qhull.org/

- Triangle, a 2D Quality Mesh Generator and Delaunay Triangulator. http://www.cs.cmu.edu/~quake/tripaper/triangle0.html

- Triangulate, an efficient algorithm for Terrain Modelling. http://local.wasp.uwa.edu.au/~pbourke/papers/triangulate/index.html

- G. Leach: Improving Worst-Case Optimal Delaunay Triangulation Algorithms. (June 1992)

- INRIA - GSH3D inserts points along edges for thin structures. http://www.inria.fr/

- David Marcum in Engineering Research Center (ERC) at Mississippi State University developed an advancing front method. Nodes are inserted incrementally, but added from the boundary towards the interior. http://www.me.msstate.edu/faculty/marcum/marcum.html

5.1.3 Advancing Front Method

In many finite element applications, there is a requirement that an existing surface triangular mesh needs to be maintained. In most Delaunay-based methods, a 3D tessellation of the geometric surface is produced before internal nodes are generated, and there is no guarantee that the given surface mesh will be maintained. The octree-based methods generally create a new mesh for the input boundary. Starting from a given boundary triangular mesh, advancing front methods [145, 254, 336] build tetrahedral elements progressively inward with the input boundary mesh preserved in the resulting tetrahedral mesh, maintaining an active front where new elements are formed. In 2D, the front advances to fill the remainder of the area with triangular elements. In 3D, we can compute an ideal location of a new fourth node for each triangle in the front. We select either the new fourth node or an existing node to form the new tetrahedron. Here some criteria like the Delaunay criterion is used to decide which one produces the best tetrahedra. A self-intersection checking is needed, and a size function can be designed to control mesh adaptation and mesh size.

There are several available codes for advancing front methods [12]; for example, the ANSYS suite of mesh generation tools [10], and Pirzadeh's VGRID software [300] which is available from TetraUSS in NASA, Langley [8]. Advancing layers are also used for boundary layer generation for computational fluid dynamics applications, which is suitable to generate anisotropic meshes.

5.2 Octree-Based Triangular and Tetrahedral Mesh Generation from Images

In scanned images such as CT and MRI, the geometric surface often exists implicitly as an isosurface or a level set. Although there have been tremendous progresses in geometric modeling and mesh generation, high-fidelity mesh generation directly from images is still a challenging problem. The volumetric imaging data V can be written as a scalar field sampled on rectilinear grids, that is, $V = \{f(i, j, k) \mid i, j, k$ are indices of x, y, z coordinates in a rectilinear grid$\}$. Within a cubic cell or voxel, we can construct a trilinear function to define a continuous field in it.

As explained in Chapter 4.5, there are two main isocontouring methods to extract isosurfaces directly from images: marching cubes [256] and dual contouring [189]. These two isocontouring methods have been extended to finite element mesh generation. The marching cubes techniques [256] construct tetrahedral elements between two isosurfaces directly from images, and the construction process was accelerated via a branch-on-need octree data structure [146]. Each cube can also be split into several tetrahedra first and then we visit each tetrahedra to generate meshes for interval volume [277]. In addition, recursive subdivision and edge-bisection methods were combined to build a multiresolution framework [444]. From slices of images, Bajaj *et al.* built triangular surface meshes [33] and also tetrahedral meshes for the volumetric region formed by the planar contours and the surface mesh [34]. We have extended the idea of dual contouring to automatic volumetric tetrahedral mesh generation for single-material [428] and multiple-material domains [432]. The constructed meshes conform to boundaries defined as level sets or isosurfaces, with correct topology and feature preservation. Our methods yield adaptive and quality meshes which can be used directly in finite element applications.

5.2.1 Preprocessing

Due to the limitation of the scanning techniques and object motion, the obtained images may have a lot of noise with poor quality. To generate high-fidelity quality finite element meshes for regions of interest, we need to apply image processing techniques to improve the quality of the images and extract useful information from them. As discussed in Chapter 3, the image processing techniques include contrast enhancement, noise removal, classification and segmentation of regions of interest, as well as alignment of images scanned at different time or using different modalities.

To improve the image contrast, an adaptive transfer function or a stretching function was designed for each voxel using its neighboring intensity field [419]. Both isotropic and anisotropic diffusion methods were developed to remove noise. Anisotropic methods preserve features like edges and corners during noise smoothing [37], while isotropic diffusion blurs the boundary edges. Classification and segmentation are usually coupled together to categorize voxels into different groups based on the intensity values, and also identify the clear boundary of each group [418]. In addition, contour spectrum [36] and contour tree [76] can assist in selecting desired isosurfaces with certain topological information.

5.2.2 3D Mesh Extraction

Our 3D mesh generation is basically an octree-based dual contouring method [428, 441]. Starting from the input image and an isovalue, a top-down scheme is first adopted to build the adaptive octree. We choose a feature-sensitive error function to decide where we should subdivide the cells and where we should keep the coarse levels. Given a voxel at level i, suppose the intensity values attached to its eight corners are f_{000}, f_{001}, f_{010}, f_{011}, f_{100}, f_{101}, f_{110} and f_{111}, respectively. Then, we can use them to build a trilinear function

$$
\begin{aligned}
f^i(x, y, z) \;=\;\; & f_{000}(1 - x)(1 - y)(1 - z) + f_{001}(1 - x)(1 - y)z \\
+ \;\; & f_{010}(1 - x)y(1 - z) + f_{011}(1 - x)yz \\
+ \;\; & f_{100}x(1 - y)(1 - z) + f_{101}x(1 - y)z \\
+ \;\; & f_{110}xy(1 - z) + f_{111}xyz.
\end{aligned} \tag{5.2}
$$

For any point within this voxel, we use this trilinear function to compute its intensity value. Here we define a set S_M to represent all the 12 edge middle points, 6 face centers and the cube body center. We use Eq. (5.2) to compute the intensity values at these points in S_M, and compare them with the data read directly from the input image which are also the intensities at level $(i+1)$ or $f^{i+1}(x, y, z)$. The feature-sensitive error function is defined as

$$
f_e^i = \sum_{j \in S_M} \frac{\left| f^{i+1}(x_j, y_j, z_j) - f^i(x_j, y_j, z_j) \right|}{\left| \nabla f^i(x_j, y_j, z_j) \right|}, \tag{5.3}
$$

which measures the difference of isosurfaces at two adjacent levels i and $(i+1)$. To generate elements with good aspect ratio, we restrict the level difference between two neighboring cells to be at most one and therefore obtain a strongly-balanced octree.

In Chapter 5.1, we discussed the dual contouring method and how to use it to generate triangular meshes for 2D planar domains. Instead of visiting each cell and connecting intersection points as in Marching Cubes, the dual contouring method visits each sign change edge and interior edge. In the adaptive octree, since each leaf cell may have neighbors at different levels, here we choose to always analyze the minimal edges. In 2D, each sign change edge is always shared by two quadtree cells. We compute a minimizer point for each cell and connect them to represent the isocontour. From volumetric images in 3D, we aim at extracting triangular isosurfaces and generating tetrahedral meshes for the volume interior to the isosurface or an interval volume between two isosurfaces. For isosurface extraction, we only need to handle sign change edges, and each sign change edge is shared by either four or three boundary cells. We obtain one minimizer point for each cell by minimizing the quadratic error function, and then connect them to generate a quadrilateral or triangle. The quadrilateral can be split into two triangles later. By visiting all the sign change edges, a triangulated isosurface is obtained.

For tetrahedral mesh extraction, we use the same way to handle sign change edge, as shown in Figure 5.7(a-b). The obtained surface quadrilaterals or triangles form a pyramid or tetrahedron together with the interior endpoint of the sign change edge. In addition, interior edges in the boundary cells and interior cells should also be analyzed. There are two methods to handle the interior edges. Similar to the 2D case in Figure 5.3(a), the first method [428] handles interior edges in boundary cells and interior cells differently. For each interior edge in a boundary cell, we first compute three or four minimizer points for its surrounding cells, and then use two adjacent minimizers and the interior edge to construct a tetrahedron. For each interior face in a boundary cell, we use the minimizer point of this cell and the interior face to build a pyramid, which can be decomposed into two tetrahedra. Finally, we split the interior cell into a set of five tetrahedra. Note that we need to choose a suitable decomposition for two adjacent cells to avoid introducing any diagonal conflict on their shared face.

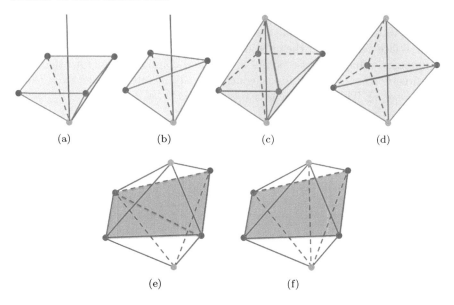

(a)   (b)   (c)   (d)

(e)   (f)

FIGURE 5.7: (a-b) The sign change edge (red edge) with four or three minimizer points (red dots); (c-d) the interior edge (red edge) with four or three minimizer points (red dots); and (e-f) two splitting methods for a diamond with different splitting edges (red dashed edge).

Similar to the 2D case in Figure 5.3(b), the second method [441] adopts the same way to handle all the interior edges, without distinguishing boundary cells or interior cells. In general, the second method yields a bit denser mesh for interior regions, but it is easier to implement and the resulting mesh has a bit better aspect ratios. For each interior edge, we find its surrounding four or three minimizer points and use them together with the two endpoints of the interior edge to construct a diamond or pyramid; see Figure 5.7(c-d). The obtained diamond and pyramid can be split into tetrahedra using two splitting methods depending on which method yields better aspect ratios. In Figure 5.7(e), the four minimizers form a quadrilateral and we can choose a diagonal to split it into two triangles first, and then each triangle together with one endpoint of the interior edge forms a tetrahedron. Differently in Figure 5.7(f), two adjacent minimizers together with the two endpoints of the interior edge form a tetrahedron.

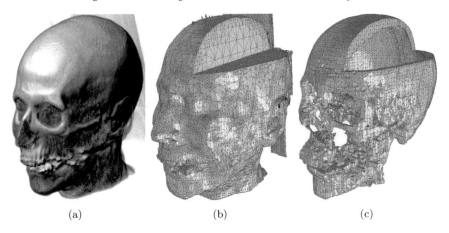

(a)   (b)   (c)

FIGURE 5.8: Adaptive tetrahedral meshes of a CT image (UNC Head, pictures from [428]). (a) The volume rendering result; (b) adaptive tetrahedral mesh interior to the skin surface; and (c) adaptive tetrahedral mesh for the skull.

Since we are limited to one minimizer point in each cell, we require that each cell can only contain at most one boundary isosurface in order to avoid generating any nonmanifolds and preserve the correct topology. For a cell containing two boundary isosurfaces, we keep subdividing it until no such situation exists. One drawback of this subdivision method is the lack of efficiency; sometimes we need to subdivide the cells many times. To efficiently resolve topology ambiguities, we allow multiple minimizer points in one cell [436]; see the detailed discussion in Chapter 5.3.

We have developed interactive software packages for 3D mesh extraction and rendering from images. Error tolerances and isovalues can be adjusted interactively from the software interface. Figure 5.8 shows the extracted meshes of the skin and the skull from CT images. The mesh adaptation is controlled flexibly by changing the error tolerance.

5.2.3 Meshing for Multiple-Material Domains

The scanned images usually contain many material regions or heterogeneous materials, such as the bone, blood and soft tissues in the human body. Such multiple-material domains have complex non-manifold boundary surfaces, and high-fidelity geometric modeling and quality meshes are need for each material region with conformal boundaries. For example, given a domain Ω consisting of n closed material regions, we denote them as Ω_0, Ω_1, ..., and Ω_{n-1}. The boundary of each region Ω_i is denoted as B_i; we then have $\Omega_i \cap \Omega_j = B_i \cap B_j$ when $i \neq j$. Note that $\cup B_i$ can be manifold or non-manifold curves or surfaces. Since in a scalar field each data point can only have one unique function value, we cannot represent the non-manifold boundaries implicitly by an isocontour. In other words, the isocontouring methods do not work directly for a domain with non-manifold boundaries.

To automatically construct high-fidelity meshes for all the material regions, here we discuss an extended dual contouring approach [432] to automatically detect all the non-manifold boundaries and mesh the domain with multiple materials simultaneously. In particular, we extend the standard dual contouring method by introducing a new type of edge, *material change edge*, whose two endpoints lie in two different material regions. During mesh generation, we replace all the sign change edges with material change edges. In addition, when we compute the minimizer point for each boundary cell by minimizing the quadratic error function in Eq. (4.209), we only considered one material region with the background or two materials. Now we may have two to eight different materials in one boundary cell with eight corners. Here we first identify all the intersection points on the cell edges, without considering which two materials share each of them. Then we include them into the summation of the quadratic error function to compute a unique minimizer point in this cell. For each interior cell, we simply choose the cell center as the minimizer.

We analyze all the material change edges to extract the non-manifold conformal boundary meshes. Each material change edge is shared by three or four boundary cells, so we obtain three or four minimizer points to form a triangle or a quadrilateral. The quadrilateral can be split into two triangles. Note that in our octree data structure, each leaf cell has a unique index and only one minimizer point is generated for each cell. In other words, each minimizer point has a unique index associated to it. This property provides us a great deal of convenience to efficiently record the minimizer points without any duplication in the resulting meshes.

Similar to meshing a single-material domain [428], we analyze both material change edges and interior edges to generate tetrahedral meshes for the multiple-material domain. For a uniform case, each edge is shared by four cells. We obtain four minimizers to build a quadrilateral. The quadrilateral and the two endpoints of the edge construct a diamond, which can be divided into four tetrahedra later. For an adaptive case, each edge is shared by four or three cells. Therefore we can obtain four or three minimizer points, yielding a quadrilateral or a triangle. The quadrilateral or triangle and the two endpoints of the edge construct a diamond or a pyramid. Finally we split it into four or two tetrahedra. The mesh adaptation can be controlled flexibly by a feature sensitive function, regions of interest, finite element solutions and user-defined error function.

As shown in Figure 5.9(a-b), we have applied our approach to the Brodmann brain atlas, a segmented volumetric map with 48 material regions (http://www.sph.sc.edu/ comd/rorden/mricro.html). Non-manifold boundary surfaces are constructed in both triangular and tetrahedral meshes. Each color represents a material region.

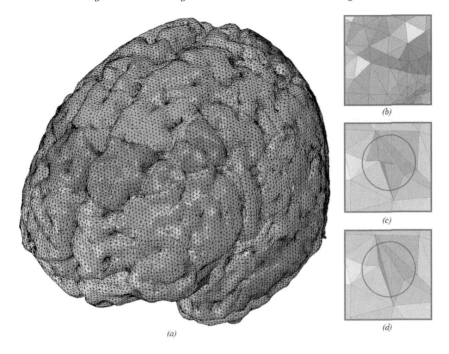

FIGURE 5.9: (a) An adaptive tetrahedral mesh of the segmented Brodmann brain atlas with 48 components (pictures from [305]). (b) The adaptive mesh of the boundary between two components; (c) zoom-in details of an interior local region with self-intersections (the red circle); and (d) the resolved local region in (c).

5.2.4 Intersection-Free Tetrahedral Meshing

Many finite element applications require an automatic and robust method to extract high-fidelity intersection-free meshes from volumetric images. The Marching Cubes (MC) technique [256] has no self-intersection problem because all the generated triangles are limited within the disjointed grid cells. However, it is poor in dealing with adaptive grids and it yields dense elements with many small angles. On the other hand, the Dual Contouring (DC) method [189] performs well in generating adaptive meshes with good aspect ratios, but it may have self-intersections due to the bad location of minimizer points within a cell. Such meshes need to be further improved by relocating minimizers or using special untangling operations [128, 204] based on smoothing and optimization. This problem can also be resolved by using pre-defined envelopes in surface meshing [190, 249]. Differently, here we extend the octree structure to a hybrid grid, based on which we develop a robust and systematic algorithm to build disjointed envelopes and generate adaptive intersection-free tetrahedral dual meshes [305].

Self-intersection situations happen when multiple elements intersect with each other, which is usually introduced by improper vertex positions or wrong geometry topologies. Our intersection-free meshing algorithm is an extension of our octree-based dual contouring method [428]. It has three main steps: constructing disjointed envelopes, identifying self-intersections and generating intersection-free meshes. We first analyze sign change edges and interior edges in adaptive grids to build tetrahedral, pyramid or diamond envelopes, which split the domain into disjointed spaces. Then we check each envelope to see which one may contain potential intersections. An intersection can be introduced by adaptive grids or concave envelopes. For envelopes with no potential intersections, we use our octree-based dual contouring method to split each envelope into tetrahedral elements. For envelopes with intersections, we identify their associated octree cells and we split each cell into 12 tetrahedral cells to build a hybrid octree and then use the envelope meshing scheme to generate its dual mesh.

Recall that MC has no self-intersections because all the elements are generated within disjointed cubic cells. Similarly in our octree-based dual contouring method, we construct disjointed envelopes around each sign-change edge and interior edge, which can be tetrahedra, pyramids or diamonds as shown in Figure 5.7(a-d). To obtain intersection-free meshes, we need to add new vertices on the envelope edges. As shown in Figure 5.10, an *envelope* is a nonoverlapping polyhedron surrounding each sign-change edge or interior edge $P_1 P_2$ (the red dashed edge). $P_1 P_2$ is shared by several cells, and we obtain one minimizer (the green dot) for each cell. Two adjacent cells share a cell face, and their minimizers form an edge which intersects with the shared cell face at a *face node* (the blue dot), which are the newly inserted vertices on the envelope. As shown in (a, c), all the minimizer points and the face nodes associated with $P_1 P_2$ form a polygon (shaded in yellow), which has eight edges for uniform cases and six edges for adaptive cases. Together with P_1 and P_2, this polygon leads to an envelope; see the blue polyhedra in (b, d). The envelope can be easily tessellated into tetrahedral elements.

In hybrid grids, where the topology ambiguity can be resolved automatically [437], the envelopes can be defined in a similar manner. We first identify those ambiguous cells and split each into 12 tetrahedral cells to form a hybrid tree consisting of both cubic and tetrahedral leaves. When we analyze all the edges, the envelopes are still formed by the sign-change edge or interior edge together with its corresponding minimizer points and face nodes. Note that some of these nodes are computed from tetrahedral cells. The constructed polygon may have 6 to 16 edges.

It has been proven that valid envelopes belonging to different sign-change edges or interior edges are disjointed, that is, they do not intersect with each other [190]. Therefore, if a triangle or tetrahedron lies completely within an envelope, it does not intersect with triangles or tetrahedra in other envelopes. The above envelope definition also works for multiple-material domains, where we need to replace each sign-change edge with material-change edge.

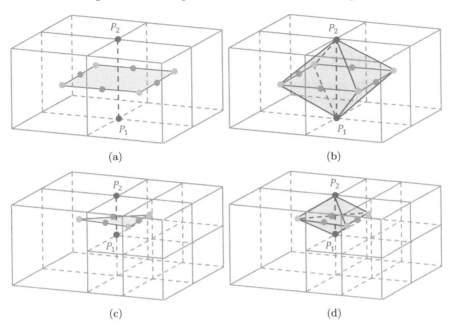

(a) (b)

(c) (d)

FIGURE 5.10: Envelopes associated with the edge $P_1 P_2$ in the uniform (a, b) and adaptive (c, d) cases. The minimizer points and the face nodes are marked in green and blue, respectively. (a, c) show the polygon (shaded in yellow) formed by these minimizer points and face nodes; and (b, d) show the resulting envelope (shaded in blue).

Although free of self-intersections, the envelope meshing technique generates many more elements compared to the naive octree-based dual contouring method [428]. Generally, there are only a few regions in a dataset with self-intersections, and we tend to use the naive meshing if there is no self-intersection detected. We first identify self-intersection situations using the necessary and sufficient conditions [190, 249, 305], and then apply the envelope meshing technique only when necessary. Self-intersections may be introduced by adaptive grids or concave envelopes, which can be identified using a two-round checking. After we identify all the self-intersections, we build the envelopes and tessellate them into intersection-free tetrahedral elements. Note that whenever a hanging node is generated in the resulting mesh, we need to split all the adjacent triangles or tetrahedra accordingly.

Our developed algorithm is automatic and robust in generating intersection-free tetrahedral meshes for single-material and multiple-material domains. The topology ambiguities can be resolved when the hybrid octree is utilized. We have applied our approach to the Brodmann brain atlas; see Figure 5.9(c, d). Our method is also very efficient, yielding about 1.6 million elements within 200 seconds.

5.3   Resolving Topology Ambiguities

High-fidelity representation of an isosurface with accurate surface and correct topology plays an important role in scientific visualization and mesh generation. Given scanned images, we aim at extracting quality and accurate triangular surfaces with correct topology. The MC technique [256] visits each voxel in the volumetric image and performs a local triangulation based on the sign configuration of the eight corner grids. The isosurface extraction process was accelerated [35, 407] to reduce the computational cost by avoiding visiting unnecessary voxel cells. The isosurface inside a voxel may have complicated geometry or topology ambiguities. *Topology ambiguity* is a situation where the same sign or material configuration may lead to more than one possible topologies in representing the isosurface. For example, as show in Figure 4.30(a, b), the same diagonally opposite cases in 2D have two equally possible interpretations. In 3D, the situations can be much more complex. Given a cubic voxel with eight corner grids, we have an intensity value attached to each corner. Suppose the isovalue is α. Figure 5.11 shows two pairs of topological ambiguities with the diagonal opposite case happening on a face or through the cubic body. The red dots represent the grids with the intensity value greater than or equal to α, while the green dots represent the grids with the intensity value less than α. We can observe that the same configuration can lead to two separate components like (a, c), one component in (b) or even a tunnel formed in the voxel like (d).

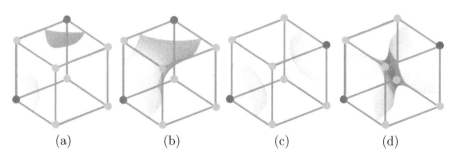

(a)   (b)   (c)   (d)

FIGURE 5.11: Two pairs of topology ambiguity in 3D. (a, b) One pair of topology ambiguity due to the diagonal opposite case on a face; and (c, d) another pair of topology ambiguity due to the diagonal opposite case through the cubic body.

To resolve these topology ambiguity problems in MC, the face and body saddle points in the cell can be used to decide the correct topology for a consistent triangulation [255, 276]. As we discussed before, MC has several drawbacks such as generating uniform and dense elements with many small-angle triangles as well as ignoring sharp features. To enable an adaptive mesh generation, we need to insert a fan of triangles surrounding the gravity center of the coarse triangles in order to avoid generating any cracks [401]. Another iso-contouring method, the octree-based DC method [189], combines SurfaceNets [155] and the extended MC [207] algorithms, which support adaptive isosur-face extraction with good aspect ratios and sharp feature preservation. These nice features make the DC method suitable for finite element mesh generation, but it has the drawback that one cell is restricted to have only one minimizer point, leading to possible non-manifold local situations and wrong topolgies. A vertex clustering algorithm [375, 425] together with several topology con-straints [63, 193, 322] was developed to address this deficiency. However, for datasets with complicated geometry and topologies, the existing DC meth-ods fail in resolving all the topology ambiguities and thus yield invalid finite element meshes.

In the following, we discuss how to address topology ambiguities in the octree-based dual contouring methods for single-material domains [436] and also multiple-material domains [437] using two different strategies.

5.3.1 Single-Material Domains

To identify topology ambiguities, we need to study the function properties within each cubic voxel cell. Given the function values attached to the eight grid points of a cell f_{ijk} (i, j, $k = 0$, 1), we represent the interior region using a trilinear interpolation $f(\xi, \eta, \zeta)$. We have

$$
\begin{aligned}
f(\xi, \eta, \zeta) =\ & f_{000}(1 - \xi)(1 - \eta)(1 - \zeta) + f_{001}(1 - \xi)(1 - \eta)\zeta \\
& + f_{010}(1 - \xi)\eta(1 - \zeta) + f_{011}(1 - \xi)\eta\zeta \\
& + f_{100}\xi(1 - \eta)(1 - \zeta) + f_{101}\xi(1 - \eta)\zeta \\
& + f_{110}\xi\eta(1 - \zeta) + f_{111}\xi\eta\zeta.
\end{aligned} \tag{5.4}
$$

Depending on the sign configurations of these eight grid points, there are 14 unique configurations for one cell; see Figure 4.32. Some of these configurations have topology ambiguities. These ambiguities can be on one face of the cell or interior to the cell. To identify them, we compute the function values at the body and face saddle points using the above trilinear function [276], and use them to help decide whether the isosurface forms a tunnel linking the two grid points or it should be separated into several components within the cell. After considering all face and interior ambiguities, we obtain a comprehensive set of 31 cases [436], which are actually a detailed version of the 14 fundamental cases. For example, Figure 5.12 shows the three possible topologies for Case 3b. The red dots represent the cube corners with the function value greater than or equal to the isovalue. The blue dots represent the minimizer points in the current cell, and the green dots are minimizers in surrounding cells. There is one ambiguity on the front face with two red dots on the opposite diagonal, and the two red grid points can be separated or connected on this face. When the two red grid points are separated like in (a) and (b), an additional interior body ambiguity arises. That is, the two positive grid points can be either separated or joined inside the cube, leading to Cases 3b-1-1 and 3b-1-2, respectively. When the two red points are connected on this face, we have only one case; see Case 3b-2 in (c). Note that the 3-index labeling of these 31 cases denotes their detailed topologies. The first index denotes the case number in the original 14 fundamental cases, the second index represents the resolution of a face ambiguity, and the last index indicates an interior body ambiguity. We can observe that we introduce multiple minimizer points (the blue dots) to a cell in order to resolve the topology ambiguities.

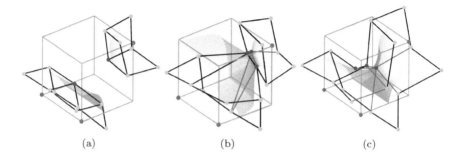

(a)   (b)   (c)

FIGURE 5.12: Case 3b with a face ambiguity (a, b) and an interior body ambiguity (c). (a) Case 3b-1-1 with two minimizer points; (b) Case 3b-1-2 with three minimizer points; and (c) Case 3b-2 with three minimizer points. The red dots represent the grid corners with the function value greater than or equal to the isovalue. The blue dots represent the introduced minimizer points in the current cell, and the green dots are minimizers in surrounding cells.

With all these ambiguities resolved, we have multiple minimizers introduced within one cell, eliminating the restriction that only one minimizer is allowed within one cell. Then, we revise the local mesh generation cell by cell by properly reconnecting these minimizer points with the neighboring cell minimizers, and obtain a triangulated isosurface with all topology ambiguities resolved. Furthermore, we adopt an advancing front method to generate tetrahedral meshes for the volume formed by these reconnected triangles. The quality of the tetrahedral mesh is finally improved via a combination of face swapping, edge removal and geometric flow-based smoothing [429]. Our algorithm [436] generates conformal meshes with good quality and correct topology. Figure 5.13 shows the generated tetrahedral meshes for a complex trabecular bone structure with all topology ambiguities resolved.

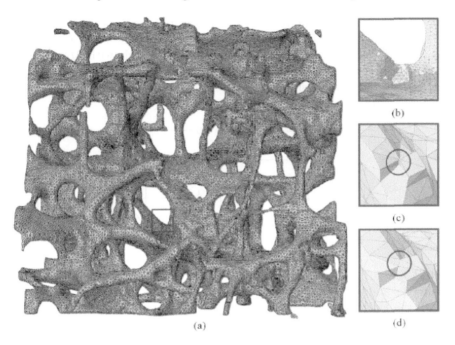

FIGURE 5.13: Resolving topology ambiguity for the trabecular bone structure (pictures from [436]). (b-d) show some details of (a).

5.3.2 Multiple-Material Domains

There are many heterogeneous materials in scanned images consisting of multiple material regions, and topology ambiguity is also an important issue for meshing such domains. In Section 5.3, we categorize all possible ambiguities into 31 cases and also modify the dual contouring mesh generation method accordingly to generate quality triangular and tetrahedral meshes with all topology ambiguities resolved. However, this method only works for single-material domains. For a cell within a multiple-material domain, its eight corner grids may belong to 1~8 different materials. The binary sign configurations do not work anymore. We cannot model the cell simply using a trilinear interpolation function as what we did for single-material domains.

Differently, here we design an indicator variable to check each material in a cell and see whether there is any topology ambiguity. Given a voxel cell with eight corners, the function value f_i ($i = 0, 1, \cdots, 7$) attached to each corner represents which material it belongs to. Suppose there are n ($1 \leq n \leq 8$) different materials in a cell, and the material indices are $M_0, M_1, \cdots, M_{n-1}$, respectively. For each grid point i and the material M_j ($j = 0, 1, \cdots, n-1$), we define the indicator variable as

$$\kappa_{ij} = \begin{cases} 1 & If \ f_i = M_j, \\ 0 & Otherwise. \end{cases} \qquad (5.5)$$

Figure 5.14 shows an example of the indicator variable. In a cubic cell, there are three materials: M_1 (red), M_2 (green) and M_3 (blue). We analyze one material at one time. For the M_1 material, we set grids with M_1 as 1 and all the other grids as 0, then we can obtain the binary configuration or the indicator variable κ_{ij} as shown in (b). Similarly, we can obtain κ_{ij} for M_2 and M_3; see (c, d).

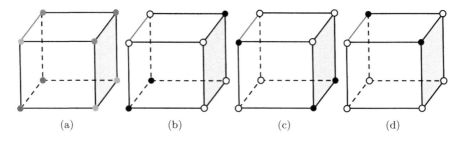

(a)   (b)   (c)   (d)

FIGURE 5.14: An example of the indicator variable. (a) A cell with three materials and their material indices are M_1 (red), M_2 (green) and M_3 (blue); (b-d) the indicator variables of material M_1, M_2 and M_3, respectively, where a black dot denotes $\kappa_{ij} = 1$ and a white dot denotes $\kappa_{ij} = 0$.

Using this indicator variable, we can construct a trilinear function for each material as we did for a single-material domain. In this way, we build three trilinear functions because this cell contains a total of three materials. Given an interior point P in a cell with n materials, its material index is simply the one with the maximum indicator variable value. For each material contained in the cell, we study the topology classification in the same way as in the single-material domain [436]. Instead of using the function value at the face or body saddle points, here we study the material indices at the face and body centers of each cell. If any of the materials in the cube has topology ambiguity, the cell is identified as an ambiguous cell.

Once an ambiguity is found, we split the cubic cell into 12 tetrahedral cells. There is no ambiguity in a tetrahedral cell because it has only four corner nodes. We then modify the dual contouring method for a hybrid octree consisting of both cubic and tetrahedral leaf cells. For each leaf cell, we insert one minimizer by either minimizing a pre-defined error function or simply choosing the cell center. Then, we classify all edges into cell edges, internal edges and face diagonals. In the following, we analyze each material change edge, whose two endpoints lie in two different materials, and use its surrounding minimizers to form a polygon during the triangular mesh generation, or a polyhedron during the tetrahedral mesh generation. For the latter, we also need to study each interior edge, whose two endpoints belong to the same material, to construct interior tetrahedra. Figure 5.15 shows five local topology ambiguities resolved using our algorithm.

After mesh generation, we classify all the vertices into several groups based

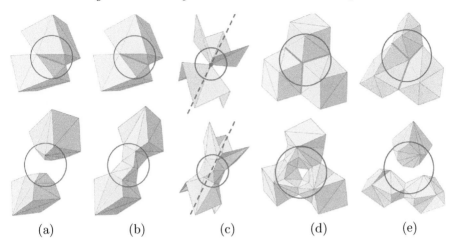

| (a) | (b) | (c) | (d) | (e) |

FIGURE 5.15: Five examples of resolved local topologies (picture from [437]). The top row shows the wrong topologies; and the bottom row shows the resolved corresponding correct topologies.

on their relative location and improve them correspondingly using different strategies; see Section 5.4 for details. Finally we improve the mesh quality with all non-manifold boundary features preserved, and also generate triangular surface meshes and tetrahedral volume meshes with all topology ambiguity resolved. Figure 5.16 shows the generated tetrahedral mesh of a 92-grain beta titanium alloy.

5.4 Quality Improvement

Poorly shaped elements influence the convergence and stability of finite element simulations, therefore it is critical to generate meshes with good quality. First we need to define proper criteria to measure the quality of various types of meshes. For triangular meshes, a popular metric is the aspect ratio defined as $2 * r_{in}/r_{out}$, where r_{in} is the radius of the inscribed circle and r_{out} is the radius of the circumcircle. For tetrahedral meshes, people choose various mesh quality functions to measure them; for example, the dihedral angle [144], the ratio of the element diameter such as the longest edge over the in-radius [150], the edge ratio of the longest edge over the shortest edge in a tetrahedron, the Joe–Liu parameter [244] and a minimum volume bound. With these measures, the mesh quality can be judged by checking the worst, mean and best mesh quality, as well as the distribution of elements in terms of their quality metrics or the histograms.

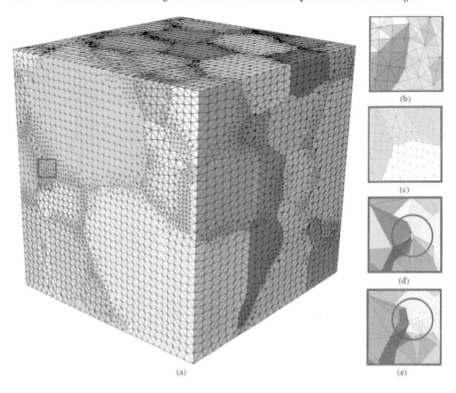

FIGURE 5.16: Resolving topology ambiguity for the 92-grain beta titanium alloy (pictures from [437]). (b-d) shows details of (a).

5.4.1 Three General Mesh Improvement Techniques

Generally the extracted meshes cannot be directly used in finite element computations due to some poorly shaped elements, and we need to improve and optimize the mesh quality. There are three kinds of mesh quality improvement techniques as reviewed in [285, 370]:

1. Local refinement/coarsening by inserting/removing mesh vertices;

2. Local remeshing by face/edge swapping; and

3. Mesh smoothing by relocating mesh vertices.

The first two techniques improve the mesh quality by making local changes to the mesh topology or connectivity, leading to shape (or aspect ratio) improvement and topology improvement, while the last smoothing technique does not. These three techniques are usually used together.

Refinement is an operation performed on the mesh that effectively reduces the local element size, which may be required in order to capture more detailed local physical phenomena or improve the local element quality. Some refinement schemes are also considered as mesh generation schemes like generating adaptive meshes. Starting from a coarse mesh, a refinement procedure is applied until a desired node resolution is achieved. There are three principal refinement methods for triangular and tetrahedral meshes: edge bisection, point insertion and templates. Edge bisection is simply to split an edge in the middle, while all the elements sharing this edge have to be split to avoid introducing any hanging nodes. Inserting a point generally requires to remesh the local region. A Delaunay approach can be used to delete several local elements and reconnect the nodes maintaining the Delaunay criterion. A template refers to a specific decomposition of the triangle or tetrahedron - like subdivision. Note that no hanging nodes are allowed in the resulting meshes.

Coarsening is an inverse operation of refinement. For example, we can reduce the worst edge ratio using edge contraction by merging an interior vertex to a boundary vertex, an interior vertex to another interior vertex, or a boundary vertex to another boundary vertex. For each iteration, we search the tetrahedron with the maximum edge ratio, and use edge contraction to remove it. We keep removing the worst tetrahedron until the maximum edge ratio is below a given threshold. Note that before performing edge contraction, we need to ensure to remove all surrounding valence-3 nodes, otherwise we may obtain overlapping elements.

For triangular meshes, simple diagonal swaps are often performed to improve the local mesh quality. For each interior edge, a check can be made to determine which edge would effectively lead to better overall or minimum aspect ratio of its two adjacent triangles. The Delaunay criterion "empty circle" is often used here. For tetrahedral meshes, a series of local transformations are designed to improve the element quality and local topology (e.g., the valence number of a single node) by swapping interior edges, such as 2-2 flip, 2-3 flip, 3-2 flip and 4-4 flip as shown in Figure 5.17. Edge swapping can also help reduce the deviation from the surface and align with the surface features in a better way. In hybrid meshes with both triangular and quadrilateral elements, a single poor quality quadrilateral can be split into two triangles with good aspect ratio.

Smoothing techniques aim at iteratively relocating individual nodes to improve the local quality of the elements. The simplest discretization of the Laplacian−Beltrami operator [38, 111] for a node is the average of all its neighbors. Laplacian smoothing is an efficient heuristic method, but it is possible to produce an invalid mesh containing inverted elements or elements with negative volume. The contribution of each neighboring node can also be weighted in the average function using edge length, area or volume [151]. Sometime, the node movement is constrained in order to avoid creating inverted elements [136, 257]. These smoothing techniques can also be applied to anisotropic meshes [162, 184, 243]. The optimization techniques measure the quality of the surrounding elements to a node and tend to optimize it [99, 320, 321], which is similar to a minimax technique used to solve circuit design problems [82]. The optimization-based smoothing yields better results

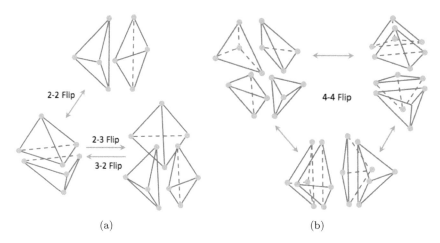

(a)   (b)

FIGURE 5.17: Swapping interior edges to improve the local quality of tetra-hedral meshes. (a) 2-2 flip, 2-3 flip and 3-2 flip; and (b) 4-4 flip.

while it is more expensive than Laplacian smoothing. Therefore, people usu-ally choose a "smart" Laplacian smoothing [73, 143, 144], which relocates the vertex only when the quality of the local mesh is improved. Physically-based smoothing techniques are used to reposition nodes based on a simulated phys-ically based attraction or repulsion force, like a system of springs interacting with each other [253]. Anisotropic meshes can be obtained from a bubble equilibrium system [60, 350, 351]. With various magnitude and direction of interparticle forces, different anisotropic characteristics and element sizes can be achieved. For high order elements like quadratic and cubic elements, repo-sitioning mid-nodes can help improve the high order element quality [317].

5.4.2 Geometric Flow-Based Smoothing

Generally in quality improvement, we need to use a combination of the above-mentioned three general techniques. Here, we adopt face swapping and edge removal coupled with geometric flow-based smoothing [437, 441]. There are two kinds of vertices in tetrahedral meshes, boundary vertices and inte-rior vertices. For each boundary vertex, we choose geometric flow to denoise the surface and improve the aspect ratio. For each interior vertex, we choose a weighted averaging method to improve the aspect ratio like the volume-weighted averaging.

Geometric flows or nonlinear geometric partial differential equations (PDEs) have been intensively used in surface and image processing [411, 412]. These nonlinear equations are discretized based on discrete differential geo-metric operators like the Laplacian−Beltrami operator over triangular meshes.

Given a vertex x on the surface M_0, a geometric flow can be defined as

$$\frac{\partial x}{\partial t} = V_n(\kappa_1, \kappa_2, x)\vec{n}(x) + v(x)\vec{T}(x), \tag{5.6}$$

where $M(0) = M_0$, $V_n(\kappa_1, \kappa_2, x)$ is the velocity along the normal direction, $\vec{n}$ is the unit surface normal vector, and $v(x)$ is the velocity on the tangent plane $\vec{T}(x)$. When $V_n(\kappa_1, \kappa_2, x)$ is defined in different ways, we obtain various geometric flows with different properties.

1. *Mean curvature flow* with the property of area shrinking:

$$V_n = -H = -(\kappa_1 + \kappa_2)/2, \tag{5.7}$$

   where H is the mean curvature;

2. *Average mean curvature flow* with the property of volume preserving and area shrinking:

$$V_n = h(t) - H(t), \tag{5.8}$$

   where

$$h(t) = \frac{\int_{M(t)} H d\sigma}{\int_{M(t)} d\sigma}; \tag{5.9}$$

3. *Surface diffusion flow* with the property of volume preserving and area shrinking:

$$V_n = \triangle H, \tag{5.10}$$

   where $\triangle$ is the Laplace–Beltrami operator; and

4. *High order flow* with the property of volume preserving when $k \geq 1$:

$$V_n = (-1)^{k+1}\triangle^k H. \tag{5.11}$$

Here we choose the surface diffusion flow to smooth the surface mesh because it preserves volume and furthermore approximates spheres accurately with a quadratic precision. A discretization scheme for the Laplace–Beltrami operator over triangles can be found in [270, 412]. We have

$$\triangle f(p_i) = \frac{1}{A(p_i)} \sum_{j \in N_1(i)} \frac{\cot\alpha_{ij} + \cot\beta_{ij}}{2} [f(p_j) - f(p_i)], \tag{5.12}$$

$$H(p_i)N(p_i) = \frac{1}{2A(p_i)} \sum_{j \in N_1(i)} \frac{\cot\alpha_{ij} + \cot\beta_{ij}}{2} (p_i - p_j), \tag{5.13}$$

$$\triangle^k f(p_i) = \triangle\left(\triangle^{k-1}f\right)(p_i)$$

$$= \frac{1}{A(p_i)} \sum_{j \in N_1(i)} \frac{\cot\alpha_{ij} + \cot\beta_{ij}}{2} \left[\triangle^{k-1}f(p_j) - \triangle^{k-1}f(p_i)\right], \tag{5.14}$$

where α_{ij} and β_{ij} are two angles opposite to the edge $p_i p_j$ as shown in Figure 5.18(a), $N_1(i)$ is a neighborhood around p_i, and $A(p_i)$ is the area defined as in Figure 5.18(b).

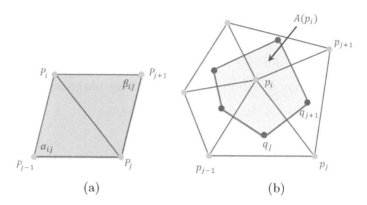

(a)   (b)

FIGURE 5.18: (a) The definition of two angles α_{ij} and β_{ij}; and (b) the definition of the area $A(p_i)$.

In the temporal space, we choose the Euler scheme. For surface vertices, the normal movement can remove noises. The surface diffusion flow preserves volume and also sphere property accurately if the initial mesh is embedded and close to a sphere. This is very suitable for biomolecular meshes because biomolecular surfaces are modeled as a union of hard spheres. For the tangential movement, we choose an area-weighted averaging method that improves the aspect ratio of surface meshes and also is suitable for adaptive meshes. Note that the tangential movement does not change the surface shape. For each interior vertex, we relocate it to the mass center of its surrounding tetrahedra. In this way, we can improve the quality of tetrahedral meshes. Figure 5.19 shows the quality improvement results of a tetrahedral mesh for a protein model (PDB ID: 2O53) using the surface diffusion flow. We can observe that the resulting mesh becomes much more regular with good aspect ratio and surface feature preserved.

5.4.3 Quality Improvement for Multiple-Material Meshes

In Section 5.2, we discussed how to generate adaptive tetrahedral meshes for multiple-material domains with non-manifold boundaries. Quality improvement for such complex meshes is more challenging than that of single-material meshes. The key point is how to improve the mesh quality while preserving all the boundary features.

All the vertices of the tetrahedral meshes can be classified into four groups: fixed vertices, curve vertices, surface vertices and interior vertices. Different kinds of vertices are handled with different strategies in order to preserve the boundary features. It is obvious that interior vertices lie inside each material

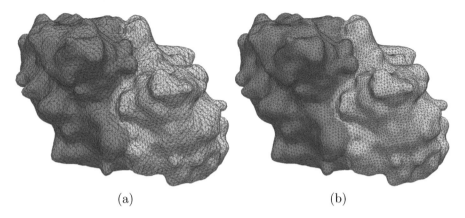

(a)   (b)

FIGURE 5.19: Quality improvement of tetrahedral meshes for a protein (2O53). (a) Before quality improvement; and (b) after quality improvement.

region. Two neighboring material regions share a boundary surface patch, and surface vertices lie on the surface patch. Two or more surface patches intersect to form boundary curves. Curve vertices lie on these boundary curves. The intersections of boundary curves are called corner vertices. These corner vertices and any other non-manifold nodes are fixed vertices, which do not move during the quality improvement.

For polycrystalline materials as shown in Figure 5.20, there is an outer boundary that is generally a cubic representative volume element (RVE). We can classify all the vertices into seven groups according to their relative locations [437].

- Group 1 — The eight corners of the RVE box, which are handled as fixed vertices in order to keep the RVE boundary.

- Group 2 — Vertices on the 12 edges of the RVE box, which only move along the edge. If they lie inside one grain (2a), they are smoothed only along the edge; if they are shared by two or more than two grains (2b), they are fixed during the improvement.

- Group 3 — Vertices on the six faces of the RVE box, which are smoothed only on the plane. If they are shared by more than two grains (3a), they are planar non-manifold points and are fixed during the improvement. If they are on the planar curve shared by two grains (3b), they are smoothed along the tangent direction of the planar curve during the improvement. Finally, if they are the vertices of one grain (3c), they are only allowed to move on the plane.

- Group 4 — Vertices inside one grain, which will be improved using the weighted-averaging or optimization method.

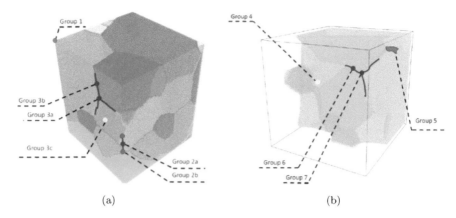

(a)   (b)

FIGURE 5.20: Vertex classification for the polycrystalline materials with the cubic representative volume element boundary (pictures from [437]).

- Group 5 – Vertices located on the grain boundary surface patches shared by two grains, which are smoothed on the tangential plane of the grain boundary during the improvement.

- Group 6 – Vertices located on interior curves, which can only move along the tangent direction of the interior curve.

- Group 7 – Interior non-manifold vertices shared by more than two interior curves, which are fixed during the improvement.

Our quality improvement algorithm for multi-material tetrahedral meshes consists of the four techniques [223], which are used together to smooth the surface, improve the aspect ratio of tetrahedral elements, as well as preserve all non-manifold geometric features.

1. Boundary curve smoothing and regularization – For curve vertices, we aim to denoise them to get a smooth boundary curve and try to regularize them. A shape-preserving curve diffusion flow is used to smooth curves, coupled with the tangential movement to achieve a uniform distribution. The curve diffusion flow can be defined as

$$\frac{dx}{dt} = -\triangle \kappa \vec{n}, \tag{5.15}$$

where $\triangle$ is the Laplace−Beltrami operator, and κ is the curve curvature computed using a local quadratic curve interpolation.

2. Boundary surface patch smoothing and regularization – The averaged mean curvature flow in Eq. (5.8) is used to smooth surface vertices along the normal direction, which has the property of volume-preserving. In

addition, triangular surface meshes are also optimized by minimizing an energy function defined based on quality metrics. For example, given surface vertices $\{x_i\}_{i=1}^N$, we can define the energy function as

$$E(x_i) = \sum_{j \in N(i)} \left(\|x_j - x_i\| - h \right)^2 , \tag{5.16}$$

where $N(i)$ is the neighborhood of x_i and h denotes the local average edge length around x_i.

3. Tetrahedral mesh regularization – An optimization-based method is applied to relocate interior vertices and improve all tetrahedra in the meshes. For example, the energy function can be defined as

$$E(x_i) = \sum_{j \in N(i)} \left(\frac{\left(\sum_{j=1}^6 e_j^2 \right)^{1.5}}{8 \times 3^{2.5} V} - q \right)^4 , \tag{5.17}$$

where $\{e_j\}_{j=1}^6$ are six edges of a tetrahedron and V is the volume, $N(i)$ is the neighborhood around x_i, and q is a threshold (we set $q = 1/0.9$, for example).

4. Topological transformation – Topological transformations like face

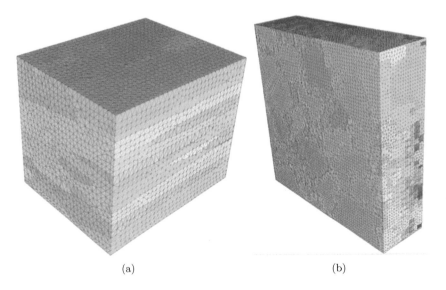

(a)   (b)

FIGURE 5.21: Tetrahedral meshes of two multiple-grain polycrystalline materials. (a) Foam; and (b) nickel. The imaging data were provided by Professors Anthony Rollett and Robert Suter at Carnegie Mellon University.

swapping (2-2 flip, 2-3 flip, 3-2 flip and 4-4 flip as shown in Figure 5.17) and edge removal are also used to improve the mesh valence and connectivity.

The first three techniques relocate vertices without changing their connectivity. Differently, the last technique of topological transformation eliminates some elements and reconnects vertices locally to improve the mesh quality. Figure 5.21 shows quality tetrahedral meshes of two multiple-grain polycrystalline materials: foam and nickel.

5.4.4 Dual Contouring-Based Meshing with Guaranteed Quality

For the octree-based methods, only a few algorithms study how to guarantee the angle range. By using the body-centered cubic tetrahedral lattice and adjusting the cutting points, an isosurface stuffing method [213] guarantees that all the dihedral angles are between $10.78°$ and $164.74°$ for a uniform boundary, and between $1.66°$ and $174.72°$ for an adaptive boundary. This method was extended to work for domains consisting of multiple materials [64]. Later, another method was developed in [386] to guarantee a minimal dihedral angle $5.71°$ for an adaptive tetrahedral mesh.

As shown in Figure 5.22, in this section we will introduce an improved dual contouring method, which can guarantee a better dihedral angle range for adaptive triangular and tetrahedral mesh generation [240]. This algorithm is based on quadtree or octree structure, and can generate interior and exterior meshes with conformal boundary. Given a planar and closed smooth curve, we have three steps to generate an adaptive triangular mesh for the interior region with guaranteed angle bounds: (1) adaptive quadtree construction; (2) grid points adjusting; and (3) improved DC method.

We first decompose the input curve into a set of non-uniform points based on its local curvature, and then use them to help build an adaptive quadtree. For each grid point i which is very close to the input curve, we use a size function $grid(i)$ and a distance function $dist(i)$ to decide how much we should move away from the curve along the normal direction. In this way, we can leave enough space between the grid point and the input curve, and then generate good-quality triangles around the boundary. In the last step, we design strategies to relocate the minimizer points and also in certain conditions we introduce one more minimizer point to guarantee generating good-quality triangles. Please refer to [240] for detailed algorithm and criteria. This three-step algorithm has been extended to 3D tetrahedral mesh generation.

We have theoretically or numerically proven that following the three steps of our algorithm, we can guarantee the obtained triangle mesh has an angle range of $(19.47°, 141.06°)$ for any given closed smooth curve, and the tetrahedral mesh has a dihedral angle range of $(12.04°, 129.25°)$ for any given closed smooth surface. These angle ranges are better than the body-centered cubic

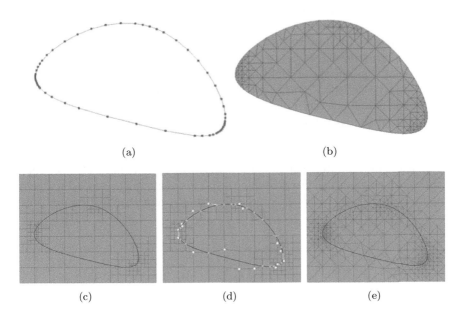

(a)   (b)

(c)   (d)   (e)

FIGURE 5.22: Flow chart of triangular meshing with guaranteed quality (pictures from [240]). (a) Input curve decomposition; (b) output triangular mesh; (c) adaptive quadtree construction; (d) grid points adjusting; and (e) improved dual contouring method. In (d), pink grid points are moved toward the curve, and yellow grid points are moved away from the curve.

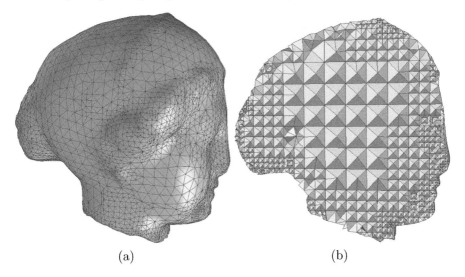

(a)   (b)

FIGURE 5.23: Tetrahedral mesh for the Igea model (pictures from [240]), with the dihedral angle range of $(11.4°, 131.9°)$ and a $0.64°$ perturbation.

methods or MC-based methods, this is mainly because in our dual contouring-based method, the minimizer points (which are also the final mesh vertices) are allowed to move on the surface inside the boundary cell. This gives us a lot of degrees of freedom to further improve the minimal and maximal angle in the resulting mesh.

In practice, since the straight line or planar cutting plane assumption can not always be satisfied strictly, we allow a small perturbation for the lower and upper bounds of the proved angle range. Figure 5.23 shows the tetrahedral mesh of the igea model, and the dihedral angle range is (11.4°, 131.9°) with a 0.64° perturbation. This algorithm provides a fundamental study on guaranteed-quality tetrahedral mesh generation, with a great scientific value for research in computer graphics or visualization. For real finite element applications, we may not need to guarantee a specific angle range, and adjustments can be made to meet various requirements from specific application problems.

Homework

Problem 5.1. In marching cubes, there are 14 unique cases. Which cases have ambiguity? List all of them using the case number 0-13.

Problem 5.2. Use the octree techniques to generate triangular meshes for the region inside the curve (Figure 5.24): (a) marching cubes − uniform quadtree; (b) marching cubes − adaptive quadtree; (c) dual contouring − uniform quadtree; and (d) dual contouring − adaptive quadtree.

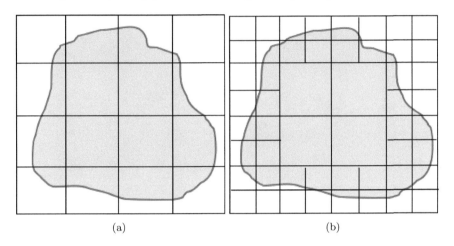

(a)   (b)

FIGURE 5.24: (a) Uniform quadtree; and (b) adaptive quadtree.

Problem 5.3. Perform a survey for unstructured triangular and tetrahedral mesh generation; discuss pros and cons for each technique. What are the challenges and what are the future directions?

Problem 5.4. Why is quality improvement important for mesh generation? Perform a literature survey and compare various quality improvement techniques for triangular and tetrahedral meshes.

Chapter 6

Image-Based Quadrilateral and Hexahedral Meshing

Quadrilateral (quad) and hexahedral (hex) meshes are superior to triangular and tetrahedral meshes in several aspects. In this chapter, we first review unstructured quad and hex mesh generation, and then talk about image-based meshing schemes. We cover several important topics on quad and hex mesh generation, including refinement templates, adaptive meshing using hybrid octree, meshing for multiple-material domains, sharp feature preservation and RD-tree based hex meshing. In the end, we also discuss various methods developed to improve the mesh quality, including pillowing, geometric flow-based smoothing and optimization for hex meshes, as well as guaranteed-quality all-quad mesh generation.

6.1   A Review of Unstructured Quadrilateral and Hexahedral Mesh Generation

Quadrilateral (quad) and hexahedral (hex) meshes have a lot of finite element applications with several superior properties over triangular and tetrahedral meshes in terms of increased accuracy, smaller element counts and improved reliability. Compared to triangular and tetrahedral mesh generation, quad and hex meshing is much more challenging because not all polygons or polyhedra can be decomposed into quads or hexes directly. In addition, unstructured hex meshes usually have complex topology, and extending 2D quad meshing to 3D hex meshing is generally not straightforward.

Mapped meshing or sweeping is a straightforward method to generate structured quad and hex meshes. It requires the opposite edges or faces of the to-be-meshed domain must have an equal number of divisions or exactly the same surface mesh. By sweeping from one boundary edge or face to its opposite, we obtain a structured quad or hex mesh. For an arbitrary geometry, the mapped meshing method may not work because we cannot guarantee the opposite faces always have the same surface mesh. Sometimes people perform user interactions to split a complex geometry into several mapped meshable regions. Each region can be meshed separately and then the resulting meshes are merged together. Note that we need to ensure a conformal boundary shared by any two adjacent regions. In the CUBIT project at Sandia National Laboratories [1], a lot of research has been carried out to automatically recognize features and decompose geometry into mapped-meshable areas or volumes, with several toolkits developed to assist the domain decomposition.

Instead of directly decomposing the complex geometry using user interactions, sometimes people use corner angles and edge directions to make an appropriate virtual decomposition [405]. Each mapped meshable region is meshed separately using a sweeping method and finally all these region meshes are merged. This method is suitable for many CAD models with blocky components, well-defined corners and cubic regions.

Given a source face and its opposite target face with the same surface quad mesh, we sweep the quad mesh along a curve to mesh the volume by creating layers of hex elements. This method is actually $2\frac{1}{2}$-D meshing. Note that during sweeping we need to avoid generating self-intersections when we locate internal nodes and cross sections. In the *Cooper Tool* [51], Ted Blacker generalized the applicability of sweeping by allowing multiple source and target surfaces, while a single sweeping direction is still required. In this method, the topology of the source and target surfaces can be different, and branches may happen along the sweeping direction [341, 404]. The Cooper Tool is available in the Fluent pre-processor named Gambit, and now it is a product of ANSYS.

6.1.1 Unstructured Quadrilateral Meshing

In unstructured meshes, the valence number of interior nodes is relaxed. This also gives us a lot of flexibility for mesh generation. There are two main methods for unstructured quad meshing: indirect methods and direct methods. In indirect methods, the domain is first meshed into triangles and then we convert each triangular element into four quads using the edge middle points and the face center. In contrast, direction methods generate quad elements directly without going through triangular meshing.

A simple indirect method is to divide each triangle into three quads. This method guarantees a resulting all-quad mesh but with many irregular (valence-3) nodes and poor-quality elements. Instead of splitting triangles, we can also combine two adjacent triangles into one quad [218]. In this method, better quality elements can be obtained, but we may have some triangles left. By maximizing the number of resulting quads, we can generate quad-dominant meshes with a small number of triangles. In addition, local element splitting and swapping can be used to increase the number and also the quality of quads in the mesh.

Given even number of edges on the boundary, an advancing front method was developed to generate an all-quad mesh [218]. The triangle edges on the boundary are defined as the initial set of front. A pair of triangles are systematically combined at the front, advancing towards the interior. Such indirect methods are generally fast, but they may introduce many irregular nodes which need to be improved via topological cleanup operations. Quad morphing [287] is another advancing front method to convert triangles into quads, which can significantly reduce the number of irregular nodes by swapping local edges and introducing additional nodes.

For direct methods, we can decompose the domain into simple regions first, and then mesh each region using templates or other methods. In the following, we will mainly review four direct methods [285, 370]: grid-based or quadtree-based methods, medial axis, advancing front and paving methods. The quadtree-based decomposition method [32] is a robust method for any complex geometry. It starts with an initial decomposition of the 2D space into a quadtree using local feature sizes. Quad elements are generated from the quadtree leaves and nodes are adjusted to conform to the boundary. The general domain decomposition can also be recursively subdivided into simple polygonal shapes, and each of them satisfies a limited number of templates for meshing [32, 366]. Good programming techniques are required to maintain compatibility of element divisions between two adjacent regions. Medial axis can be used to decompose the domain [28, 367]. The medial axis is a union of centers of the maximal circles as they roll through the area. The advancing front method [445] starts with an initial placement of nodes on the boundary, and then elements are formed by projecting edges towards the interior. Sometimes two triangles are created first using the traditional advancing front methods, which are combined to form a quad. Similarly, the paving technique

[52] starts from the boundary and works in by forming complete rows of elements. In paving, projecting nodes, handling special geometric situations and intersection of opposing fronts are all important issues. The paving technique is available in CUBIT software, MSC Patran and Fluent's Gambit software.

6.1.2 Unstructured Hexahedral Meshing

Starting from a tetrahedral mesh, Eppstein [127] split each tetrahedron into four hexahedra using edge middle points, face center points and the body center point. This method is straightforward and guarantees a resulting all-hex mesh, but it yields a large number of elements and many elements have poor quality. In addition, it also produces many irregular (valence-4) vertices. Another indirect method is to combine five or more tetrahedra to form a single hexahedron. Due to the complex 3D topology in hex meshes, this method is still not tractable so far.

Direct unstructured all-hex mesh generation [285, 370] is more preferable, and there are mainly five methods developed over the years: grid-based or octree-based, medial surface, plastering, whisker weaving and the recently developed vector field-based methods. The grid-based approach yields a fitted 3D grid of hex elements in the interior of the volume [325, 327]. Hex elements are inserted at the boundaries to fill the gaps between the regular grid of hexes and the boundary surface. Local modification and smoothing are needed in order to create topologically correct hex elements. This method is robust for complex geometry, but it creates poorly shaped elements around the boundary. Hex elements are in general not conformal to the boundary, especially for specific regions with sharp features. Sharp feature preservation in all-hex meshing is still a challenging problem. For grid-based or octree-based methods, the orientation of the resulting meshes depends on the direction of the interior grids. These methods are available in the Hexar software from Cray Research [406] and MARC's Mentat software.

We have developed octree-based quad and hex meshing schemes using the dual contouring idea to handle conformal boundaries [427, 428], which will be discussed in detail later in this chapter. Each interior grid point is shared by four cells in 2D for uniform cases. We compute one minimizer point for each cell and connect them to form a quad element. After constructing uniform quad meshes, we then use templates to locally refine them and obtain adaptive quad elements. The newly generated points are projected onto the boundary surface. In 3D, each interior grid point is shared by eight cells and we obtain eight minimizer points to form a hex element. This method is very suitable to create interior and exterior meshes with the conformal boundary between them. Adaptive hex meshes can be obtained by using template-based local refinement and look-up tables. The mesh adaptation can be controlled flexibly using a feature-sensitive error function, areas of interest, finite element solutions or user-defined error function. This idea has been extended to mesh domains with multiple materials [432].

Medial surface methods [302, 303] start with an initial decomposition of the volume using medial surfaces, which are surfaces formed by the center of a maximal sphere as it rolls all over the volume. The volume is decomposed to generate several mapped meshable regions, and each region is meshed using a series of pre-defined templates. Medial surface methods have been incorporated into the FEG's CADFix hex mesh generator and Solidpoint's Turbomesh software.

Plastering [53, 72] is basically an advancing front hex meshing technique, which extends the 2D paving algorithm to 3D. It first places elements along the boundary and then advances towards the interior of the volume. Note that we need to detect intersecting faces and determine how to use the existing nodes and when to seam faces. In addition, complex interior voids may appear, which sometimes cannot be meshed with all-hex elements. In these situations, we have to modify the surrounding existing elements to facilitate placement of hexes towards the interior. The plastering techniques work for many objects, however it has not been proven to be robust and reliable for any complex domain. The whisker-weaving hex meshing [368] was developed based on the concept of the spatial twist continuum (STC), which is a dual of the hex mesh formed by intersecting surfaces bisecting hex elements in three directions. In this method, the STC is constructed first, and then it is used as a guide to fill the volume with hex elements. Like the plastering technique, the whisker-weaving technique has not been proven to be robust for arbitrary geometry.

The vector field-based methods are techniques developed in recent years. CubeCover [278] generates uniform all-hex meshes with the guidance of a valid 3D frame field via a global parameterization. This method was later extended to singularity-restricted field guided volume parameterization [236]. Although these methods can generate all-hex meshes with high quality, they consider uniform hex meshes only. In addition, the robustness and reliability of this kind of method are still questionable; it may not work for any complex geometry. All-hex meshing is challenging, therefore some people investigated hex-dominant meshing methods to create hybrid meshes [413]. Note that to connect tetrahedral elements with hexes, we need some transition elements like pyramid- or prism-shaped elements.

In the following, we will discuss several octree-based dual contouring schemes for automatic all-quad and all-hex mesh generation from scanned images, including meshing for multi-material domains, hybrid octree-based adaptive meshing, sharp feature preservation, RD-tree based meshing and quality improvement.

6.2   Octree-Based Quadrilateral and Hexahedral Mesh Generation from Images

Unstructured quad and hex mesh generation attracts many researchers' interests due to its broad applications in a lot of finite element simulations. However, it is still a challenging task to generate adaptive and quality all-quad and all-hex finite element meshes directly from volumetric images, such as CT and MRI. In Chapter 5.2, we have discussed how to use the dual contouring method coupled with the octree data structure to generate triangular and tetrahedral meshes from images by analyzing sign change edges and interior edges [428, 441]. In this section, we will discuss in detail how to generate all-quad and all-hex meshes by analyzing each interior grid point in the octree-based dual contouring method [427, 428].

To yield uniform quad and hex meshes with correct topology from images, we adopt a suitable starting octree level using a bottom-up surface topology preserving scheme. In this scheme, the key idea is to check whether a fine isocontour is topologically equivalent to a coarse one. Given the volumetric data and an isovalue, we convert each intensity value to a binary sign by comparing it with the isovalue. In other words, we assign "1" to a grid point if its intensity value is greater than or equal to the isovalue; otherwise the sign is "0". According to [189], given an edge (or face or cube), the fine and coarse isocontour is topologically equivalent to each other if and only if the sign of its middle vertex is identical to the sign of at least one endpoint of that edge (or face or cube). Generally we choose a suitable starting octree level to guarantee the correct topology for uniform mesh generation, and then the topology will be preserved in the following adaptive mesh refinement.

In uniform cases, each grid point within the to-be-meshed region is always shared by four cells in 2D and eight cells in 3D. We calculate a minimizer point for each boundary cell, and simply set the cell center as the minimizer point for each interior cell. These four or eight minimizers form a quad in 2D and hex in 3D. Figure 6.1 shows the interior and exterior all-hex meshes for the human head model. They share the conformal surface represented implicitly as an isosurface in the volumetric data. Note that the mesh quality needs to be improved before they can be used in finite element applications.

To generate adaptive quad and hex meshes, we use refinement templates [304, 427, 435] to locally refine the obtained uniform meshes. Alternatively, we can build a generalized hybrid octree structure [172] and use it to generate adaptive quad and hex meshes. We will talk about these two adaptive meshing techniques as follows.

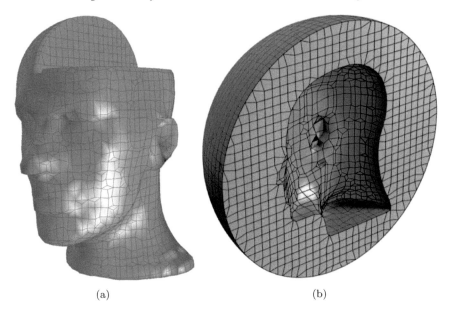

(a)   (b)

FIGURE 6.1: Interior and exterior all-hex meshes of the human head with a shared conformal surface (pictures from [428]). (a) The interior hex mesh; and (b) the exterior hex mesh within a bounding sphere.

6.2.1 Refinement Templates

The extracted uniform quad and hex meshes can be made adaptive using refinement templates, with the newly inserted vertices projected onto the boundary. There are two kinds of refinement templates: 3-refinement and 2-refinement. 3-Refinement templates basically split an edge into three smaller edges equally; while 2-refinement templates split an edge into two smaller edges. In the following, let us discuss these two templates in detail.

3-Refinement Templates. Figure 6.2 shows the 2D and 3D 3-refinement templates for adaptive quad and hex mesh generation. The red dots represent the minimizer points of those to-be-refined quadtree or octree cells. The designed templates should satisfy the following five basis requirements [427]:

- All the resulting elements are quads in 2D and hexes in 3D;

- No hanging node is allowed;

- The resulting mesh represents the object surface accurately;

- The resulting elements have good aspect ratio; and

- The resulting mesh introduces a small number of new elements and vertices.

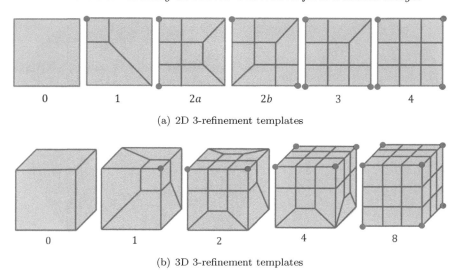

(a) 2D 3-refinement templates

(b) 3D 3-refinement templates

FIGURE 6.2: 2D and 3D 3-refinement templates for quad and hex meshing. (a) Six 2D templates; and (b) five 3D templates. The number below each template indicates the number of minimizers (the red dots) generated from to-be-refined octree cells.

From Figure 6.2, we can observe that there are a total of six templates in 2D. While in 3D, we have only five templates because not all the configurations can be divided into all-hex elements. We set a sign for each corner vertex of the hex element or the associated cell in the uniform level, indicating if this cell needs to be refined or not. Then we check each hex element to see if it matches with any of the five basic templates. If not, we use a pre-defined look-up table [427] to update the related signs until all the configurations fall within these five templates. Finally, we apply the 3-refinement templates to generate adaptive hex meshes. The mesh adaptation can be controlled flexibly using various ways, including the feature-sensitive error function to detect curvature-related surface features, areas of interest, simulation results and user-defined error function.

Figure 6.3 shows an adaptive hex mesh of a complex protein called mouse acetylcholinesterase (mache), which exists in our neuromuscular junction system in the tetramer format. The active site or the cavity is detected and also refined locally to capture more detailed surface features. It is also the main structure deciding the constant diffusion rate of this protein.

2-Refinement Templates. Compared to 3-refinement templates, it is much more complex to implement 2-refinement templates for adaptive all-hex mesh generation. Due to this reason, very little research has been done and it has been restricted to structured meshes only [125, 181, 328]. When they are applied to an unstructured mesh, an overall refinement of all elements is

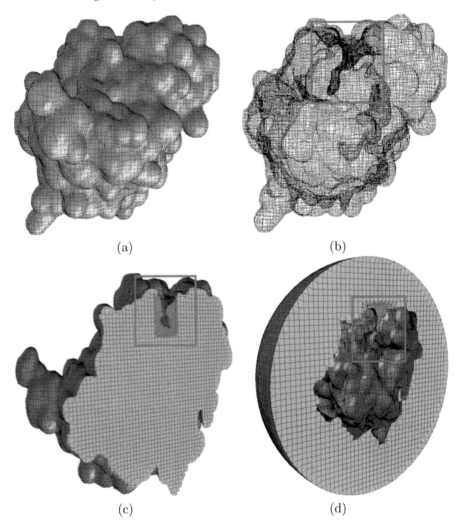

(a)   (b)

(c)   (d)

FIGURE 6.3: Adaptive all-quad and all-hex meshes of a protein called mouse acetylcholinesterase (mache, pictures from [427]). (a) Surface quad hex; (b) wireframe of (a); (c) a cross section of the interior all-hex mesh; and (d) a cross section of the exterior all-hex mesh. The red boxes show the locally refined elements around the active site or the cavity.

usually required beforehand, which increases the number of elements rapidly by seven times [122]. In the following, we discuss a new 2-refinement implementation [304] for any unstructured all-hex meshes, which removes all the hanging nodes with a limited local propagation.

The main difficulty of implementing 2-refinement templates is how to automatically remove all the hanging nodes without introducing too much prop-

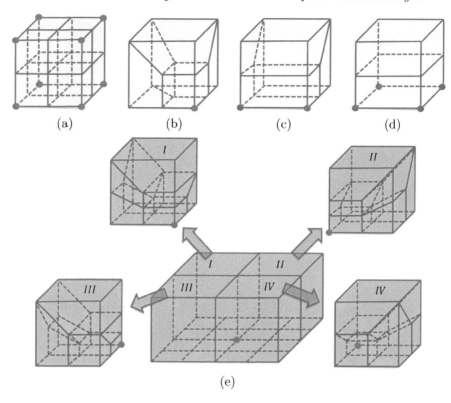

FIGURE 6.4: 2-Refinement templates. The red dots represent the nodes in the refined region. Template (e) must be used in a block of four transition elements surrounding a valid transition block center (the red dot).

agation. Figure 6.4 shows the five templates [326, 328] used in our implementation. We categorize all the hex elements into three regions: the to-be-refined region, the transition region and the unchanged regions. A *to-be-refined region* is formed by all the elements that need to be refined using Template (a); a *transition region* is formed by those elements adjacent to the to-be-refined region, which need to be refined using Templates (b-e); and finally the *unchanged regions* are elements far away from the to-be-refined region and they stay unchanged. During our 2-refinement process, we first refine all the elements in the to-be-refined region following Template (a). As a result, hanging nodes are introduced on the boundary of the to-be-refined region. Then we identify the adjacent transition region and eliminate all the hanging nodes using Templates (b-e). The detailed steps [304] are discussed as follows.

1. Define a valid to-be-refined region and refine it (Step 1) – There are three requirements to define a valid to-be-refined region: (a) only regular nodes are allowed to be on the boundary of the to-be-refined region;

(b) all the transition blocks must have a valid center as shown in Template (e); and (c) all the elements in the transition region need to be nonconcave. If any of the above three requirements is violated, we add the related elements into the to-be-refined region until all these three requirements are satisfied. Then, we obtain a valid to-be-refined region and refine each element in it following Template (a). Hanging nodes are therefore created, which will be removed later.

2. Create two layers in the transition region (Step 2) − From Step 1, we obtain the transition region with all the elements nonconcave. They have only one node, edge or face lying on the boundary of the to-be-refined region. Using the three templates in Figure 6.4(b-d), we create two layers for the transition region.

3. Remove all the hanging nodes (Step 3) − In the transition region, we first identify all the valid transition block centers and then apply Template (e) to them directly. In the to-be-refined region, we mark all the nodes connecting to the valid transition block centers. Then, we identify all the elements in the transition region with one marked node and apply Template (b) to refine them.

Discussion. In 2-refinement, the element size is reduced by a factor of 2, while in 3-refinement the element size is reduced by a factor of 3. Therefore 2-refinement schemes usually yield fewer elements with a more gradual,

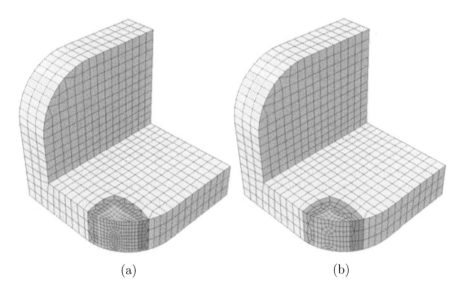

(a)   (b)

FIGURE 6.5: A comparison of the 3-refinement (a) and 2-refinement (b) schemes (pictures from [304]). The to-be-refined regions are marked in blue; the transition regions are in pink and the unchanged regions are in yellow.

smoother transition and better aspect ratio than 3-refinement schemes. The implementation of 2-refinement is much more complex, where we need to deal with adjacent elements in pairs, while 3-refinement templates can be applied individually. Compared with the literature, our 2-refinement implementation is robust for any unstructured hex meshes and we eliminate all the hanging nodes with a limited propagation. Figure 6.5 shows a comparison of the 3-refinement and 2-refinement schemes. We can observe that 2-refinement and 3-refinement have similar propagation, but 2-refinement yields a smoother transition, better aspect ratio and fewer elements.

6.2.2 Hybrid Octree for Adaptive Meshing

Instead of using local refinement templates as we discussed above, we can also generalize the octree structure to a hybrid one [172] and use it to create adaptive quad and hex dual meshes. In the hybrid octree, the leaf cells can be more general polyhedra besides cubes.

In our octree-based meshing methods, we build a strongly-balanced octree using the balancing and pairing rules. The balancing rule restricts the level difference between two neighboring octants to be at most one. The pairing rule is embedded within the balancing rule. It requires all the siblings of an octant (the other seven octants from the same parent) to be subdivided if this octant is subdivided in order to comply to the balancing rule. To build a hybrid octree, a new cutting operation is developed to eliminate all the hanging nodes and at the same time guarantee that each grid point is always shared by eight leaf cells. Note that each leaf cell can be a cube or a general polyhedron. In our dual contouring-based meshing methods, we compute one minimizer point for each leaf cell and obtain eight minimizers surrounding each interior grid point to form a hex element.

Let us first take a look at the cutting operation in 2D. In the strongly balanced quadtree, four sibling quadrants with the same parent form a quadtree block. In Figure 6.6(a), there are two blocks: the blue one and the yellow one. When one of the adjacent blocks (the yellow one) is subdivided, we obtain a transition edge (the green edge) with two hanging nodes (the red dots). To eliminate hanging nodes, we move the transition edge toward the subdivided block and cut the transition cells as shown in Figure 6.6(b). We can observe that one new grid point (the blue dot) is introduced in the transition region (the orange region) and the resulting hybrid quadtree cell is not always a square; it can be a general polygonal cell. Note that our cutting operation ensures each grid point is always shared by four cells, guaranteeing to yield an all-quad dual mesh later. Different from [265], our cutting method eliminates hanging nodes in the subdivided block with less propagation. In addition, when two adjacent transition edges follow two different directions, we can apply the cutting operation in each direction independently; see the transition region and two transition edges in Figure 6.6(d).

In the strongly-balanced octree in 3D, eight octants with the same parent

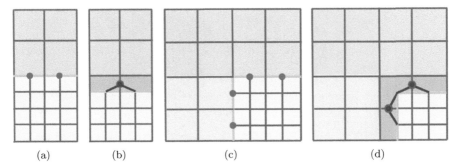

(a) (b) (c) (d)

FIGURE 6.6: The cutting operation in 2D. (a, b) Two neighboring quadtree blocks with one transition edge and the cutting result; and (c, d) four neighboring quadtree blocks with two transition edges along two directions and the cutting result. The red dots are hanging nodes, the green edges are transition edges, the blue dots are newly inserted grid points, and the blue edges are cutting edges. In (b, d), the orange region is the transition region.

form an octree block. A transition case happens when adjacent blocks are at different levels. There are five unique transition cases, including one transition case on face and four transition cases on edge. Figure 6.7 shows all these transition cases and their corresponding cutting results. To eliminate hanging nodes, our cutting operation cuts in the subdivided blocks to guarantee all grid points in the resulting hybrid octree are always shared by eight polyhedral cells. According to the pairing rule, we have a limited number of configurations and the transition along each direction can be carried out independently.

After obtaining the hybrid octree, we can extract an all-hex dual mesh. Each grid point is shared by eight polyhedral cells, we compute one minimizer point for each cell and connect them to build a dual hex element. For interior cells, we simply choose the cell center as the minimizer point. All grid points in the hybrid octree are always shared by eight cells, therefore the resulting dual mesh contains all-hex elements. Figure 6.8 shows two hex meshes generated using this hybrid octree method.

6.2.3 Meshing for Multiple-Material Domains

The octree-based dual contouring method [427, 428] has been extended to generate all-quad and all-hex meshes for multiple-material domains [432]. Instead of analyzing edges such as material change edges and interior edges for triangular and tetrahedral meshing, we first choose a starting octree level and analyze each interior grid point to construct a uniform all-quad or all-hex mesh. In a uniform case in 2D, as each interior grid point is shared by four quadree cells, we can calculate four minimizer points to construct a quad. In

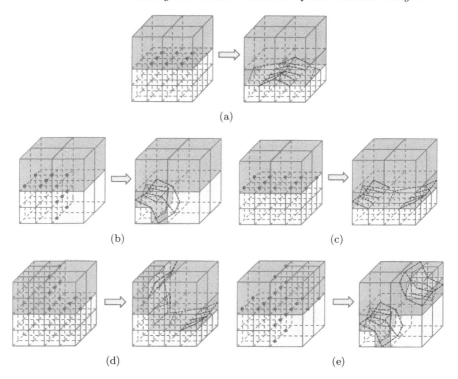

FIGURE 6.7: The cutting operation in 3D. (a) One transition case on face for two blocks; and (b-e) four transition cases on edge for four blocks. The red dots are the handing nodes, the green planes are transition planes, the red and blue dashed lines are the cutting edges in two different directions.

3D, each grid point is shared by eight octree cells, so we can obtain eight minimizer points to construct a hex element.

An error function is calculated for each octree cell and is compared with a given threshold to decide the sign configuration of the minimizer point for this cell. All configurations can be converted into the basic 3-refinement or 2-refinement templates [304, 427, 435] to refine the uniform mesh adaptively without introducing any hanging nodes. The 3-refinement templates satisfy one criterion: in all templates, the resulting refinement around any minimizer point, edge or face with the same sign configuration is the same. This criterion guarantees that no hanging nodes are introduced during the process of mesh refinement. The 2-refinement scheme needs a bit more propagation, but it yields a more gradual and smoother transition. The hybrid quadtree and octree [172] can also be used here to construct adaptive quad and hex meshes.

Figure 6.9 shows the all-quad and all-hex meshes constructed for three areas (Areas 19, 37 and 39) in the segmented Brodmann brain atlas, with one area represented by a different color. We can observe that the multi-material

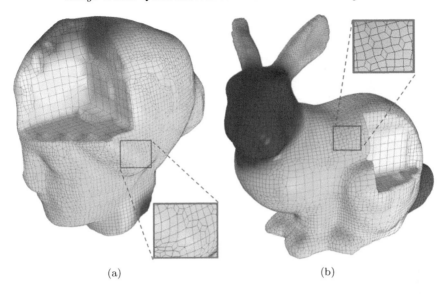

(a) (b)

FIGURE 6.8: The adaptive hex meshes generated using the hybrid octree. (a) The igea model; and (b) the bunny model. The color represents a temperature distribution from finite element analysis.

domain usually has non-manifold boundaries and our created meshes have conformal boundaries between any two adjacent areas.

6.3 Sharp Feature Preservation for CAD Assemblies

Automatic mesh generation for CAD models plays an important role in many engineering simulations. These complicated 3D domains like multiple-component CAD assemblies need to be discretized into piecewise-linear tetrahedral or hex elements, while all-hex meshes are more preferred in a lot of finite element applications. Generally CAD models contain many sharp features such as designed curves, corners and surface patches. Starting from B-Reps (boundary-representations), we need to incorporate them into mesh generation in order to preserve all the given boundary features.

Although some traditional mesh generation methods such as sweeping [51, 203, 214, 341, 358, 404], plastering [53, 72, 359] and whisker weaving [139, 368] have attained much success, robust unstructured all-hex mesh generation with sharp feature preservation is still a challenge. Among the recent developments, the grid-based or octree-based methods have been proven to be robust for complex geometry and topology. The octree-based isocontour-

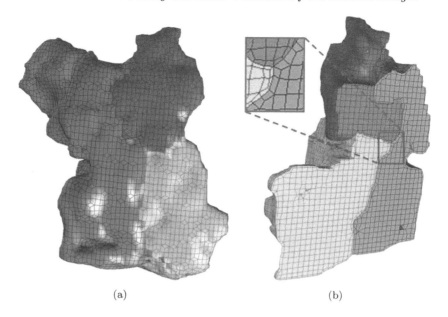

(a)   (b)

FIGURE 6.9: The all-quad and all-hex meshing results for three areas (Areas 19, 37 and 39) in the segmented Brodmann brain atlas (pictures from [432]). (a) Quad mesh; and (b) hex mesh with one cross section.

ing methods [427, 428, 432] analyze each interior grid point and generate a dual mesh of the background grids for any complicated single-material and multiple-material domains. The feature embedding methods [265, 286, 340] embed the geometric topology into a base mesh and then map the base mesh onto the given surface geometry.

In the following, we talk about an automatic and robust method to generate conformal all-hex meshes with sharp feature preservation for multiple-component CAD assemblies [304].

6.3.1 Meshing Pipeline

Figure 6.10 shows the pipeline of our all-hex mesh generation algorithm for CAD assemblies. Given B-Reps, we first extract NURBS (non-uniform rational B-splines) curves and surfaces, and triangulate the entire surface using NURBS sampling. Then, we embed the surface model within grids and generate the binary grid data to distinguish the interior and exterior regions for the object. For assemblies with multiple components, the binary sign attached to each grid point represents the component index. From the constructed grid data, we use the octree-based dual contouring method [427, 428, 432] to construct a uniform all-hex base mesh. One minimizer point is calculated for each cell. For binary grid data, we simply choose the cell center as the minimizer point. We analyze each interior grid point to generate a dual hex mesh of the background grids.

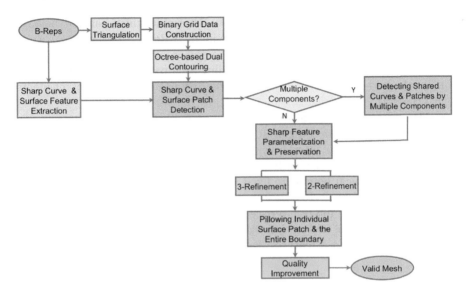

FIGURE 6.10: All-hex meshing pipeline for CAD assemblies. The modules in pink are critical steps involving sharp feature detection and preservation.

After that, we identify and distinguish the shared feature curves and surface patches by multiple components, and embed all the sharp features into the built base mesh. Based on the given geometry topology, all the mesh vertices are categorized into four groups. To preserve all the sharp features, we relocate each group of vertices to their appropriate positions via a curve and surface parameterization. The users can choose 3-refinement or 2-refinement templates to generate adaptive all-hex meshes. To remove triangle-shaped quads along the feature curves and "doublets," we first pillow each surface patch individually and then pillow the entire boundary surface. Finally, a combination of smoothing and optimization is used to improve the mesh quality. In the following, let us talk about sharp feature preservation and the two-step pillowing technique in detail.

6.3.2 Sharp Feature Detection and Preservation

From the given B-Reps, we need to detect and preserve two kinds of sharp features: boundary curves and surface patches. For multiple-component CAD assemblies, a boundary curve may be shared by multiple surface patches, and a surface patch may be shared by two different components.

To identify each boundary curve, we first detect its one endpoint and then trace the mesh edges to find the complete curve trajectory path. Note that for a closed curve, its two endpoints are identical. To guarantee the correct curve topology, we restrict each edge segment can only belong to one specific curve. In addition to boundary curves, we also need to identify each surface patch. We check each boundary quad element to determine which patch it belongs to by comparing its four corners with the input B-Reps. Alternatively, we can also detect boundary quads using a propagation scheme. From the B-Reps we have obtained all the curves surrounding each patch. We start with those quads along the detected curves, and then propagate to adjacent elements until the entire set of quads is identified.

After identifying all the vertices on sharp curves and surface patches, we classify them into four categories: the critical points, the curve vertices, the surface vertices and the interior vertices. To preserve all the boundary features, we relocate and regularize each vertex via a curve and surface parameterization scheme. Critical points are fixed all the time. Curve and surface vertices are relocated and projected back onto the given NURBS curves and surface patches. Finally, the interior vertices are relocated to their volume center via a weighted-averaging and optimization method. In this way, we can preserve all the boundary sharp features from the input B-Reps in our all-hex mesh generation.

6.3.3 Two-Step Pillowing for Nonmanifold Domains

After sharp feature preservation, the constructed all-hex mesh conforms to the given B-Reps. However, some boundary quads may have two adjacent edges lying on the same curve, leading to triangle-shaped quads along the curves. In addition, some hex elements may have two or three adjacent faces lying on the same surface patch, leading to large dihedral angles or bad Jacobian. These two situations destroy the mesh quality and they can be resolved using the pillowing technique [271, 340, 342].

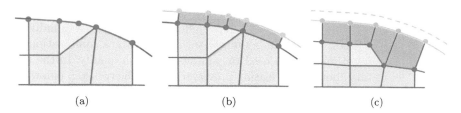

| (a) | (b) | (c) |

FIGURE 6.11: Steps to generate a pillowed layer in 2D. (a) The original quad mesh with the red boundary to be pillowed; (b) a duplicated boundary (the green curve) with a new layer of hexes (orange); and (c) the smoothing result. Note that the yellow element is improved a lot.

Pillowing is a sheet insertion technique that refines the mesh boundary. Figure 6.11 shows the steps to generate one pillowed layer in 2D. We identify the boundary surface, duplicate it and connect the corresponding points to construct a layer of hexes. After that, we move the duplicated nodes to the surface and use smoothing to improve the element aspect ratio. Note that the yellow element in (a) has three nodes on the boundary, and it has been improved a lot as shown in (c).

Pillowing can eliminate the situations where two adjacent hexes share two faces, called a "doublet" [271]. The doublet needs to be eliminated because the dihedral angle between these two adjacent faces is usually large and it is impossible to remove all negative Jacobians with only smoothing. The pillowing method inserts a new layer of elements at the boundary of the selected region called "shrink set," and splits the doublet hexahedron into several elements. This method is not only useful in removing doublets but also has proven to be a powerful tool to insert layers of elements in existing meshes [340, 342].

In the following, we present a two-step pillowing technique to improve the mesh for multi-component CAD assemblies. In the first step, we pillow each individual surface patch; and in the second step, the overall boundary surface of each component is pillowed. Note that the surface patch contained in one single component and the surface patch shared by two components should be pillowed in different ways.

In particular, for surface patches contained in one single component, the first pillowing step tends to eliminate the triangle-shaped boundary quads along curves. The second pillowing step aims to eliminate the hex elements with multiple quads on the same surface patch. For CAD assemblies with multiple components, patch matching is very important especially for patches shared by two adjacent components, and here we adopt different matching strategies for these two pillowing steps. There are two kinds of surface patches in CAD assemblies: patches contained in one component and the shared patches between two components. During the first pillowing step, we first mark the sets of shared patches. Then we pillow the faces contained in only one component, and insert one layer of elements at one time. Whenever we meet a shared patch, we insert two instead of one layer of hexes. For the second pillowing step, patch matching is straightforward. When pillowing the entire surface for one component, we shrink the original nodes inside while keeping the newly inserted nodes at the boundary. Therefore, all the nodes in the adjacent components connecting to the original boundary should be reconnected to the new boundary. In this way, the resulting meshes of adjacent components automatically match with each other at the shared patch. Figure 6.12 shows the resulting hex meshes of a two-cylinder CAD assembly with two pillowed layers.

After preserving sharp features, we locally refine the mesh using templates and pillow two layers. The resulting all-hex meshes conform to the given B-Reps and have no triangle-shaped quads or doublets. A smoothing technique coupled with optimization is then applied to improve the mesh quality; see Section 6.5 for details. Figure 6.13 shows the generated adaptive and quality

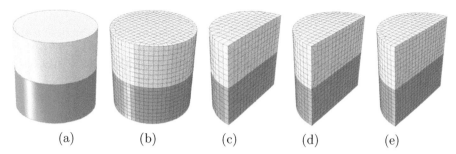

(a)	(b)	(c)	(d)	(e)

FIGURE 6.12: (a, b) A two-cylinder CAD assembly and the resulting mesh after pillowing; and (c-e) one cross section of the original mesh, the mesh with one pillowed layer, and the mesh with two pillowed layers, respectively. (Pictures from [304].)

all-hex meshes of an eddy plate assembly and a gear assembly with all the sharp features preserved.

6.4 Octree vs. RD-Tree Based Adaptive Hexahedral Meshing

In the previous section, we discussed octree-based dual contouring methods for adaptive all-quad and all-hex mesh generation with 2-refinement and 3-refinement schemes. Here we will investigate a new 2-refinement scheme which works for two different tree structures, namely, the octree and the rhombic dodecahedral (RD) tree [435]. In particular, the RD tree is a new tree structure we study which can be used for all-hex meshing. Moreover, this new scheme requires very little propagation compared to the 2-refinement scheme discussed in Chapter 6.2.

6.4.1 Octree-Based Adaptive Hexahedral Meshing

Starting from a closed smooth surface mesh, we design four steps to construct adaptive all-hex meshes based on an octree data structure: adaptive octree construction, hanging node elimination via 2-refinement, buffer zone clearance and projection.

In the first step, the given surface mesh is embedded in a bounding cube, which performs as the root of the octree and we mark it Level-0. To build the hierarchical octree structure, a Level-i cell is subdivided into eight identical Level-$(i + 1)$ cells. Here we choose the feature-sensitive error function [428] to estimate the difference of the isosurface between two neighboring levels,

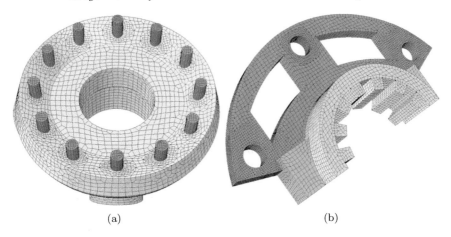

(a)   (b)

FIGURE 6.13: Adaptive and quality all-hex meshes of an eddy plate assembly and a gear assembly (pictures from [304]). (a) The eddy plate with 2-refinement applied; and (b) the gear with 3-refinement applied to the middle hole while 2-refinement applied to the other two holes.

and build the adaptive octree. To generate meshes with good aspect ratio, we restrict the level difference between two adjacent cells to be at most one. In the end, we obtain a strongly-balanced octree.

Hanging nodes are introduced in the adaptive octree when the adjacent cells are at different levels. In the following we introduce a new 2-refinement algorithm to remove hanging nodes with very little propagation. Here we apply the two 2-refinement templates shown in Figure 6.4(b, e) to unstructured all-hex meshes. Before that, we need to address the following two important transition issues.

1. The coupling of transition elements - As shown in Figure 6.4(e), this 2-refinement template needs to be applied as a block of four transition elements sharing an edge. Therefore, we need to ensure that there is a valid block we can find for each transition element, especially for unstructured meshes with arbitrary topology (or valence number). In addition, an efficient implementation is needed sometimes for such a complex four-element block with flipping; and

2. The concavity in the octree with elements containing more than one transition face - There is no template working for such transition elements so far. This problem also exists in 3-refinement methods.

The first issue can be easily addressed by subdividing all the elements surrounding each irregular transition node in the adaptive octree. However, the second one is hard to address. As shown in Figure 6.14, we can use the pillowing technique to isolate the refined region, that is, we duplicate the

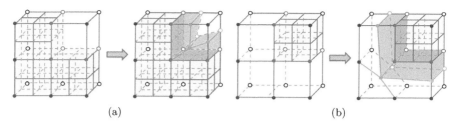

(a)   (b)

FIGURE 6.14: Pillowing for concave (a) and convex (b) cases. The yellow regions are the refined regions with solid dots on the boundary and empty circle dots interior. All the red dots (both solid and empty circle) are transition nodes. The pillowed layers are marked in orange with the green duplicated nodes (both solid and empty circle).

corresponding transition nodes for the transition elements and also insert new layers of hexes. In this way, each transition element is limited to have only one transition face. After isolating the refined regions, we can apply the template in 6.4(e) directly to eliminate hanging nodes.

During pillowing, we may have non-manifold transition regions, which happen in an eight-element block with each element at Level-i (not refined) or Level-$(i + 1)$ (refined). We have a total of $2^8 = 256$ possible configurations. Among these configurations, non-manifold cases happen in the blocks with two to six refined elements, which can be resolved by duplicating the non-manifold nodes [435]. After generating the adaptive octree with all the hanging nodes removed, we clear the buffer zone by deleting outside elements and elements close to the surface to obtain a hex core mesh. To generate good-quality elements around the boundary, we also delete elements sharing a single point, edge or face with others, as well as elements with non-manifold connectivity on the boundary.

In the end, we project all the boundary nodes of the core mesh onto the surface, and create a buffer layer. To obtain elements with better quality, the buffer layer can be split into two layers such that more flexibility is provided for the following quality optimization.

6.4.2 RD-Tree Based Adaptive Hexahedral Meshing

In addition to cubes, the rhombic dodecahedron (RD) in Figure 6.15(a) can also be used to tessellate 3D domains. There are some RD structures naturally existing in the world. For example, honeycomb is tessellated with many identical cells, and each cell is basically a hexagonal prism capped with a half-RD element. Minerals such as garnet form an RD crystal habit, and the RD structure exists in the unit cells of diamond. Such honeycomb-like structure has been used in scientific and engineering applications. An RD element has 14 nodes, 24 equilong edges and 12 rhombic faces with the 120°

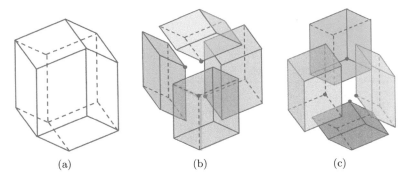

FIGURE 6.15: An RD element (a) and its two decomposition formats (b, c). The four red nodes in (b, c) are identical, which splits the RD into four hex elements.

dihedral angle. By adding a center point, we can split an RD element into four identical rhombic hexes in two ways; see Figure 6.15(b, c).

Different from the octree-based method, here we build a uniform RD tree by packing RD elements to cover the entire to-be-meshed domain. Then we split all the RD elements into hexes and convert the RD tree to a hex tree. To optimize the overall valence number of the hex tree, during splitting we check the valence number of each vertex and choose the proper template from Figure 6.15(b, c). Note that the resulting hex tree contains many irregular points with a valence other than eight.

For the adaptive tree based on RD, we first locally refine each rhombic hex, and then use pillowing and 2-refinement to eliminate hanging nodes. Since the RD tree is unstructured with many irregular nodes, we cannot apply the pillowing technique directly. We classify all the elements at the lower level surrounding a transition node into different patches according to their connectivity, and then we apply local refinement and propagation to make all the patches valid for pillowing. After pillowing, we apply the 2-refinement templates to eliminate all the hanging nodes. The following buffer zone clearance and projection steps are similar to the octree-based method.

Figure 6.16 shows the adaptive all-hex meshes of the igea model using both the octree and RD-tree based methods. Our 2-refinement algorithm works robustly for these two methods, and the resulting interior elements follow the octree or RD-tree orientations. Compared with the octree-based method, we can observe that the RD-tree method yields many more unstructured elements with a lot of irregular nodes due to its special tree structure.

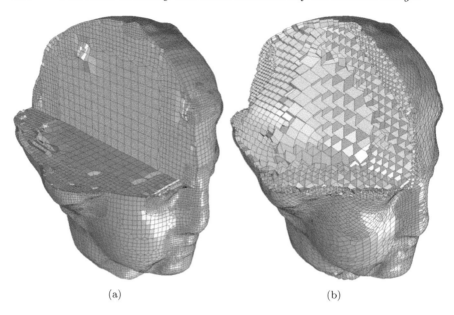

(a)   (b)

FIGURE 6.16: Adaptive all-hex meshes of the igea model (pictures from [435]). (a) The octree-based method; and (b) the RD-tree based method.

6.5   Quality Improvement

Quality improvement plays an important role in finite element mesh generation. First we need to define quality metrics to measure the quality of quad and hex meshes. Here we choose the scaled Jacobian, the condition number of the Jacobian matrix and the Oddy metric [281] as our metrics [205, 206, 208]. Given a vertex $x \in \Re^3$ in a quad or hex element, suppose $x_i \in \Re^3$ ($i = 1, \cdots, m$) are its adjacent vertices, where $m = 2$ for a quad and $m = 3$ for a hex. We obtain the edge vectors $e_i = x_i - x$, which can be used to build the Jacobian matrix $J = [e_1, \cdots, e_m]$. The determinant of the Jacobian matrix is called the *Jacobian*, or the *scaled Jacobian* when the normalized edge vectors are used. An element is "inverted" if any of its Jacobians ≤ 0. Here we adopt the Frobenius norm as a matrix norm, $|J| = \sqrt{tr(J^T J)}$. The condition number of the Jacobian matrix can be written as $\kappa(J) = |J||J^{-1}|$, where $|J^{-1}| = \frac{|J|}{det(J)}$. Then, we obtain the three quality metrics for a vertex x in a quad or hex.

$$Jacobian(x) \quad = \quad det(J), \tag{6.1}$$

$$\kappa(x) \quad = \quad \frac{1}{m}|J^{-1}||J|, \tag{6.2}$$

$$Oddy(x) \quad = \quad \frac{|J^T J|^2 - \frac{1}{m}|J|^4}{det(J)^{4/m}}, \tag{6.3}$$

where $m = 2$ for quad meshes and $m = 3$ for hex meshes. Instead of using edges vectors, the Jacobian matrix can also be defined using finite element basis functions [178, 282]. We have

$$J = \begin{bmatrix} \frac{\partial x}{\partial \xi} & \frac{\partial x}{\partial \eta} & \frac{\partial x}{\partial \zeta} \\ \frac{\partial y}{\partial \xi} & \frac{\partial y}{\partial \eta} & \frac{\partial y}{\partial \zeta} \\ \frac{\partial z}{\partial \xi} & \frac{\partial z}{\partial \eta} & \frac{\partial z}{\partial \zeta} \end{bmatrix}. \tag{6.4}$$

As discussed in Section 6.2, template-based local refinement [304, 427, 435] can be applied to improve the local node density and mesh accuracy. The pillowing technique [271, 340, 342] is also an effective quality improvement method for quad and hex meshes. It guarantees that for each element there is at most one face lying on the boundary surface. This gives us a lot of flexibility to further improve the aspect ratio of the hex mesh.

Smoothing and optimization are two important quality improvement methods for quad and hex meshes. They relocate vertices without changing the mesh connectivity and topology. Similar to triangular and tetrahedral meshes, we can use the Laplacian smoothing and weighted-averaging methods. Optimization-based smoothing techniques measure the quality of the surrounding elements about a node and tend to optimize the local region with respect to the node location. The node is moved along the gradient direction until an optimum is reached. The conjugate gradient method is used here to solve the optimization problem. Optimization-based smoothing techniques are generally time-consuming. Therefore, people use a combined Laplacian and optimization approach, that is, the Laplacian smoothing is applied most of the time, switching to the optimization-based smoothing only when needed. To improve the quad and hex mesh quality, generally a combination of pillowing, smoothing and optimization techniques is adopted.

In the following, we will discuss geometric flow-based smoothing [429], quality improvement for meshes of multiple-material domains [222, 306], as well as guarantee-quality all-quad meshing [237, 238].

6.5.1 Geometric Flow-Based Smoothing

In Section 5.4, we discussed various geometric flows and how to use them for triangular and tetrahedral mesh quality improvement, based on a discretized format of the Laplacian-Beltrami operator (LBO) over triangular meshes [270, 412]. Similarly, in this section we discretize the LB operator over quad meshes, and discuss how to apply the discretizated format to surface smoothing and quality improvement of quad and hex meshes [429]. There are mainly three steps to smooth the surface and improve the quality of quad and hex meshes.

1. Discretize the LBO and smooth the surface mesh, which is basically to relocate vertices along the normal direction with volume preservation;

2. Improve the aspect ratio of the surface quad mesh, which is to adjust vertices in the tangent plane with feature preservation; and

3. Improve the aspect ratio of the hex mesh, which is to relocate vertices inside the volume.

For quad meshes, we only need Steps 1 and 2; while for hex mesh quality improvement, we need all these three steps.

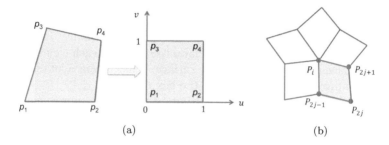

(a)   (b)

FIGURE 6.17: (a) A quad is mapped onto a bilinear parametric plane; and (b) the node labeling of a quad surrounding the vertex p_i.

To smooth surface quad meshes, we derive a discretized format of the LBO and apply the surface diffusion flow by relocating vertices along their normal directions. The surface diffusion flow has the property of volume-preserving, and it also preserves the sphere shape accurately if the initial mesh is embedded and close to a sphere. Here we use the bilinear interpolation to calculate the area of a quad, and then use it to derive the discretized format. Given a quad shown in Figure 6.17(a), we define a bilinear parametric surface $S(u, v)$ which interpolates four vertices of the quad. We have

$$S(u, v) = (1 - u)(1 - v)p_1 + u(1 - v)p_2 + (1 - u)vp_3 + uvp_4. \tag{6.5}$$

The area of the surface $S(u, v)$ for $(u, v) \in [0, 1]^2$ is defined as

$$
\begin{aligned}
A &= \int_0^1 \int_0^1 \sqrt{||S_u \times S_v||^2} du dv \\
&= \int_0^1 \int_0^1 \sqrt{||S_u||^2 ||S_v||^2 - (S_u, S_v)^2} du dv, \tag{6.6}
\end{aligned}
$$

which can be computed using the four-point Gaussian quadrature. We can also obtain ∇A from the above equation:

$$\nabla A = \alpha_2(p_2 - p_1) + \alpha_3(p_3 - p_1) + \alpha_4(p_4 - p_1), \tag{6.7}$$

where α_2, α_3 and α_4 are coefficients; see [429] for details.

As shown in Figure 6.17(b), given a node p_i with the valence of n on the surface and a neighborhood region with the diameter D, we can compute the mean curvature at p_i [270] and obtain

$$H(p_i) = \lim_{D \to 0} \frac{2\nabla A}{A}. \tag{6.8}$$

We denote the coefficients of the quad $[p_i p_{2j-1} p_{2j} p_{2j+1}]$ as α_j^i, β_j^i and γ_j^i. Then, the discretized mean curvature can be written as

$$H(p_i) = \lim_{D \to 0} \frac{2\nabla A}{A} \approx \sum_{k=1}^{2n} w_k^i (p_k - p_i), \tag{6.9}$$

where

$$w_{2j}^i = \frac{2\gamma_j^i}{A(p_i)}, \quad w_{2j-1}^i = \frac{2(\alpha_j^i + \beta_{j-1}^i)}{A(p_i)}, \quad w_{2j+1}^i = \frac{2(\alpha_{j+1}^i + \beta_j^i)}{A(p_i)}. \tag{6.10}$$

According to the relationship $\triangle x = 2H(p_i)$, we can obtain the discretized LB operator

$$\triangle f(p_i) \approx 2 \sum_{k=1}^{2n} w_k^i \left(f(p_k) - f(p_i) \right). \tag{6.11}$$

Therefore, we have

$$\triangle H(p_i) n(p_i) \approx 2 \sum_{k=1}^{2n} w_k^i \left(H(p_k) - H(p_i) \right) n(p_i)$$

$$= 2 \sum_{k=1}^{2n} w_k^i \left[n(p_i) n^T(p_k) H(p_k) - H(p_i) \right]. \tag{6.12}$$

The aspect ratio of the surface mesh can be improved by adjusting vertices in the tangent plane, and surface features are preserved since the movement in the tangent plane does not change the surface shape. For each vertex p_i, its mass center $m(p_i)$ is calculated to find its new position on the tangent plane. We have

$$m(p_i) = \sum_{j=1}^{n} \left(\frac{p_i + p_{2j-1} + p_{2j} + p_{2j+1}}{4} A_j \right) \Big/ A_{total}^i, \tag{6.13}$$

where A_{total}^i is the total quad area around p_i. The vertex tangent movement is basically an area-weighted relaxation method, therefore it is also suitable for adaptive quad meshes.

Besides the movement of surface vertices, interior vertices also need to be relocated in order to improve the aspect ratio of hex meshes. Given an interior vertex p_i, its mass center is calculated as

$$m(p_i) = \sum_{j \in N(i)} \left(\frac{1}{8} \sum_{j=1}^{8} p_j V_j \right) \bigg/ V_{total}^i, \qquad (6.14)$$

where $N(i)$ denotes the 1-ring neighborhood of p_i and V_{total}^i is the total volume of all elements around p_i.

Figure 6.18(a, b) shows the improved all-quad meshes of a vascular structure and a bladder model using the geometric flow-based smoothing method. Pillowing, geometric flow-based smoothing and optimization are used together to improve the quality of hex meshes; see the improved hex mesh of a brain model with a tumor in Figure 6.18(c, d).

6.5.2 Quality Improvement for Multiple-Material Meshes

As discussed in Section 5.4, we need to classify all the vertices in the mesh into different groups for multiple-material domains. For example, for the representative volume element (RVE) dataset of scanned polycrystalline materials in Figure 5.20, vertices are categorized into seven groups according to their relative locations in the individual grain and in the entire RVE data. With this vertex classification, we apply different schemes to each group to improve the mesh quality [222, 306].

The hex meshes generated from the 3D volumetric data can be noisy, have triangle-shaped elements or "doublets" on the boundary, or even have negative Jacobians. These will result in a poorly conditioned stiffness matrix, and affect the stability, convergence and accuracy of the finite element analysis. Therefore, quality improvement is a critical step after mesh generation. Here we talk about a comprehensive approach based on a modified pillowing technique, relaxation-based smoothing and optimization [222, 306]. There are four main steps in our algorithm.

1. A relaxation-based smoothing is implemented to smooth the grain boundaries while preserving surface features;

2. To eliminate doublets, a modified pillowing technique is developed to create a boundary layer;

3. An additional smoothing scheme is implemented to drag the shrink set inside and improve the mesh quality; and finally

4. An optimization is carried out to improve the worst Jacobian.

During the relaxation-based smoothing, different schemes are applied to various groups of vertices. We use B-spline interpolation and resampling to regularize vertices on the planar and interior curves. Each grain boundary surface

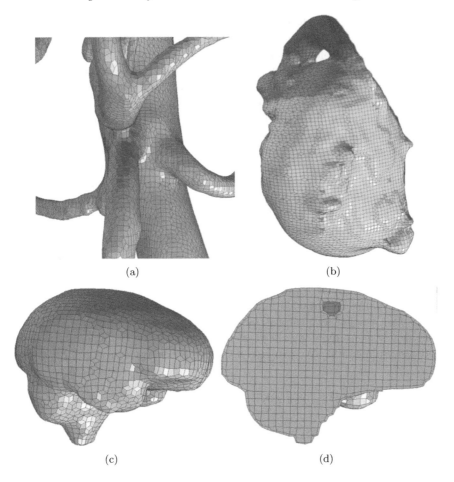

(a) (b)

(c) (d)

FIGURE 6.18: (a, b) Quality all-quad meshes of a vascular structure and a bladder model; and (c, d) quality all-hex mesh of the brain model with a tumor. The blue region in (d) is the tumor.

is smoothed and regularized using the tangential movements, and interior vertices are smoothed using the weighted-averaging coupled with optimization.

In the hex mesh of microstructure materials, triangle-shaped elements or "doublets" may happen on the shared boundary between two different grains. The pillowing technique was developed to resolve such situation [271, 340, 342]. For the grain containing such element, we pillow one layer along the boundary, and then use smoothing to improve the mesh quality. Note that when triangle-shaped elements or doublets appear on both sides of the grain boundaries, two sheets need to be inserted. The key problem in pillowing is how to automatically and efficiently define the shrink set.

In our algorithm, the pillowing process is applied grain by grain. If the grain completely embeds within the RVE volume, we set the entire grain boundary as the shrink set. Sometimes the grain boundary may end at the RVE boundary, therefore its shrink set can also be an open surface. In addition, one grain boundary may have local non-manifolds. For this situation, we detect these non-manifold edges, disable all the surrounding elements, and set the grain boundary except the disabled elements as the shrink set. After pillowing, we add the disabled elements back to the mesh, and also update the vertex classification for the grain boundary.

There are many boundary curves in the multi-material data, which start and end at non-manifold points. As noises exist on the curve, a curve fairing method based on the mean curvature flow was carried out to smooth both the planar and interior curves. In [222], a shape-preserving curve diffusion flow is designed to evolve the curve. Given a boundary curve $[x_0 x_1 \cdots x_n]$ with two fixed endpoints x_0 and x_n, we have

$$\frac{dx}{dt} = -\left[(\triangle \kappa)^T n\right] n, \tag{6.15}$$

where $\triangle$ is the Laplace operator, κ is the curvature and n is the normal.

To regularize curves while preserving geometric features at the same time, we fix the starting and ending non-manifold points of the curves, and move interior vertices only along the tangent direction using B-spline interpolation and resampling [306]. We can also minimize an energy function [222] defined as

$$E_{curve} = \frac{1}{2} \sum_{i=1}^{n} \left(\|x_i - x_{i-1}\| - h \right)^2, \tag{6.16}$$

where h is the average length of each two neighboring vertices. The variational method together with an L^2-gradient flow can be used to minimize this energy function.

Vertices on grain boundary surface patches can be improved along the normal direction using various geometric flows like the surface diffusion flow. For tangential movements to regularize the boundary quad meshes, we design an energy function and minimize it. Given surface vertices $\{x_i\}_{i=1}^N$, suppose m_i is the valence number of vertex x_i. The neighboring vertices are labeled following Figure 6.17(b). We have

$$E_{quad} = \frac{1}{2} \sum_{i=1}^{N} \left(E_1(x_i) + \lambda E_2(x_i) \right), \tag{6.17}$$

where

$$E_1 = \sum_{j=1}^{m_i} \left(\|x_{2j-1} - x_i\| - h \right)^2 + \sum_{j=1}^{m_i} \left(\|x_{2j} - x_i\| - \sqrt{2}h \right)^2, \tag{6.18}$$

$$E_2 = \sum_{j=1}^{m_i} \left(det\left(x_{2j-1} - x_i, x_{2j+1} - x_i, n_i\right) - J_i \right)^2. \tag{6.19}$$

Here we choose $h = \sqrt{A_m}$. A_m is the average area of the quads surrounding x_i, J_i is the average Jacobian and n_i is the normal at x_i. It is obvious that E_1 is globally minimized when the length of each edge connecting to x_i equals to h and the diagonal of each quad is close to $\sqrt{2}h$. The E_2 term aims to optimize the Jacobian defined by the two edge vectors.

Similarly, we can design and minimize the regularization energy function for the interior vertices of the hex meshes [222]. In addition to forcing the edge and the diagonal of each quad to be close to the local average length, we can also force the diagonal of each hex to be close to $\sqrt{3}h$, where $h = \sqrt[3]{V_m}$ and V_m is the average volume of hex elements in one grain. Figure 6.19 shows the improved hex mesh of the 92-grain beta titanium data, as well as a local mesh for three adjacent grains. We can observe that one boundary layer is created for each grain.

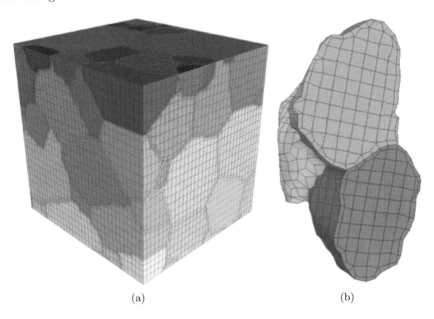

(a)   (b)

FIGURE 6.19: (a) The improved hex mesh for the 92-grain beta titanium data; and (b) a local mesh for three grains. (Pictures from [306].)

6.5.3 Guaranteed-Quality All-Quadrilateral Mesh Generation

Provably good-quality triangular mesh generation has been studied a lot for both planar and curved surfaces. However, fewer algorithms exist in the literature for all-quad mesh generation, and most of them are heuristic. There are only a few studies on guaranteeing a certain angle range for all-quad meshes. As proved in [50], any planar n-gon can be meshed by $O(n)$ quads with all the angles between $45° - \epsilon$ and $135° + \epsilon$. Circle-packing techniques

[48] were developed to generate quads with an upper angle bound like $120°$, but no lower bound was provided. A balanced quadtree was later utilized to generate a quad mesh with a lower angle bound ($18.43°$), but the upper bound is as large as $180°$ [30].

In this section, we discuss two approaches to generate guaranteed-quality all-quad meshes for given points cloud or planar curves, using quadtree and hexagon tree. In the quadtree-based method [237], all the angles of any element are within $[45° - \epsilon, 135° + \epsilon]$, where $\epsilon \leq 5°$, except badly shaped elements required by the sharp angles in the input geometry. In the hexagon tree-based method [238], we achieve an even better angle range, $[60° - \epsilon, 120° + \epsilon]$.

Guaranteed-Quality Quadtree-Based All-Quad Meshing. Our quadtree-based algorithm yields unstructured adaptive all-quad meshes [237]. Given N points in a planar domain, we define a size function, and then use strongly-balanced quadtree together with 2-refinement templates to produce all-quad meshes conforming to the given points cloud, with all the element angles within $[45°, 135°]$.

For a closed smooth curve C in a planar domain represented by cubic splines, a set of points or polygons, we design six steps to construct guaranteed-quality all-quad meshes. We first decompose each curve into a set of piecewise-linear edge segments based on its local curvature. Then, we build a strongly balanced quadtree and remove all hanging nodes using 2-refinement templates. The element size is adapted to the boundary curvature and narrow regions. We also remove elements outside and around the boundary to create the quadtree core mesh and a buffer zone. Next, we design four categories of templates to adjust the boundary edges, and finally create two buffer layers between the core mesh and the input boundary. We have proved that for any planar smooth curves, all the angles in the resulting all-quad mesh are within $[45° - \epsilon, 135° + \epsilon]$ ($\epsilon \leq 5°$). In other words, all the scaled Jacobians defined by two edge vectors are in the range of $[sin(45° - \epsilon), sin90°]$, or $[0.64, 1.0]$. A few bad angles may be required to preserve sharp features like small angles on the boundary. In addition, multiple boundary layers can be generated by splitting the second buffer layer.

Our algorithm [237] provides a theoretical basis of guaranteed-quality all-quad mesh generation. To guarantee the angle range, sometimes the algorithm needs an aggressive refinement, leading to many elements in the final result. This may not be practical for real applications. Moreover, we did not apply quality improvement techniques in our algorithm because it would make the angle guarantee unpredictable. For practical purposes, smoothing and optimization could further improve the overall mesh quality.

Matching Interior and Exterior All-Quad Meshes with Guaranteed Angle Bounds. Using the above method , we can generate an individual interior or exterior mesh at one time, but they do not match at the boundary. As a follow-up, we develop a new algorithm [239] to match the generated inte-

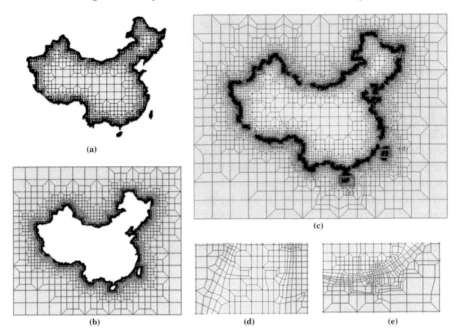

FIGURE 6.20: All-quad meshes of the China map with guaranteed angle range (pictures from [239]). (a) An interior mesh; (b) an exterior mesh; (c) the matching result with conformal boundary, all the angles are within [43°, 138°]; and (d, e) zoom-in pictures of (c).

rior and exterior meshes with conformal boundary, with the guaranteed angle bounds preserved. In addition, we introduce a sharp feature layer to preserve sharp features.

Furthermore, the quadtree-based algorithm constructs lots of quad elements with similar shapes, such as square element and right-angle trapezoids. Such special property can be used in finite element analysis to speed up the stiffness matrix construction. In our algorithm, all the quad elements are classified into six element types, and most element types only need a few flops in constructing the element stiffness matrix. Figure 6.20 shows the interior and exterior all-quad meshes with conformal boundary, and all the angles in the mesh are within [43°, 138°].

Guaranteed-Quality Hexagon Tree-Based All-Quad Meshing. In addition to quadtree, an adaptive hexagonal subdivision scheme [363, 364] can also be used to generate all-quad meshes. Here, we discuss a new hexagon-tree based method [238] for any complicated smooth curves, which is superior over the quadtree-based algorithm [237] with a better guaranteed angle range, $[60° - \epsilon, 120° + \epsilon]$ ($\epsilon \leq 5°$). Moreover, this angle range can also be guaranteed for geometry with sharp features, with the exception of small angles in the input geometry.

For any smooth planar curves, we also use six steps to construct all-quad meshes with guaranteed angle bounds. We first decompose each curve into a set of edge segments. Instead of using the quadtree, we construct an adaptive hexagon-tree, which does not contain any hanging nodes and also provides a better angle range. A root hexagon is generated to enclose all the points. Then each cell in the hexagon-tree is recursively refined using two refinement templates; see Figure 6.21(a, b). After that, each pair of semi-hexagons can be grouped into one hexagon as shown in Figure 6.21(c). For each refinement or grouping, we update the related cell levels. Note that hexagons are in even levels and semi-hexagons are in odd levels. Here we also apply the strongly balanced criterion. For those cells violating this criterion, we split and group them following Figure 6.21(d). Our algorithm supports local refinement, coarsening and grouping. In the end, we split each hexagon to two semi-hexagons and obtain all-quad meshes.

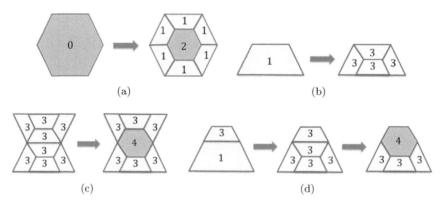

FIGURE 6.21: Adaptive hexagon-tree construction. (a, b) Two refinement templates; (c) grouping; and (d) strongly balanced tree construction. The numbers represent the cell levels. Note that hexagons are in even levels, and semi-hexagons are in odd levels.

After building the hexagon-tree and generating all-quad elements, we remove elements outside and around the boundary to create a hexagonal core mesh and a buffer zone. Several templates are designed to adjust the boundary edges and then two buffer layers are created between the core mesh and the input boundary. We have proved that for any planar smooth curves, all the angles in the resulting mesh are within $[60° - \epsilon, 120° + \epsilon]$ ($\epsilon \leq 5°$). In addition, all the scaled Jacobians defined by two edge vectors are in the range of $[sin(60° - \epsilon), sin90°]$, or $[0.82, 1.0]$. Our algorithm can also handle curves with sharp features and match interior and exterior meshes with a relaxed angle range. Figure 6.22 shows the guarantee-quality all-quad mesh obtained from a hexagon tree; all the angles are within $[57°, 124°]$.

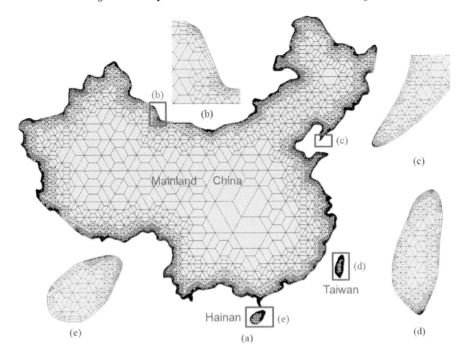

FIGURE 6.22: (a) All-quad mesh of the China map, all the angles are within [57°, 124°]; and (b-e) zoom-in pictures. (Pictures from [238].)

Similar to the quadtree-based method, we also categorize all the quads in the generated meshes into five element types. In finite element analysis, the stiffness matrix of four element types can be precomputed, which can significantly reduce the computational time, save memory and provide more precise results.

Discussion. Compared to the quadtree method, the hexagon-tree approach yields better angle bounds and no hanging nodes. However, the hexagon-tree produces many irregular nodes and nonsquare elements in the core mesh. In addition, the quadtree-based method can be easily extended to 3D, while extending the hexagon-tree based method to 3D is not straightforward. Due to the complex topology of hex meshes, guaranteed-quality all-hex meshing for arbitrary geometry is still an open problem.

Homework

Problem 6.1. Use the quadtree-based isocontouring techniques to generate quadrilateral meshes for the region inside the curve (Figure 6.23): (a) uniform quad meshing, and (b) adaptive quad meshing.

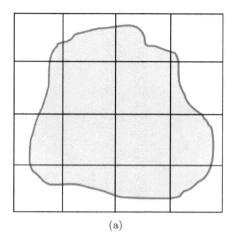

(a)

FIGURE 6.23: Uniform and adaptive quadrilateral meshing.

Problem 6.2. Perform a survey for unstructured quadrilateral and hexahedral mesh generation; discuss pros and cons for each technique. What are the challenges and what are the future directions?

Problem 6.3. Perform a survey and compare various quality improvement techniques for quadrilateral and hexahedral meshes.

Chapter 7

Volumetric T-Spline Modeling

With the rapid development of isogeometric analysis in recent years, there is an urgent need for volumetric parameterization such as volumetric T-spline model construction. In this chapter, we will first review NURBS (non-uniform rational B-spline) and T-spline, and then discuss several volumetric T-spline modeling techniques we developed, including converting any quad/hex meshes to standard and rational T-splines, polycube-based parametric mapping, feature preservation using eigenfunctions, Boolean operations and skeletons, truncated hierarchical Catmull–Clark subdivision, weighted T-splines, conformal

T-spline modeling, as well as incorporating T-splines into commercial CAD and FEA software.

7.1   Introduction

With the rapid development of 3D scanning data acquisition and automatic geometric modeling techniques, the obtained geometries are usually represented in the form of polygonal meshes. Note that the scanned data can be slices of images like CT and MRI as we discussed in the previous chapters, or they can be points cloud acquired from Reverse Engineering. Besides polygonal meshes, spline models are popularly used to represent free-form surfaces, especially in computer aided design (CAD), computer aided manufacturing (CAM), and computer aided engineering (CAE). Unlike piecewise-linear elements, splines can accurately represent a wide range of geometric objects with high order continuity. In recent years, a spline-based analysis method named "isogeometric analysis" was developed [42, 104, 179], which integrates design with analysis directly and has many advantages over the traditional finite element method (FEM). Isogeometric analysis utilizes the same basis functions to construct the geometry and the solution space, it basically incorporates the isoparametric idea into FEM. To improve geometric accuracy and surface continuity, and to achieve compatibility with CAD systems and isogeometric analysis, one of the most significant challenges facing isogeometric analysis researchers is developing analysis-suitable volumetric spline parameterizations such as solid NURBS and T-splines from boundary representations.

In the following, let us first review NURBS and T-spline, and then talk about wavelets-based NURBS simplification and fairing, as well as solid NURBS construction using offsets.

7.1.1 A Review of NURBS and T-Spline

To represent free-form surfaces precisely for design and manufacturing, NURBS was developed starting from the 1950s [133, 297]. Currently the NURBS surfaces are widely used in aircraft, shipbuilding and automobile industry. For example, in the design of aircraft wings, ship hulls, propellers, turbines and car bodies. Based on the Bernstein polynomials, a French engineer named Pierre Bézier invented the Bézier method [100, 297]. Roughly at the same time Paul de Casteljau proposed a numerical method to evaluate Bézier curves [133], which was named the "de Casteljau algorithm." This is really a milestone for surface description, translating the representation of mechanical parts from blueprints to computers. To overcome the shortcomings of Bézier, Carl R. de Boor replaced a single segment with piecewise polynomials and generalized the Bézier method to B-spline [133]. In addition, several knot insertion and degree elevation schemes were developed for B-splines, including Boehms knot insertion algorithm [55], the Oslo algorithm [259], the degree elevation algorithm [297] and the blossoming algorithm [308]. As a generalization of B-spline, NURBS was invented in the 1960s to unify the geometry representation of standard analytical shapes such as conic curves and free-form shapes. Currently NURBS has become the standard form for CAD, CAM and CAE.

A NURBS surface is defined by its degree, two global knot vectors and control points [297]. A NURBS surface can be represented mathematically as

$$S_N(u,v) = \frac{\sum_{i=0}^{m}\sum_{j=0}^{n} C_{ij}w_{ij}N_{i,d}(u)N_{j,d}(v)}{\sum_{i=0}^{m}\sum_{j=0}^{n} w_{ij}N_{i,d}(u)N_{j,d}(v)}, \qquad (7.1)$$

where C_{ij} represents a control point, and w_{ij} is the weight associated to C_{ij}. The NURBS control points must lie topologically in a rectangular $m \times n$ grid or mesh. $N_{i,d}(u)$ and $N_{j,d}(v)$ are B-spline basis functions defined by two global knot vectors, $u = [u_0, u_1, \cdots, u_{m+d}, u_{m+d+1}]$ and $v = [v_0, v_1, \cdots, v_{n+d}, v_{n+d+1}]$, and d is the degree. NURBS provides a unified geometry representation for conic sections and free-form shapes. They have been widely used in engineering design and manufacturing.

Invented by Thomas W. Sederberg in 2003, T-spline [334] is a generalization of NURBS, and also a mathematical tool for geometric modeling. Unlike NURBS with global knot vectors, T-splines allow T-junctions in their control grids to support local refinement. A T-junction terminates a row or column of control points in the control grids. T-splines have been applied to solve several important problems in computer-aided geometric design (CAGD) such as local refinement [332], NURBS merging [334], control mesh simplification [395] and trimmed NURBS conversion [333]. In particular, T-splines can conveniently model geometry with arbitrary topological genus [334] using a single patch no matter how complex the topology is. A T-spline surface can be written as

$$S_T(s,t) = \frac{\sum_{i=0}^{n} w_i C_i B_i(s,t)}{\sum_{i=0}^{n} w_i B_i(s,t)}, \qquad \forall (s,t) \in \Omega, \qquad (7.2)$$

where w_i is the weight associated to the control point C_i, $B_i(s,t) = N_i(s)N_i(t)$. $N_i(s)$ and $N_i(t)$ are B-spline basis functions defined by two local knot vectors, $s = [s_0, s_1, s_2, s_3, s_4]$ and $t = [t_0, t_1, t_2, t_3, t_4]$ when degree $d = 3$. Ω is the parameter domain of the T-spline surface.

To define a T-spline surface, we need both the degree and the control mesh (or the T-mesh). Different from NURBS, T-spline control points may not connect in the form of a rectangular grid. A T-mesh is introduced to record the connectivity of all the control points; see a local T-mesh shown in Figure 7.1. In the T-mesh, each edge connecting two control points is labeled with a *knot interval*, representing its parametric length. In the parametric space, each edge must follow one parametric direction, either s or t. In other words, its two end nodes share the same t or s value. Such an edge is called an s- or t-edge. The local knot vectors for each control node can be inferred from the T-mesh. Note that for a NURBS surface patch, there are two *global knot vectors*, u and v; while for a T-spline, each control node C_i has its own *local knot vectors*, s and t, which are inferred locally from the T-mesh. A T-mesh contains regular nodes, extraordinary nodes and T-junctions. A *regular node* has a valence of four, and an *extraordinary node* has a valence other than four and is not a T-junction. NURBS can be considered as a special case of T-splines, where the T-mesh is simply a rectangular grid with no T-junctions. T-splines use local parametric coordinate systems, especially for the region surrounding extraordinary nodes.

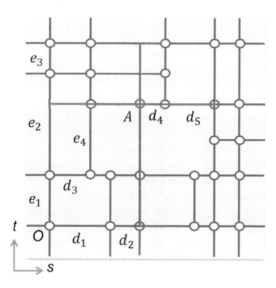

FIGURE 7.1: A local region of a T-mesh. O is the parametric origin, d_i are the s-edge knot intervals, and e_i are the t-edge knot intervals. The red lines are two rays we shoot to infer the local knot vector.

Given a regular node or T-junction C_i in the T-mesh, we can infer its two local knot vectors from its local T-mesh neighborhood using *the knot vector inference rule* [332]. Supposing its parametric coordinates are (s_2, t_2), we shoot a ray along each parametric direction to compute its two local knot vectors. Consider a ray along the s direction, $R(\alpha) = (s_2 + \alpha, t_2)$. It intersects with t-edges on the left and on the right. Note that the ray is shot to the right when $\alpha > 0$ and to the left when $\alpha > 0$. We find out its first two intersection points on the left (s_0 and s_1) and also on the right (s_3 and s_4). Then we build the local s-knot vector and obtain $s = [s_0, s_1, s_2, s_3, s_4]$. Similarly, we shoot another ray along the t direction, $R(\beta) = (s_2, t_2 + \beta)$, and find the first two intersection points with s-edges below and above C_i. Then we build the other local knot vector and obtain $t = [t_0, t_1, t_2, t_3, t_4]$.

Now let us take node A in Figure 7.1 as an example to see how to infer its two local knot vectors s_A and t_A. Supposing O is the parametric origin, we can easily obtain A's coordinates $(d_1 + d_2, e_1 + e_4)$. To infer s_A, we shoot a ray starting from A along the s direction, and then we have $R(\alpha) = (d_1 + d_2 + \alpha, e_1 + e_4)$. We find out the first two intersection points with t-edges on the left and also on the right, and then use them to build the local s-knot vector, $s_A = [0, d_3, d_1 + d_2, d_1 + d_2 + d_4, d_1 + d_2 + d_4 + d_5]$. Similarly, we shoot a ray along the t direction and obtain the local t-knot vector, $t_A = [0, e_1, e_1 + e_4, e_1 + e_2, e_1 + e_2 + e_3]$.

In addition, Sederberg *et al.* [332] also defined three types of T-splines by checking partition of unity, including standard, semi-standard and nonstandard T-splines. The blending functions of a *standard T-spline* satisfy

$$\sum_i B_i(s, t) = 1, \quad \forall (s, t) \in \Omega. \tag{7.3}$$

Including the weights into the above summation, a *semi-standard T-spline* satisfies

$$\sum_i w_i B_i(s, t) = 1, \quad \forall (s, t) \in \Omega, \tag{7.4}$$

where not all the weights $w_i = 1$. A *non-standard T-spline* is one for which we cannot find any set of weights for the blending functions to satisfy $\sum_i w_i B_i(s, t) = 1$.

7.1.2 Wavelets-Based NURBS Simplification and Fairing

In Reverse Engineering (RE), the NURBS data are generated from points cloud or CAD systems, and they are usually very dense. To improve the computational speed and shortening the transmission time, we need to simplify the geometry. In addition, the obtained NURBS curves and surfaces are sometimes unsatisfactory due to the inevitable modeling error and existing noises. The fairness and smoothness of curves and surfaces are important in modeling, manufacturing, finite element analysis [178, 282] and isogeometric analysis [104, 179]. For example, in industry they affect the quality, processability and utilization characteristics of products.

Most existing simplification methods were designed for polygonal surfaces, and very little research was conducted on NURBS simplification [147, 260]. For NURBS fairing, several approaches were developed in the literature. Based on the knot removal and reinsertion method, bad knots were identified and removed to smooth curves and surfaces [132, 202]. This method works well for planar curves, but is not satisfactory for space curves or surfaces. The fairing process can also be modeled as an energy minimization problem [123], which was derived from the balance of a thin elastic beam or thin plate with minimal strain energy or bending energy under a certain load [319]. This is basically a nonlinear optimization problem and numerically expensive. In recent years, the wavelet-based methods were developed to improve the fairness of curves and surfaces [22, 93]. These methods are based on the uniform or quasi-uniform wavelet transform, and they cannot directly work for non-uniform B-spline curves and surfaces.

In this section, we talk about NURBS simplification and fairing schemes based on an orthogonal non-uniform spline wavelet transform [390]. Here, the compression nature of wavelets is used for NURBS simplification. To preserve the continuity after simplification, we provide sufficient conditions for C^0- and G^1-continuity, and adjust control points at the shared boundary between adjacent patches. To filter out high frequency noises and preserve geometric features and continuity, we decompose the NURBS curves and surfaces and then selectively reconstruct them with boundary constraints.

Non-Uniform Wavelet Transform. The wavelet transform is basically a mathematical tool to hierarchically decompose a given function into a coarse overall structure with details. That is, the wavelet transform projects a given function in space V^L onto its two subspaces, V^{L-1} and W^{L-1}. Generally, the given function can be a signal, image, curve or surface. In the classical wavelet theory, the "coarser" level scaling and wavelet functions are obtained from the "finer" level scaling and wavelet functions by dilation. We only require that each scaling function in V^{L-1} can be represented as a linear combination of the scaling functions in V^L, with wavelets defined on a bounded interval.

The wavelet transform starts with a nested set of spaces $\cdots V^{L-1} \subset V^L \subset V^{L+1} \cdots$. A NURBS curve can be written as

$$\bar{\gamma}^L(u) = \sum_{i=0}^{n} \bar{C}_i^L N_{i,d}^L(u), \tag{7.5}$$

or in the matrix form

$$\bar{\gamma}^L(u) = \phi^L \bar{C}^L, \tag{7.6}$$

where $\phi^L = \left[N_{0,d}^L, \cdots, N_{n^L,d}^L \right]$ is a set of B-spline basis functions of degree d, and $\bar{C}^L = \left[\bar{C}_0^L, \cdots, \bar{C}_{n^L}^L \right]$ is a set of homogeneous control points. Suppose the basis functions of V^{L-1}, W^{L-1} are ϕ^{L-1} and ψ^{L-1}, respectively. Note that V^{L-1} and W^{L-1} are the two subspaces of V^L, therefore we can find two matrices P and Q such that $\phi^{L-1} = \phi^L P^L$ and $\psi^{L-1} = \phi^L Q^L$. Then we have

$$\begin{aligned} \phi^L \bar{C}^L &= \phi^{L-1} \bar{C}^{L-1} + \psi^{L-1} \bar{D}^{L-1} \\ &= \phi^L P^L \bar{C}^{L-1} + \phi^L Q^L \bar{D}^{L-1}, \end{aligned} \tag{7.7}$$

and the wavelet transform of a curve can be written as

$$\begin{aligned} \bar{C}^L &= P^L \bar{C}^{L-1} + Q^L \bar{D}^{L-1} \\ &= \begin{bmatrix} P^L & | & Q^L \end{bmatrix} \begin{bmatrix} \bar{C}^{L-1} \\ \bar{D}^{L-1} \end{bmatrix}. \end{aligned} \tag{7.8}$$

Similarly, we represent a B-spline surface as

$$\bar{S}^L(u,v) = \sum_{i=1}^{m^L} \sum_{j=1}^{n^L} \bar{C}_{ij}^L N_{i,d,u}^L(u) N_{j,d,v}^L(v), \tag{7.9}$$

where $N_{i,d,u}^L(u)$ and $N_{j,d,v}^L(v)$ are the B-spline basis functions. It can also be represented in the matrix form

$$\bar{S}^L = \phi_u^L \bar{C}^L \left[\phi_v^L\right]^T, \tag{7.10}$$

where $\phi_u^L = \left[N_{0,d,u}^L, \cdots, N_{m^L,d,u}^L\right]$ and $\phi_v^L = \left[N_{0,d,v}^L, \cdots, N_{n^L,d,v}^L\right]$. Similar to the B-spline curve, we have

$$\begin{aligned}
\bar{C}^L &= P^{Lu}\bar{C}^{L-1}\left[P^{Lv}\right]^T + P^{Lu}\bar{D}_{\phi^{Lu-1}\psi^{Lv-1}}\left[Q^{Lv}\right]^T \\
&+ Q^{Lu}\bar{D}_{\psi^{Lu-1}\phi^{Lv-1}}\left[P^{Lv}\right]^T + Q^{Lu}\bar{D}_{\psi^{Lu-1}\psi^{Lv-1}}\left[Q^{Lv}\right]^T. \tag{7.11}
\end{aligned}$$

We can also rewrite it in the matrix form and obtain

$$\bar{C}^L = \left[\begin{array}{c|c} P^{Lu} & Q^{Lu} \end{array}\right] \left[\begin{array}{cc} \bar{C}^{L-1} & \bar{D}_{\phi^{Lu-1}\psi^{Lv-1}} \\ \bar{D}_{\psi^{Lu-1}\phi^{Lv-1}} & \bar{D}_{\psi^{Lu-1}\psi^{Lv-1}} \end{array}\right] \left[\begin{array}{c} \left[P^{Lv}\right]^T \\ \left[Q^{Lv}\right]^T \end{array}\right]. \tag{7.12}$$

Note that Matrix P can be calculated using the Oslo knot insertion algorithm [259], and Matrix Q can be obtained using the minimal support and orthogonal condition [261]. It is obvious that the wavelet transform has been converted into a linear system.

Simplification. Given a NURBS curve $\gamma^L(u)$ or surface $S^L(u,v)$ in V^L and a threshold ϵ ($\epsilon > 0$), we aim at finding a subspace V^{L-1} of V^L in the lowest possible dimension and a new curve $\gamma^{L-1}(u)$ or surface $S^{L-1}(u,v)$ in V^{L-1}, keeping the simplification error within the threshold ϵ. There are two main problems for simplification or compression: (1) how to choose the subspace; and (2) how to preserve the continuity at the boundary after simplification.

During the subspace selection, we tend to remove as many knots as possible from the knot vector(s) while keeping the approximation error within ϵ. For a NURBS curve, the simplification error is defined as

$$e_c = \frac{max_i \left\{dist\left(\bar{C}_i, \left[P^L\bar{C}^{L-1}\right]_i\right)\right\}}{average\left\{dist\left(\bar{C}_{i+1}, \bar{C}_i\right)\right\}}. \tag{7.13}$$

Similarly for a surface, the error or weight is defined as

$$e_s = \frac{1}{\lambda}max_{i,j}\left\{dist\left(\bar{C}_{i,j}, \left[P^{Lu}\bar{C}^{L-1}\left[P^{Lv}\right]^T\right]_{i,j}\right)\right\}, \tag{7.14}$$

where

$$\lambda = \frac{1}{2}\left(average\left\{dist\left(\bar{C}_{i+1,j}, \bar{C}_{i,j}\right)\right\} + average\left\{dist\left(\bar{C}_{i,j+1}, \bar{C}_{i,j}\right)\right\}\right). \tag{7.15}$$

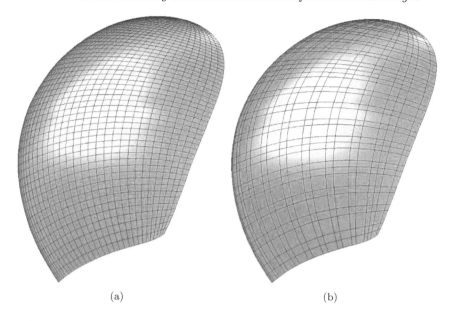

FIGURE 7.2: The simplification result of a propeller model (pictures from [390]). (a) The original model with two patches and 3,990 control points; and (b) the compression result with 1,584 control points. The simplification error is 2.13×10^{-4}, and the computational time is 95.64 seconds.

We remove the knots whose weights are smaller than the given tolerance ϵ. Note that the error will accumulate if one removes two or more neighboring knots. Therefore when two or more adjacent knots all have weights less than ϵ, we keep every other one.

After choosing the subspace, we can implement the wavelet decomposition for NURBS simplification. For curves, we set the first and the last row of Matrix Q to be zero. This can keep the end control points unchanged after decomposition. For surfaces, the neighboring patch should use the same subspace for the shared parametric direction. After compression, we adjust the control points at the shared boundary to preserve C^0- or G^1-continuity. Figure 7.2 shows the simplification result of a propeller model.

Fairing. The wavelet transform decomposes a given function into different frequency components, which could be used to detect and remove high-frequency noises while preserving detailed features. There are two criteria for fairing: a curve is fair if (1) its curvature plot is continuous; and (2) it has few monotone pieces. Similarly, a surface is fair if (1) the isoparametric curves along the u and v directions are fair; and (2) the intersectant curves of the surface and a set of parallel planes are fair.

In our fairing algorithm, decomposition and selective reconstruction are used to remove high frequency noise in two steps with two subspaces. One contains all the odd inner knots, and the other one contains all the even inner knots. First, we remove noises associated to all the even inner knots using the first subspace, and then we remove noises associated to all the odd inner knots using the second subspace. In this way, our algorithm removes high frequency noises while preserving geometric features at the same time. Note that our algorithm avoids the complicated steps to compute the curvature and its derivatives to find "bad" knots, and in addition, it does not need an iteration. The boundary continuity is automatically preserved during fairing. Figure 7.3 shows the fairing result of a thoracic aorta from CT images.

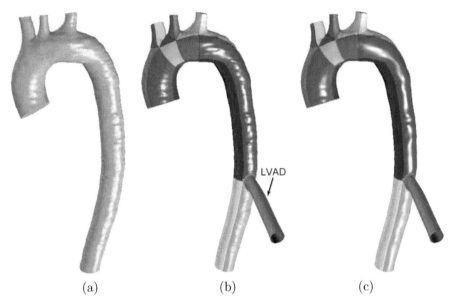

(a) (b) (c)

FIGURE 7.3: The fairing result of a thoracic aorta (pictures from [390]). (a) The triangle mesh generated from CT imaging data; (b) the NURBS model generated from the triangle mesh with 23 NURBS patches, 21,297 control points and a left ventricular assist device (LVAD) inserted; and (c) the NURBS surface after fairing. It is obvious that the surface becomes much smoother after fairing. The computational time is 87.40 seconds.

7.1.3 Solid NURBS Construction via Offsetting

Given NURBS surface patches, we can obtain solid NURBS by offsetting it along the normal direction. This is a very useful technique for building thin or shell structures for manufacturing and computation. Starting from the original curve or surface, an offset is defined as the locus of all the points with a constant distance along the normal direction. Given a NURBS curve $C(u)$ or surface $S(u, v)$ and a distance d, the offset operation is to find $C_d(u)$ or

$S_d(u,v)$ such that $C_d(u) = C(u) + d \times N(u)$ or $S_d(u,v) = S(u,v) + d \times N(u,v)$, where $N(u)$ and $N(u,v)$ are the unit normal vectors. Although creating an offset for curves and surfaces is mathematically known for many years and the offset for a line, plane, circle or sphere is very obvious, unfortunately for free-form curves and surfaces it is not straightforward. Note that due to the operation of square root when calculating the unit normal, the exact offset of NURBS cannot be represented as a NURBS curve or surface anymore. In computation or manufacturing, what we use for the offset of a NURBS curve and surface is generally an approximation.

Driven by the application of offset in tool path generation and solid modeling, many techniques have been developed to generate the offset of free-form curve and surface. There are two main categories of NURBS offsetting methods, the sample-based method and the control points based method. The former samples the original curve or surface based on the local feature. Then we compute the offset of the obtained sample points via curve or surface fitting [212, 298]. The latter calculates the offset by directly manipulating the control points. In [126], the offset was obtained via offsetting the control points and then perturbing the control points to minimize the offset error.

Starting from NURBS surface, we have developed an error-bounded offsetting method to construct solid NURBS for isogeometric analysis [392]. To improve the offset accuracy, we first refine the original control mesh using subdivision. We insert one knot for each knot interval except the two ends, and calculate the new control points based on the Oslo knot insertion algorithm. Then, we calculate the normal for each control point, and for boundary control points we take the average of the normal from all neighboring patches. For each control point P_{ij}, we calculate the corresponding parameter value (u_{ij}, v_{ij}) at which the two basis functions get the maximum. After that, we calculate the offset for the refined control mesh using the obtained normal. When the offset distance exceeds the local curvature radius, the offset surface may have self-intersections. In this situation, we consider the control mesh as a quad mesh and use the mesh smoothing technique to remove them. Next, we use the NURBS wavelet decomposition to remove all the inserted knots. Finally, we check the error and iterate the entire procedure until the error is within a given error bound.

Figure 7.4 shows the solid NURBS results of two complex Navy structures using the above error-bounded offsetting method, a boat with the propeller shaft and rudder and a submarine.

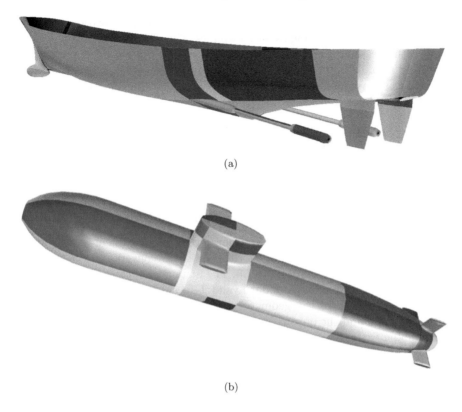

(a)

(b)

FIGURE 7.4: (a) The constructed solid NURBS model of the boat with the propeller shaft and rudder (74 patches, 6,412 control points); and (b) the constructed solid NURBS model of the submarine with 66 patches and 2,592 control points.

7.2 Converting Unstructured Quadrilateral and Hexahedral Meshes to T-Splines

Over the entire history of finite element development and broad applications, many unstructured bilinear quad and trilinear hex meshes have been accumulated and employed in the traditional finite element analysis. To improve the geometry accuracy and surface continuity, and to integrate CAD systems with isogeometric analysis, we need to convert these polygonal meshes into continuous, high-order spline parametric representations.

Considerable work has been conducted to convert triangles to 2D spline representations, but very few research was conducted to convert quads and hexes into splines. In [24], NURBS surfaces are constructed from triangular meshes for organic structures. Periodic Global Parameterization was later utilized for converting triangles into T-splines [232, 233]. Starting from a poly-cube with the same topology as the input geometry, a parametric map was built and also utilized to construct T-splines [382]. In [430], a skeleton-based method was developed to construct solid NURBS for isogeometric analysis of blood flow in human vascular structures.

In the following, we will discuss how to convert an arbitrary unstructured quad and hex mesh into a standard or rational T-spline with C^2-continuity everywhere on the surface except at local regions surrounding extraordinary nodes [393, 394].

7.2.1 Converting Quadrilateral Meshes to Standard T-Splines

Given an unstructured quad and hex mesh, we take them as the initial T-mesh and adopt a two-stage algorithm [393] to convert it to a standard T-spline; see the pipeline in Figure 7.5. The main goal of the topology stage is to construct a gap-free and standard valid T-mesh, preserving sharp features in the input model. In the geometry stage, we minimize the error between the input mesh and the output T-spline surface, and adjust the control points surrounding extraordinary nodes to improve the local surface continuity.

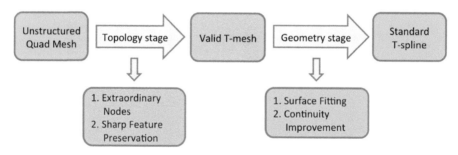

FIGURE 7.5: An overview of the two-stage algorithm to convert an unstructured quad mesh to a standard T-spline.

To obtain a valid T-mesh, the key problem is how to handle extraordinary nodes. If the input mesh contains regular nodes only, the initial T-mesh is topologically correct and automatically valid. However, extraordinary nodes are unavoidable for complex geometry, and they may introduce gaps on the T-spline surface if they are not handled properly. Moreover, the surface continuity is no longer C^2 around extraordinary nodes [334]. To generate a gap-free valid T-mesh, we first categorize all the quad configurations based on how many extraordinary nodes they have, and then design templates to handle each of them.

There are a total of six types of quads in the initial T-mesh: quads with zero, one, two (neighboring or diagonal), three and four extraordinary nodes.

In our designed templates as shown in Figure 7.6, both the number of newly inserted nodes and the region influenced by the extraordinary nodes are minimized. All of these templates have a conforming 1-ring neighborhood, and they satisfy the gap-free requirement. Using this template set, the resulting surface continuity is C^0 at the extraordinary point and across the edge within the 3-ring neighborhood, and C^1 outside the 3-ring neighborhood until the 4-ring neighborhood.

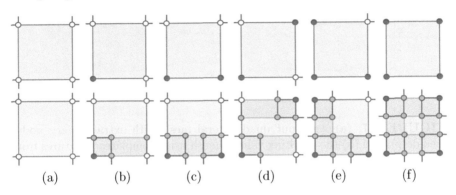

(a)   (b)   (c)   (d)   (e)   (f)

FIGURE 7.6: Standard T-spline templates for six element types, including quads with zero (a), one (b), two (c, d), three (e) and four (f) extraordinary nodes. The red dots are extraordinary nodes, the green dots are newly inserted nodes, and the red short edges are zero-length edges.

During template development, we may obtain semi-standard or non-standard T-meshes in some local regions. To standardize these regions, we apply edge extension to ensure the related nodes have the same knot vector along one specific parametric direction. Figure 7.7 shows the valid T-mesh generated using our template set. The blue edges are generated during the template development, and the green edges are extended edges during T-mesh standardization.

After obtaining a valid T-mesh, we infer the local knot vectors for each control node to build cubic T-splines. For a regular node or a T-junction, its two knot vectors can be inferred using the knot vector inference rule [332]. As we discussed before, we shoot a ray along the s (or t) direction, and then find two intersections on the left (or the bottom) and two other intersections on the right (or the top). These four intersections together with the current knot form the local knot vector. For nodes adjacent to an extraordinary node, we follow the same rule except that whenever the ray meets the extraordinary node, we repeat the knot associated to this extraordinary node.

CAD models contain many sharp corners and sharp edges. To preserve these sharp features in the input geometry, we insert zero-length edges across sharp edges and also around sharp corners in the T-mesh. Since each sharp edge is shared by two surface patches, we duplicate sharp edges on each patch and assign a zero knot interval to all the transverse edges. By repeating knots,

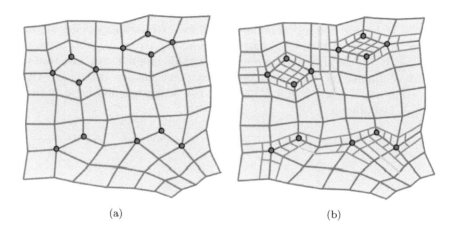

(a)   (b)

FIGURE 7.7: (a) An input quadrilateral mesh with extraordinary nodes (red dots); and (b) the resulting valid T-mesh using templates. (Pictures from [393].)

we decrease the surface continuity to C^0 across the sharp edges and corners and as a result, we preserve all sharp features.

From the topology stage, we obtain a standard, gap-free and valid T-mesh. In the geometry stage, we use an efficient surface fitting technique to improve the geometric accuracy by adjusting control points. We can guarantee that the output T-spline surface interpolates or passes through all the nodes in the input mesh. Then the control nodes around each extraordinary node are also adjusted to improve the local surface continuity. After that, a standard cubic T-spline is constructed from the T-mesh.

From the constructed T-spline, we then extract Bézier elements for isogeometric analysis applications [58, 329]. For a quasi-uniform T-mesh, each nonzero-area parametric domain corresponds to a Bézier element. Therefore, the number of Bézier elements in the resulting T-spline is exactly the same as the number of quad elements in the input mesh. For each nonzero-area parametric domain, we first identify all the nodes with nonzero basis function values, and then compute the transformation matrix M^e between the T-spline basis functions and the Bézier basis functions. We have the relationship

$$B_t^e = M^e B_b^e, \tag{7.16}$$

where

$$B_t^e = \left[B_0^e, B_1^e, \cdots, B_{n^e-1}^e\right]^T \tag{7.17}$$

is the vector formed by the T-spline basis functions with nonzero function values, and

$$
B_b^e =
\begin{bmatrix}
N[0,0,0,0,1](s)N[0,0,0,0,1](t) \\
N[0,0,0,1,1](s)N[0,0,0,0,1](t) \\
\vdots \\
N[0,1,1,1,1](s)N[0,1,1,1,1](t)
\end{bmatrix}
\tag{7.18}
$$

is the vector formed by the Bézier basis functions. Here, n^e represents the number of nodes with nonzero basis function values in this domain, and M^e can be computed using the Oslo knot insertion algorithm [156]. Figure 7.8 shows the constructed T-spline and the extracted Bézier elements for a face model. Note that the number of Bézier elements in (c) is exactly the same as the number of quads in (a).

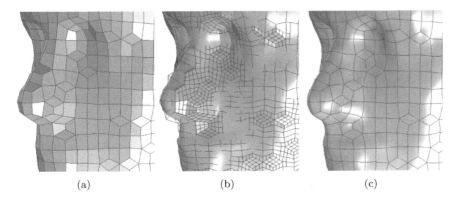

(a)   (b)   (c)

FIGURE 7.8: (a) The input quad mesh; (b) the T-mesh with its defined T-spline surface; and (c) Bézier elements extracted from the T-spline. (Pictures from [393].)

7.2.2 Converting Unstructured Meshes to Rational T-Splines

When we convert unstructured quad meshes to a standard T-spline [393] as we discussed earlier, many nodes and edges need to be inserted in order to ensure the resulting T-spline standard. To reduce the number of newly inserted nodes and edges, as a follow-up here we generalize the T-spline definition to the rational T-spline [394]. The new rational T-spline basis functions satisfy partition of unity by definition. Here, we focus on converting an arbitrary unstructured quad or hex mesh to a rational bicubic T-spline surface or tricubic T-spline solid.

As reviewed in Section 7.1, there are three types of T-splines depending on whether their basis functions or their weighted basis functions satisfy a partition of unity, including standard, semi-standard and non-standard T-splines. T-splines support local refinement and provide a lot of flexibility for modeling. On the other hand, they are also limited due to several open problems. For example, how to characterize T-mesh configurations for a standard, semi-standard, or non-standard T-spline and how to calculate the associated weights for a semi-standard T-spline are still open problems.

To enable partition of unity in a more general way, here we choose the rational basis functions to construct T-splines. The rational T-spline surface is defined as

$$S(\xi, \eta) = \frac{\sum_{i=0}^{n} w_i C_i R_i(\xi, \eta)}{\sum_{i=0}^{n} w_i R_i(\xi, \eta)}, \quad \forall (\xi, \eta) \in \Omega, \tag{7.19}$$

where C_i is the control node, w_i is the weight (here we set $w_i = 1$) and

$$R_i(\xi, \eta) = \frac{N_i^{\xi}(\xi) N_i^{\eta}(\eta)}{\sum_{j=0}^{n} N_j^{\xi}(\xi) N_j^{\eta}(\eta)} \tag{7.20}$$

is the newly defined rational B-spline basis function. $N_i^{\xi}(\xi)$ and $N_i^{\eta}(\eta)$ are B-spline basis functions defined by the local knot vectors at C_i, $\xi_i = [\xi_{i0}, \cdots, \xi_{i4}]$ and $\eta_i = [\eta_{i0}, \cdots, \eta_{i4}]$ when degree $d = 3$. Similarly, a rational solid T-spline is defined as

$$S(\xi, \eta, \zeta) = \frac{\sum_{i=0}^{n} w_i C_i R_i(\xi, \eta, \zeta)}{\sum_{i=0}^{n} w_i R_i(\xi, \eta, \zeta)}, \quad \forall (\xi, \eta, \zeta) \in \Omega, \tag{7.21}$$

where

$$R_i(\xi, \eta, \zeta) = \frac{N_i^{\xi}(\xi) N_i^{\eta}(\eta) N_i^{\zeta}(\zeta)}{\sum_{j=0}^{n} N_j^{\xi}(\xi) N_j^{\eta}(\eta) N_j^{\zeta}(\zeta)} \tag{7.22}$$

is the newly defined rational B-spline basis function, $N_i^{\xi}(\xi)$, $N_i^{\eta}(\eta)$ and $N_i^{\zeta}(\zeta)$ are B-spline basis functions defined by the local knot vectors at node C_i, $\xi_i = [\xi_{i0}, \cdots, \xi_{i4}]$, $\eta_i = [\eta_{i0}, \cdots, \eta_{i4}]$ and $\zeta_i = [\zeta_{i0}, \cdots, \zeta_{i4}]$ when degree $d = 3$. It is obvious that the rational B-spline basis functions automatically satisfy $\sum_{i=0}^{n} R_i = 1$ by definition, for any (ξ, η) in 2D and (ξ, η, ζ) in 3D. In this way, we obtain a set of basis functions satisfying partition of unity even for non-standard T-splines, and thus successfully skip the difficulty of checking the type of T-mesh configurations and calculating the associated weights for semi-standard cases.

As shown in Figure 7.9(a-c), in 2D we can simply distinguish regular nodes and extraordinary nodes using the valence number. A regular node has the valence of four, and a node with a valence of other than four and not a T-junction is an extraordinary node. The situations in 3D are much more complex than 2D. We cannot rely on the valence number only to judge if a node is regular or extraordinary. For example, node A has a valence of 8 in both (d) and (g), but it is a regular node in (d) while an extraordinary node in (g). Let us take a look at (e). There is a yellow plane separating all the elements sharing node A into two sets, one set is above the yellow plane and the other set is below the yellow plane. These two sets are symmetric with respect to the yellow plane. In such a situation, we call node A a *partial extraordinary node*. If we cannot find any symmetry in all three directions, the node is called an extraordinary node; see (f) and (g).

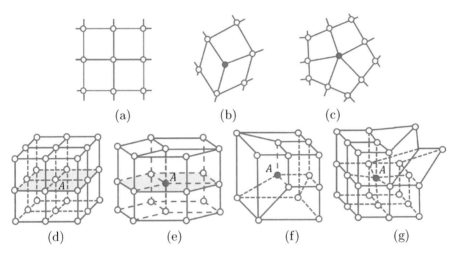

FIGURE 7.9: Regular node, partial extraordinary node and extraordinary node. (a) A regular region in 2D; (b, c) extraordinary nodes (the red dots) in 2D; (d) a regular region in 3D; (e) a partial extraordinary node (the red dot) in 3D; and (f, g) extraordinary nodes (the red dots) in 3D.

With rational T-spline basis functions, we can further simplify the designed templates for extraordinary nodes; see Figure 7.10. Compared to standard T-spline templates in Figure 7.6, we introduce fewer new points and fewer zero-length edges, especially in templates (b) and (d). In addition, no edge extension is needed since partition of unity is satisfied by definition. Figure 7.11 shows a comparison of the resulting standard T-mesh and rational T-mesh from the same input. It is obvious that (b) contains many fewer nodes and zero-length edges, leading to better surface continuity.

In addition to regular and extraordinary nodes, hex meshes also contain partial extraordinary nodes. Here, we first design general templates for the partial extraordinary and extraordinary node, and then apply them to their surrounding elements. Figure 7.12 shows a general template for a partial extraordinary node and a general template for an extraordinary node existing in hex meshes.

To preserve sharp edges in solid T-splines, we insert an edge parallel to the sharp edge on each adjacent boundary face. This step guarantees that there are two zero-length edges across the sharp edge. In other words, there are three

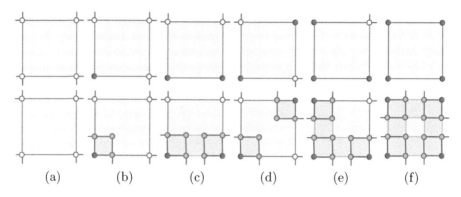

FIGURE 7.10: Rational T-spline templates for six quad element types, including quads with zero (a), one (b), two (c, d), three (e) and four (f) extraordinary nodes. The red dots are extraordinary nodes, the green dots are newly inserted nodes, and the red short edges are zero-length edges.

duplicated knots and the surface continuity is degenerated to C^0 across the sharp edge. Different from sharp feature preservation for 2D T-spline surfaces, we have newly inserted faces when we handle partial extraordinary nodes and extraordinary nodes in 3D. We also need to insert zero-length edges on them. This step makes the boundary face and the layer right below it have exactly the same topology. To preserve sharp corners in solid T-splines, we treat each

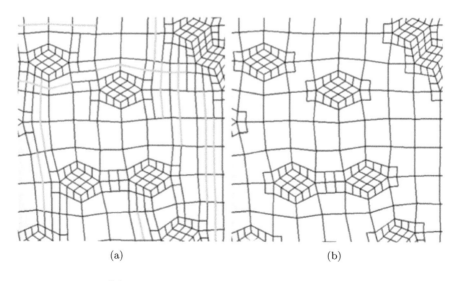

FIGURE 7.11: (a) The standard T-mesh after applying templates in Figure 7.6; and (b) the rational T-mesh after applying templates in Figure 7.10. (Pictures from [394].)

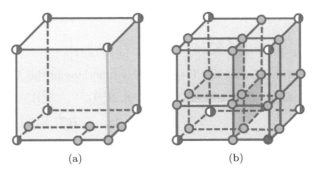

FIGURE 7.12: General templates for 3D hex meshes. (a) A general template for a partial extraordinary node; and (b) a general template for an extraordinary node.

sharp corner node as a partial extraordinary node. Again, we need to apply all the operations conducted on the boundary face to the newly inserted faces, such as zero-length edge insertion.

From the constructed T-spline surface, we can extract Bézier elements following Eqns. (7.16)-(7.18). Similarly for a solid T-spline, we have

$$B_t^e = \left[N_0^\xi N_0^\eta N_0^\zeta, N_1^\xi N_1^\eta N_1^\zeta, \cdots, N_{n^e-1}^\xi N_{n^e-1}^\eta N_{n^e-1}^\zeta \right]^T \quad (7.23)$$

and

$$B_b^e = \begin{bmatrix} N[0,0,0,0,1](\xi)N[0,0,0,0,1](\eta)N[0,0,0,0,1](\zeta) \\ N[0,0,0,1,1](\xi)N[0,0,0,0,1](\eta)N[0,0,0,0,1](\zeta) \\ \vdots \\ N[0,1,1,1,1](\xi)N[0,1,1,1,1](\eta)N[0,1,1,1,1](\zeta) \end{bmatrix}. \quad (7.24)$$

Note that we can also use the Bézier extraction technique to study the linear independence of a given T-mesh. In [235], the nullity of the T-spline-to-NURBS transformation matrix is computed to determine the linear independence of a T-spline surface model. Unfortunately, this method is not suitable for a T-spline with extraordinary nodes because converting such a T-spline to NURBS will end up with multiple NURBS patches. In our T-meshes, generally we have a large percentage of extraordinary nodes and we also need to handle 3D solid T-splines. To study the linear independence of solid T-splines with extraordinary nodes, here we adopt a different scheme. We assemble the transformation matrices of all the nonzero domains or Bézier elements following the same way as what we do for matrix assembly in the finite element method, and obtain a global transformation matrix. In other words, we compute the global transformation matrix, K, for the entire model from T-spline to Bézier elements. Then we have

$$B_t = KB_b, \quad (7.25)$$

where

$$B_t = \left[N_0^\xi N_0^\eta, N_1^\xi N_1^\eta, \cdots, N_{n-1}^\xi N_{n-1}^\eta \right]^T \tag{7.26}$$

is the vector formed by all the T-spline basis functions in the T-mesh, and

$$B_b = \left[B_b^0[0], B_b^0[1], \cdots, B_b^0[15], B_b^1[0], B_b^1[1], \cdots, B_b^1[15], \right.$$
$$\left. \cdots, B_b^{m-1}[0], B_b^{m-1}[1], \cdots, B_b^{m-1}[15] \right]^T \tag{7.27}$$

is formed by all the Bézier basis functions of the model. Here, n is the number of control nodes in the T-mesh and m is the number of nonzero domains or Bézier elements for the entire model. The matrix K is the global transformation matrix from T-spline to Bézier, with a size of $n \times 16m$ in 2D. Similarly in 3D, we have

$$B_t = \left[N_0^\xi N_0^\eta N_0^\zeta, N_1^\xi N_1^\eta N_1^\zeta, \cdots, N_{n-1}^\xi N_{n-1}^\eta N_{n-1}^\zeta \right]^T \tag{7.28}$$

and

$$B_b = \left[B_b^0[0], B_b^0[1], \cdots, B_b^0[63], B_b^1[0], B_b^1[1], \cdots, B_b^1[63], \right.$$
$$\left. \cdots, B_b^{m-1}[0], B_b^{m-1}[1], \cdots, B_b^{m-1}[63] \right]^T. \tag{7.29}$$

The matrix K is the global transformation matrix from solid T-spline to solid Bézier and the size of K is $n \times 64m$. Given a T-mesh, all the T-spline basis functions form a linear space. If all these functions are linearly independent, they form a basis of the space with a dimension of n. The following proposition provides a necessary and sufficient condition for the linear independence of a T-spline.

Proposition (Necessary and sufficient condition for linear independence). *A necessary and sufficient condition for a T-spline surface or solid to be linearly independent is that the global transformation matrix from T-spline to Bézier is in full rank.*

Figure 7.13 shows the converted rational solid T-spline from an unstructured hex mesh. Note that the number of Bézier elements is the same as the number of elements in the input mesh. The reason for this is that all the edges we inserted have a zero knot interval. The resulting solid T-spline is tricubic and C^2-continuous except in the vicinity of partial extraordinary and extraordinary nodes.

7.3   Polycube-Based Parametric Mapping Methods

In solid modeling and computer aided design, boundary representations are widely used and a 3D "solid" geometry is generally the volume bounded by

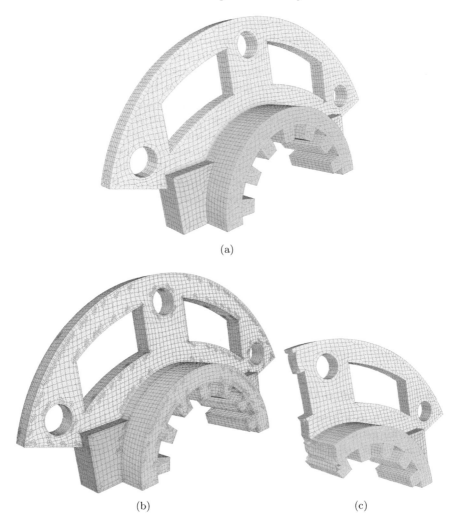

(a)

(b)   (c)

FIGURE 7.13: A CAD assembly model. (a) The input unstructured hex mesh; (b) the constructed rational solid T-spline and T-mesh; and (c) the extracted solid Bézier elements with some elements removed to show the interior structure.

closed surfaces. Despite the importance of boundary representations, interior volume carries a lot of other information such as material properties, density and physical domain distribution. In particular, finite element and isogeometric analysis [42, 104, 179] require a volumetric representation in simulating physical or biological phenomena. To integrate engineering design and analysis, a fundamental step is to construct such volumetric representation from boundary surfaces, like trivariate spline models.

Some research has been devoted to solid spline construction from boundary representations. By sweeping a closed curve, NURBS parameterizations of swept volumes were built and used in isogeometric analysis [19]. A skeleton-based sweeping method was developed to construct solid NURBS for patient-specific vascular structures from medical images for blood flow simulation [430]. Based on discrete volumetric harmonic functions, a volumetric parameterization was used to construct a single trivariate B-spline [268]. A global one-piece trivariate spline scheme was presented in [228], using the generalized polycube parameterization [382]. Harmonic functions were also used in building volumetric parameterizations [161, 234]. Based on an adaptive tetrahedral meshing and mesh untangling technique, an algorithm was developed to construct a trivariate T-spline representation for genus-zero solids [129].

In the following, we will describe a different method [438] to construct solid rational T-splines for complex genus-zero geometry from the boundary surface triangulation, with C^2-continuity everywhere over the boundary surface except for the local region of eight corner nodes and no any negative Jacobian. The definition of the rational T-spline [394] is used here to obtain partition of unity basis functions for an arbitrary gap-free T-mesh. We build a parametric mapping between the input surface and the boundary of the parametric domain, a unit cube. In addition, we also use harmonic fields to build polycubes and trivariate T-splines for high-genus objects [391].

7.3.1 Genus-Zero Objects

As shown in Figure 7.14, there are two main stages to construct a solid T-spline for genus-zero objects from a given boundary triangle mesh [438]. We build a valid T-mesh in the first stage and then construct solid T-splines in the second stage. T-mesh construction consists of four steps. We first build a parametric mapping between the input triangle mesh and a unit cube, which also serves as the parameter domain for the solid T-spline. Then we apply the strongly balanced octree subdivision to the parameter domain and each node on the cube boundary is projected onto the input surface. After that, we pillow all the boundary nodes and improve the T-mesh quality via smoothing and optimization. Finally to obtain a gap-free and valid T-mesh, we apply templates to each extraordinary node and partial extraordinary node.

With a valid T-mesh, we construct the solid rational T-spline in the second stage. For each node in the T-mesh, we shoot rays and traverse T-mesh faces and edges to infer its local knot vectors along the three parametric directions [332]. For each local domain, the nodes with nonzero basis functions are identified and the solid T-spline is computed based on the rational T-spline definition [394]. Finally, Bézier elements are extracted from the built solid T-spline to facilitate isogeometric analysis. In the following, let us discuss details of these steps in the pipeline.

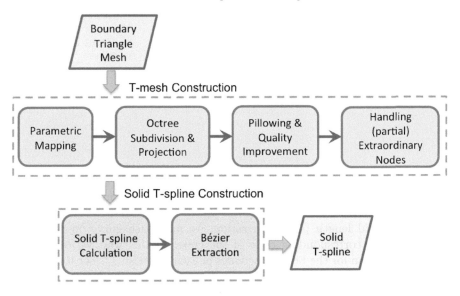

FIGURE 7.14: An overview of the solid T-spline construction algorithm from the given boundary triangle mesh with genus-zero topology.

Parametric Mapping. This step aims at creating a parametric mapping between the boundary triangle mesh T and the boundary of a unit cube C, which performs as the parameter domain of the solid T-spline. To achieve this, we first select eight vertices in T and denote them as V_i ($i = 0, \cdots, 7$), which correspond to the eight corners of the cube C. Then we find twelve curves by calculating the shortest distance between each pair of the adjacent vertices $V_i V_j$ using the Dijkstras algorithm [112]. The obtained twelve paths partition the input surface mesh into six submeshes, T^i ($i = 0, \cdots, 5$), and each submesh corresponds to one face of the cube, F_C^i ($i = 0, \cdots, 5$). Then the following work is to map each submesh T^i onto a planar unit square F_C^i via a surface parameterization.

Surface parameterization tends to build a one-to-one mapping f from a given surface $S \subset \Re^3$ to a parameter domain $\bar{S}$, and we have $f(p) = q$ where $p \in S$ and $q \in \bar{S}$. Surface parameterization has been employed a lot in texture mapping, morphing, remeshing and data fitting [137, 138, 337], and various techniques have been developed to minimize the distortion of angles and areas during the mapping. For a triangle mesh parameterization, our goal is to build a map between the mesh and a triangulation of a planar domain. In other words, given a disk-like triangular mesh $T \subset \Re^3$, we aim to find the correspondence between T and a simply connected planar region like the unit disk or a rectangle. For planar domain parameterization [137], we first map the surface boundary onto the boundary of the parametric domain, and then the parameterization of interior vertices can be obtained by solving the following linear system,

$$\sum_{j \in N_i} w_{ij} \left(f(V_j) - f(V_i) \right) = 0, \tag{7.30}$$

where $V_i \in S$, w_{ij} is the associated weight, and N_i is the first-ring neighborhood surrounding V_i. Different parameterization methods utilize various weights w_{ij} for each edge. Here, we simply choose the harmonic weights,

$$w_{ij} = cot\alpha_{ij} + cot\beta_{ij}, \tag{7.31}$$

where α_{ij} and β_{ij} are two opposite angles in the two triangles sharing the edge $V_i V_j$. These weights were derived from a discrete harmonic map to reduce the angular distortion.

Via a chord length parameterization, we map each curve on the input mesh onto its corresponding edge of the cube. In this way, we obtain the parameterization for the boundary nodes of each submesh T^i. For the interior vertices, we compute their parameterization by solving the linear system in Eq. (7.30). Note that the associated matrix is symmetric and positive definite, therefore the linear system can be solved efficiently.

Figure 7.15 shows the parametric mapping results of a duck model. (a) shows the computed twelve edges (the red edges) on the input triangular mesh, which split the surface into six submeshes as shown in (b). Each color represents a different patch. (c) is the mapping result of the duck surface on a unit cube and the mapping is bijective. When the mapping is not bijective, the obtained T-mesh may have overlapping, which can be resolved later via smoothing and optimization.

Adaptive Octree Subdivision and Projection. Based on the above

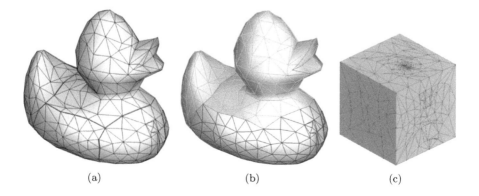

(a)   (b)   (c)

FIGURE 7.15: The mapping results for the duck model (pictures from [438]). (a) The input triangle mesh and the obtained twelve paths (the red edges); (b) six submeshes with each color representing a different submesh; and (c) the result of mapping the triangle mesh onto a unit cube.

parametric mapping results, we apply an adaptive octree subdivision to the unit cube C and generate an adaptive initial T-mesh. By subdividing one element into eight smaller ones recursively, we obtain the refined T-mesh. For each element on the cube boundary, we compute its local distance to the input triangular mesh. If the distance is greater than a given threshold ϵ, we subdivide this element. The octree subdivision is performed until the local distance is less than ϵ. Here, we still adopt the strongly balanced octree subdivision. Note that for each node on the cube boundary, we can easily obtain its parametric coordinates, and during the mapping we use the barycentric coordinates to find its physical coordinates, which are the mapped location on the input triangular mesh.

From octree subdivision, we obtain a multiresolution solid T-mesh by choosing different octree levels. Figure 7.16(a) shows an adaptive octree subdivision result for the duck model in the parametric domain.

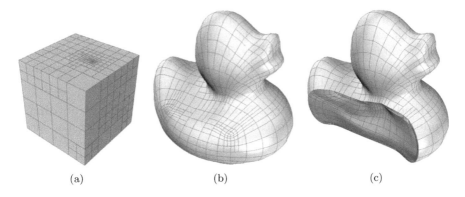

(a)   (b)   (c)

FIGURE 7.16: The duck model (pictures from [438]). (a) The subdivision result in the parametric domain; (b) the resulting solid Bézier elements; and (c) some elements are removed from the Bézier model to show the interior mesh (blue) with one pillowed layer (magenta).

Pillowing and Quality Improvement. Pillowing is a sheet-insertion operation that inserts one or multiple layers of elements around the specified region like the boundary. Here we apply this pillowing operation to our T-mesh because it provides us more flexibility to improve the T-mesh quality, and it can also help improve the surface continuity of solid T-splines.

By pillowing the boundary, we can restrict each element has at most one face lying on the boundary, which gives us more flexibility to further improve the mesh quality. We duplicate each boundary face to create a pillowed element, and assign a constant as the knot interval for the edges along the pillowing direction. The knot intervals for the other two directions stay the same. After pillowing, the eight corners in the initial T-mesh become interior extraordinary nodes and the nodes on the twelve edges become interior partial extraordinary nodes. Therefore, there are only eight partial extraordinary nodes on the resulting new boundary surface, which correspond to the eight original corners. Note that during pillowing we introduce some extraordinary nodes and partial extraordinary nodes. Instead of using the global parameterization, here we adopt a local parameterization in the following steps.

Smoothing and optimization techniques are used to remove tangling elements and improve the T-mesh quality by relocating T-mesh nodes. In smoothing, we relocate each node towards the mass center of its neighboring elements. In optimization, we use finite element or Bézier basis functions to define the Jacobian, and compute an optimal position for each node to maximize the worst Jacobian. Note that generally the smoothing and optimization techniques only work for unstructured hex meshes without hanging nodes or T-junctions. To improve the T-mesh quality around T-junctions, we split elements and convert the local T-mesh to an unstructured all-hex mesh virtually. For an edge T-junction node, we split the element into two hex elements. For a face T-junction node, we split the element into four hex elements. After that we can apply the smoothing and optimization techniques just as what we do for unstructured meshes. That is, for each node we check all its adjacent virtual "hex" elements to identify which one has the worst Jacobian, and then we relocate the node to maximize the worst Jacobian.

Handling Extraordinary Nodes and Partial Extraordinary Nodes. The extraordinary nodes or partial extraordinary nodes may introduce gaps to the resulting solid T-spline if they are not handled properly. This step aims at creating an initial gap-free T-mesh by applying the general templates in Figure 7.12. In the interior region of the obtained T-mesh, there are eight extraordinary nodes and the nodes lying on the twelve edges of the cube are partial extraordinary nodes. The solid T-spline is C^0-continuous around these eight internal extraordinary nodes and across these twelve internal edges. On the resulting boundary surface, we have C^2-continuous everywhere except the vicinity surrounding eight partial extraordinary nodes, where the continuity is C^0.

Solid T-Spline Construction and Bézier Extraction. From the constructed valid T-mesh, we can infer the knot vectors for each node by traversing T-mesh faces and edges, and then build the rational basis functions using these knot vectors and local parameterization. The entire solid T-spline model is built by looping over all the local domains and constructing the local solid T-spline elements.

After that, we extract Bézier elements from the constructed solid T-splines, which is also the primary computational objects in isogeometric analysis. Generally, a T-mesh element may contain multiple Bézier elements and the T-mesh does not contain all the reduced continuity lines or knot lines in the parametric space. To extract the Bézier elements, we first need to compute the "Bézier mesh" with all the knot lines captured. Note that all the missing knot lines in the T-mesh are introduced by T-junctions or L-junctions, therefore we can capture them using the knot vector inference lines. The Bézier mesh is thus obtained by including all these knot vector inference lines for each T-junction and L-junction.

Similarly for each T-junction or L-junction in 3D T-meshes, we use the three isoparametric planes passing through the knot vector inference lines and bounded by the end knots to refine the T-mesh, and then we obtain the Bézier mesh. All these three planes are the reduced continuity surfaces which are not contained in the T-mesh, and we obtain the Bézier mesh by adding them to the T-mesh. Each element in the Bézier mesh is called a Bézier element. Finally for each Bézier element, we compute a transformation matrix between the T-spline basis functions and the Bézier basis functions.

Figures 7.16(b, c) and 7.17 show the solid T-spline results of a duck model and a cow model, respectively. The output solid T-spline is tricubic and C^2-continuous except in the vicinity of partial extraordinary and extraordinary nodes. From the Bézier extraction results, we can observe that the surface is smooth except at the eight corner nodes.

7.3.2 High-Genus Objects

In Section 7.3, we discussed a parametric mapping based method to construct rational trivariate solid T-splines for genus-zero objects from the boundary surface triangulation [438]. To extend this algorithm to high-genus objects [391], here we first compute a harmonic field, and then use its critical points to extract the topology of the input geometry. After that, we build a polycube which is topologically equivalent to the given object. Due to its regular structure, the polycube serves as the parametric domain of the tensor-product spline representations. A trivariate solid T-spline can be built directly upon the generated polycube, instead of using the generalized polycube parameterization [228] or extending the polycube to a box domain and then restricting this domain [387]. Similar to genus-zero objects, a parametric mapping is built between the triangulation and the polycube boundary. An octree subdivision is applied to the polycube, which is then mapped onto the physical space to create the initial T-mesh. The subdivision continues until the surface approximation error is within a given tolerance. After that, we obtain a valid T-mesh via pillowing, T-mesh quality improvement, and applying templates to handle extraordinary nodes and partial extraordinary nodes. Finally, we extract Bézier elements with all positive Jacobians, which can be used in isogeometric analysis applications.

The key problem here is how to build polycubes automatically from boundary representations. Here our polycube construction method has two steps: (1) harmonic field calculation; and (2) dealing with critical points. We first compute a smooth harmonic scalar field over the input triangular mesh using the discretized Laplacian−Beltrami operator. We also compute its gradient field

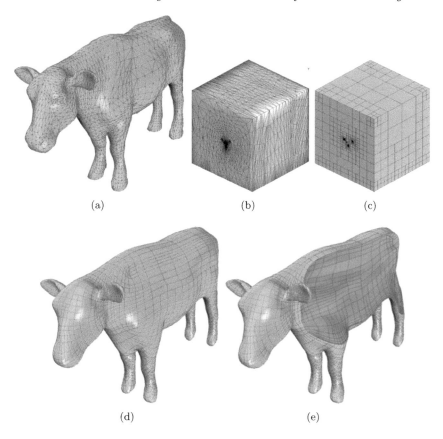

(a)   (b)   (c)

(d)   (e)

FIGURE 7.17: The cow model (pictures from [438]). (a) The input boundary triangle mesh; (b) the mapping result; (c) the subdivision result for the parametric domain; (d) the extracted solid Bézier elements; and (e) some elements removed to show the interior mesh (blue) and one pillowed layer (magenta).

and its orthogonal vector field. After that, from the harmonic field we compute all the critical points by setting the first partial derivatives to be zero, including the minimal, maximal and saddle points. To build the polycube, we treat each type of the critical points differently. The polycube edges are defined by tracing along the gradient and isocontour directions. We then obtain the T-mesh from the constructed polycube, which is also the parametric domain for the trivariate solid T-spline construction.

Harmonic Field Calculation. A harmonic field can be computed by solving the Laplace equation with certain boundary conditions, defining a scalar or vector field over the domain. For a given triangulated surface, the discretized Laplacian–Beltrami operator is used to construct a linear system of algebraic equations with imposed boundary conditions. The harmonic field has the nice property of smoothness and free of extraneous critical points, therefore it has been applied to solving a number of geometry processing problems. For example, the integral lines were traced through the gradient and orthogonal vector fields of a harmonic field to remesh arbitrary manifolds with quadrilaterals [117]. A harmonic map was utilized for 3D geometric metamorphosis between two objects which are topologically equivalent to a sphere or a disk [192]. As generalized barycentric coordinates, harmonic coordinates were utilized in volume deformation [187].

Here we use the harmonic field to help build polycubes for high-genus objects. Given an input mesh $T \subset \Re^3$, we tend to extract the topological structure of the object by computing a harmonic function $f : T \to \Re$, such that $\triangle f = 0$ with pre-defined minimum and maximum values S_{min} and S_{max} at certain vertices. Basically, computing such a harmonic function yields a scalar field with each vertex in T assigned a scalar value. For a triangle surface mesh, we use the discretization format of the Laplacian−Beltrami operator [138] to construct a linear system, which is then solved to yield the scalar function f.

After that, we compute the gradient $g_1 = \nabla f$ and its orthogonal vector field g_2 (or the isocontour field) from the scalar field f. Due to the nice properties of the harmonic scalar field, the resulting direction fields are smooth and free of extraneous critical points. For one triangle (V_i, V_j, V_k) on the surface with the unit normal vector $\vec{n}$, we obtain the gradient vector $g_1 = \nabla f$ by solving the linear system [117],

$$
\begin{bmatrix} V_j - V_i \\ V_k - V_j \\ \vec{n} \end{bmatrix} [g_1] = \begin{bmatrix} f_j - f_i \\ f_k - f_j \\ 0 \end{bmatrix}. \tag{7.32}
$$

The orthogonal vector field g_2 follows the isocontour directions of f, which can be obtained easily via isocontouring for any given isovalue. With two orthogonal vector fields g_1 and g_2 obtained, we then trace along the flow lines, which are piecewise linear curves defined over the triangular mesh following one of the vector fields. We need to handle two cases in tracing the gradient flow: the regular case and the edge case. Note that we may need to add new vertices to advance the flow lines.

Based on the obtained scalar field, we also compute all the critical points of f by setting its partial derivatives to be zero. In addition to the minimal and maximal points, we obtain two kinds of saddle points: the splitting saddle points and the merging saddle points.

Dealing with Critical Points. To build polycube, we need to deal with various critical points (the minimal, maximal, splitting saddle and merging saddle points) and use them to segment the surface into polycube structures. We first compute all the isocontours interpolating critical points. Suppose f_i is the isovalue of L_i and C_i is the corresponding isocontour. If level L_i

interpolates a saddle point, we compute two sets of isocontours, C_i^- and C_i^+, using the isovalue f_i with a small perturbation δ. For the minimal or the maximal level, we only need to compute one isocontour.

Supposing level L_i interpolates a minimal point, we choose four seed points on each closed curve, which correspond to the four lower corners of a cube C_i with the unit parametric length. For other vertices lying on this isocontour, their parametric values can be computed via the chord-length parameterization. Starting from these seed points, we trace the gradient flow line until it meets the isocontour C_{i+1}^- of the next level L_{i+1}. The four intersection vertices correspond to the four upper corners of the cube C_i. The four traced curves are then mapped onto the four vertical edges of the cube. After that, we advance the polycube construction process to the next level.

For each splitting saddle point, we find one isoparametric line to split one cube into two. Similarly, for each merging saddle point, we use the same procedure to find one isoparametric line to merge two neighboring cubes and ensure they match with each other seamlessly. By dealing with all the critical points, we construct the polycube level by level. Note that the isocontours are always mapped onto the horizontal isoparametric edges of the polycube and the gradient flow lines are mapped onto the vertical edges.

After we obtain the polycube, we follow the same manner in [438] to build the valid T-mesh, including parametric mapping, adaptive octree subdivision and projection, pillowing and quality improvement, and using templates to handle extraordinary and partial extraordinary nodes. Note that T-junctions are allowed in the shared faces between two neighboring cubes. Finally, we infer the local knot vectors for each node, build the rational trivariate solid T-spline from the T-mesh and extract embedded Bézier elements. Figure 7.18 shows the solid T-spline results of a genus-1 kitten model.

7.3.3 Polycube Construction Using Boolean Operations

Given CAD models with complicated geometry and topology, how to automatically and robustly split complex geometry into different components and transfer the input geometric information to the desired volumetric models is still an open problem. It is well known that constructive solid geometry Boolean operations have been widely used in CAD [17, 353]. Inspired by this, in this section we talk about an approach [245] for polycube construction using two Boolean operations: union and difference. In particular, we first compute a harmonic field on the given surface and use it together with the boundary information to partition the domain. After that, two primitives (cube and torus) and Boolean operations are applied to generate polycubes, which are then deformed to align with the input boundary representation to create volumetric T-splines via parametric mapping.

As shown in Figure 7.19, our algorithm consists of three main stages to construct a trivariate solid T-spline from the given CAD model, including

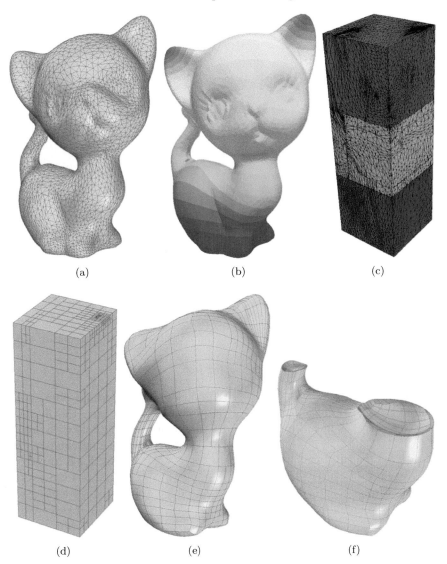

FIGURE 7.18: The kitten model (pictures from [391]). (a) The input boundary triangle mesh; (b) the harmonic field; (c) the mapping result; (d) the subdivision result for the parametric domain; (e) the extracted solid Bézier elements; and (f) some elements are removed to show the interior mesh (blue) and one pillowed layer (magenta).

curve extraction, domain decomposition and Boolean operations, as well as volumetric T-spline construction. First, we initialize all the necessary boundary information, classify the curve information from the CAD model into two

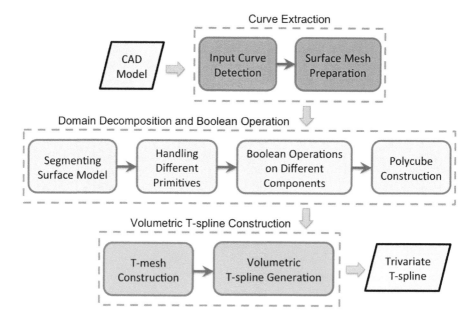

FIGURE 7.19: Algorithm overview of volumetric T-spline construction using Boolean operations.

groups, and then generate the surface mesh. Second, we perform domain decomposition and Boolean operations to generate polycubes using the obtained curve information and surface mesh. We compute a harmonic field with proper boundary conditions and then partition the surface into different components, which are topologically equivalent to either a cube or a torus. Each torus consists of four consecutive cubes with the two end faces merged to form a loop. All cubes from domain decomposition are then combined together using the union operation and holes (represented topologically as cubes) are created by the difference or subtraction operation. The resulting configuration gives us a polycube, and the input CAD surface is then mapped onto the polycube boundary. Finally, we perform an octree subdivision on the polycube to create volumetric T-splines. Here, we adopt a separate octree for each cube and require two adjacent cubes to have the same parameterization at the shared boundary. We also preserve all the detected sharp features in this step. After that, pillowing, smoothing and optimization techniques are used to improve the quality of the T-mesh. To generate a gap-free T-mesh, we apply templates [393, 394] to each extraordinary node and partial extraordinary node in the T-mesh. Finally, we obtain volumetric T-splines and extract Bézier elements for isogeometric analysis applications. In the following, let us take a look at each stage in detail.

Curve Extraction. Sharp edges or features are very common in most

CAD models, and we need to detect these features and map them to edges of the polycube. Since we have two Boolean operations, we also need to distinguish which curves are best represented as polycube edges during the union operation (we call them *feature curves*), and which curves are represented as polycube edges during the difference operation (we call them *difference curves*).

The input boundary information can be classified into three groups: corners, curves and patches. All the surface models can be represented by these three groups. Curves are a sequence of parametric boundary line segments on the surface. Corners are the intersection points of two or more curves, which correspond to the corners of the polycube. In the designed CAD models, we also need to detect sharp features such as sharp curves and sharp corners. The surface continuity across sharp curves is C^0, and the sharp corners are the intersection points of two or more sharp curves.

Domain Decomposition and Boolean Operations. For simple CAD models, the detected feature curves can be directly used to create polycube edges and we can use the difference curves to define virtual components. Created by the CSG difference operation in design, a *virtual component* such as a hole does not exist in the real model, but we can deduce it from the design and the given boundary information.

Instead of partitioning surfaces using pants decomposition [164] with a strong rely on the topology of the models, here we adopt the harmonic fields to segment a complex geometry into coherent regions [391]. For example, we can solve the Laplace equation to obtain the steady-state temperature distribution over surfaces, which is a harmonic field. Generally, the boundary conditions are set by assigning high and low temperature values to two different vertices on the model.

In computer aided design and geometric modeling, we need to define primitives for Boolean operations. Cuboids, cylinders, prisms, pyramids, spheres and cones are commonly used primitives in CSG. In our polycube construction, here we only use two primitives: the cube and the torus. Unlike conventional CSG, we use our primitives in a topological sense. Figure 7.20 shows these two primitives mapping from the physical space to the parametric space. A torus primitive is represented by four consecutive unit cubes, with the first face and the last face (the two yellow faces) merged to form a loop.

In our polycube generation, we adopt two basic Boolean operations: union and difference. We also investigate various means for the union of two cubes, the difference of two cubes, the union of a cube and a torus, as well as the difference of a cube and a torus. For example, due to different sizes and relative position, we have multiple cases to handle the union and difference operations between two cubes. For the difference operation, we build virtual components using the detected difference curves. After building T-meshes, we delete elements inside the filled holes by using the difference or subtraction operation.

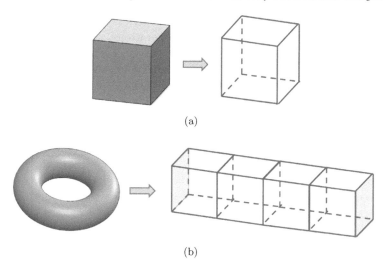

(a)

(b)

FIGURE 7.20: Two primitives mapping from the physical space to the parametric space. (a) Cube; and (b) torus.

Volumetric T-Spline Construction. The T-spline control mesh or T-mesh generation is a key step in constructing volumetric T-splines. There are five main steps in this stage, including adaptive octree subdivision and mapping, sharp feature preservation, pillowing and quality improvement, handling extraordinary and partial extraordinary nodes, trivariate T-spline construction and Bézier extraction. After applying the templates, we obtain C^0-continuity around the extraordinary and partial extraordinary nodes. For each irregular node on the surface, we relocate the Bézier control nodes in its two-ring neighborhood via applying a constrained optimization [331], yielding a continuity elevation from C^0 to G^1 for its local vicinity.

Figure 7.21 shows the resulting volumetric T-splines of a rod model using Boolean operations. A harmonic field was computed to split the torus or hole region. For the other regions, the shortest distance among the corners was traced to segment the model. In addition to the union of cubes, we also applied the torus primitive and the difference operation, and obtained high-quality elements with good surface continuity.

7.3.4 Polycube Construction Using Skeletons

Given a 3D object, its topology can be represented by 1D skeletons or medial axes. They contain important geometrical information for volumetric parameterization and we can use them to build polycubes. In this section, we describe an approach [246] of taking the skeleton as a guide for polycube construction, and then building T-meshes following the geometry topology

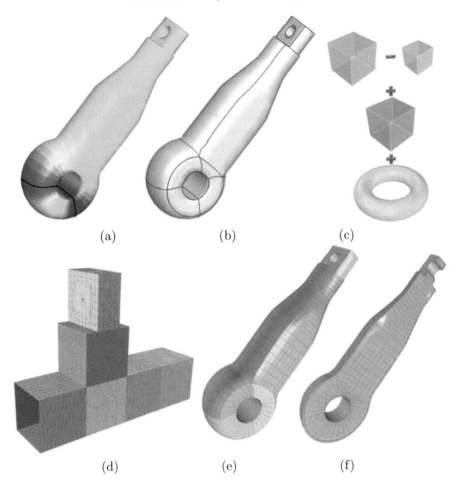

(a) (b) (c)

(d) (e) (f)

FIGURE 7.21: Rod model (pictures from [245]). (a) One temperature field to segment the bottom torus region; (b) the segmentation result; (c) Boolean operations; (d) parametric mapping result (the torus primitive is used in the bottom component, and the difference operation is used to create the small hole in the top component); (e) solid T-spline with Bézier elements; (f) some elements are removed to show the interior of (e).

with detailed features preserved. In the resulting polycubes, we enforce each cube has at most one patch on the boundary, providing the necessary condition to yield good-quality elements. In this way, we can easily obtain the singularity graph of the constructed T-mesh, which follows the edges in the interior cubes. As a result, we obtain a nice singularity distribution in the constructed solid T-splines, having only a few singular points and with no any singular edges lying on the surface. In addition, we classify all the surface features into three groups: open curves, closed curves and singular features. Various schemes are developed to preserve them. In the following, let us take a look at the detailed algorithm.

Skeleton-Based Polycube Construction. Starting from surface representations, there are many algorithms developed in the literature to extract skeletons, including mesh contraction [31], mean curvature flow [365] and the generalized sweeping method [267]. Here, we adopt the mean curvature flow method given in [365] to create the skeleton, and then split it into different branches. After that, we define a B-spline curve for each branch, and calculate the tangent direction at each point on the curve. We may need some user interactions to clean the skeleton in this step, such as deleting small branches, merging adjacent bifurcations to trifurcations, or making the local branching region coplanar.

In the next step, we construct a generalized cube using six planar or curved Coons patches for each skeleton branch [100]. To connect these generalized cubes from different branches together at the bifurcation or trifurcation, we split the cube patches into half-planes and combine them together following the templates given in [430]. We may introduce singularities to the polycubes along the shared edges of the half-planes. After we connect all the interior cubes properly, we enlarge each cross-section of these interior cubes to adapt to the input surface, and then create boundary cubes by projecting the patches of the interior cubes onto the surface.

We now partition the model into different subdomains using the interior and boundary cubes. It is obvious that such domain decomposition result follows the skeleton of the input model, and thus we obtain a T-mesh following the topology of the geometry. This method allows us to flexibly change the orientation or the number of cubes by simply modifying the skeleton at the beginning. To yield better parameterization results, we can also optimize the location of the projected cube corners on the surface.

In the constructed polycubes, we have some singular edges that can be easily detected. For example, an interior cube edge is a singular edge if it is shared by other than four cubes. All the control nodes lying on a singular edge are singular nodes. The singular graph of the T-mesh is the graph containing all the singular edges or connecting all the singular nodes. Note that the singular graph of a hexahedral mesh should not start or end in the interior of the volume [278, 303], in other words, it can only start or end on the boundary. After building the polycube, we obtain the fixed singular graph, from which we can easily predict the positions of singular points generated from octree subdivision. Figure 7.22 shows the resulting polycubes of the bunny model using skeletons as the guide.

Feature Preservation. Surface features play an important role in representing the surface details, such as smooth curves, sharp curves and singular points. In our algorithm, we perform feature preservation during the T-mesh construction process. For each cube, we map it onto a unit cube

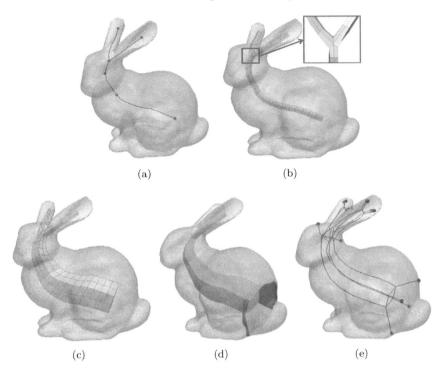

(a) (b)

(c) (d) (e)

FIGURE 7.22: The bunny model (pictures from [246]). (a) Skeleton splitting results; (b) generating interior cubes by shifting the skeleton branches; (c) updated interior cubes by enlarging the cross sections and smoothing; (d) four connecting patches of the bunny model after optimization (blue and green patches); and (e) its singular graph with red dots representing singular points on the surface.

in the parametric space and apply a strongly-balanced octree subdivision to build the T-mesh [391, 438]. The resulting T-mesh contains all the topological information represented in skeletons. Different from the previous research on skeleton-based volumetric composition and structured grid generation [103], our T-mesh generation method allows T-junctions and we do not have any singular edge lying on the surface. We classify all the surface features into three different groups: open curves, closed curves and singularity points, and then handle them using different approaches.

For open curves, we use two different methods to preserve them. One is parametric mapping and the other one is volumetric parameterization from a 3D frame field. The former method directly aligns the open curve to a certain parametric line, which may generate distorted T-mesh elements even with smoothing performed. The latter method can produce a better result with a smooth and gradual transition. A volume parameterization of geometry V from a frame field is basically an atlas of maps $f : V \to \Re^3, p \mapsto (u, v, w)^T$, where f is a piecewise linear field in each input tetrahedral element. The integer grids in $\Re^3$ can yield a hex tessellation of the entire geometry. The volume parameterization from the frame field [278] is performed by solving a minimization problem

$$min \sum_t vol \cdot D_t, \qquad (7.33)$$

where vol is the volume of a tetrahedron,

$$D_t = ||c\nabla f(u) - U_t||^2 + ||c\nabla f(v) - V_t||^2 + ||c\nabla f(w) - W_t||^2, \qquad (7.34)$$

c is the length scale of parameterization and $\{U_t, V_t, W_t\}$ are the initialized frame field.

The domain decomposition and polycube construction result depend on the extracted skeleton. In other words, we can change the ways of domain splitting and the design of polycube by modifying the skeleton. Note that we project each patch of the interior cubes onto the surface to create boundary cubes, therefore for a closed curve on the surface, we can define its enclosed region as the boundary patch and then generate a boundary cube. That is, we should add a new branch to the skeleton in order to build a boundary cube for the closed curve and preserve such a surface feature.

In addition, we design some templates to preserve surface singularities without changing the polycube, following the property of singularity distribution in the hex T-meshes. As stated in [278], the singular graph should not start or end in the interior of the hex meshes. Therefore, we need to provide a singular edge path connecting the desired surface singularity to the existing singular graph in the interior when we design templates. Figure 7.23 shows three templates [246] to insert surface singularities, which can be applied to boundary cubes or elements containing the desired singularities. With these templates, we split the cube or element into smaller elements and introduce new singularities on the surface and also in the interior.

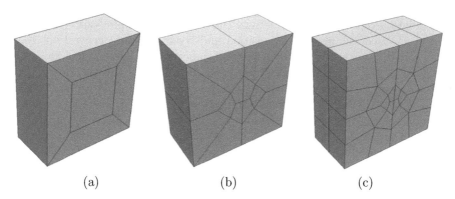

(a)   (b)   (c)

FIGURE 7.23: Three templates designed to modify singularities.

With the constructed T-mesh, we build rational volumetric T-splines. The rational T-spline basis functions satisfy the property of partition of unity by definition, which makes it suitable for geometric design and isogeometric analysis. We apply templates to deal with irregular nodes in the T-mesh, to enable a valid gap-free solid T-spline construction. From the valid T-mesh, we extract the local knot vectors and build rational solid T-splines. We use the Oslo knot insertion algorithm [259] to calculate the transformation matrix from rational T-spline basis functions to Bézier basis functions, which is then used to extract Bézier elements from T-splines for isogeometric analysis applications.

Compared to the result in [391], we have preserved detailed feature information in the kitten model as shown in Figure 7.24. For example, we preserved the feature lines of the mouth by aligning them to certain parametric lines. For the two eyes, we used two different templates to preserve the circular features. The singularity information in the left eye region is constrained in that local area, while in the right eye region it is propagated outward a bit.

7.3.5 Conformal T-Spline Modeling for CAD Models

As we all know, the designed CAD models are usually represented in splines such as NURBS or T-splines. To integrate design with analysis, the root idea of isogeometric analysis [42, 104, 179], it is critical to automatically create a conformal solid spline model with the given boundary spline representations preserved exactly.

Developing a general methodology for conformal solid T-spline construction from boundary NURBS or T-spline representations is very challenging because the input boundary representation may have very complicated geometry and topology, the T-meshes have complex connectivity, and there are various types of surface parameterizations. It is very hard to preserve the conformal property when converting a bivariate surface model to a trivariate solid model. In this section, we only consider geometry with genus-zero topology, and require the input NURBS or T-spline surface is topologically equivalent to a cube, containing eight extraordinary points only. We aim to construct rational solid T-splines, with the input boundary NURBS or T-spline representation preserved exactly [439]. We first adopt a cube as the parametric domain, and apply an adaptive subdivision to the cube. Then we project the boundary nodes from the subdivision onto the input surface via surface parameterization. To preserve the input boundary representation, we create two boundary layers between the input boundary surface and the boundary of the subdivision result. Templates are applied to irregular nodes to create a valid gap-free T-mesh. Finally, we build rational solid T-splines and extract Bézier elements.

In our conformal solid T-spline construction algorithm, we require the input surface should form a genus-zero object and a unit cube is adopted as the parametric domain of the solid T-spline, with eight valence-3 corner nodes,

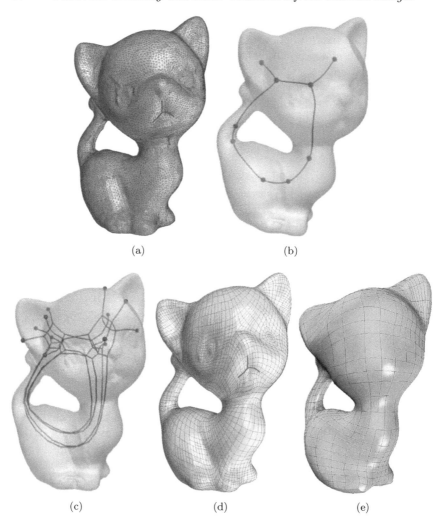

FIGURE 7.24: The kitten model (pictures from [246]). (a) The input surface triangle mesh; (b) skeleton splitting result; (c) the singular graph; (d) the extracted solid Bézier elements with detailed feature preservation; and (e) the extracted solid Bézier elements from [391].

twelve edges and six faces. We insert two boundary layers to connect the subdivision result with the input boundary surface, enabling a conformal boundary for the resulting solid T-spline. Figure 7.25 illustrates the construction process of two boundary layers. To improve the T-mesh quality, we use the smoothing and optimization techniques to relocate the interior nodes.

Figure 7.26 shows the result after inserting two boundary layers for a mouse model. In (c), the subdivision result is marked in pink, the first boundary layer

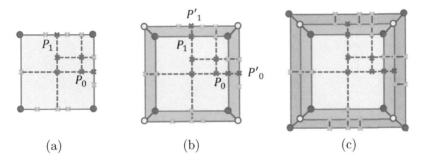

FIGURE 7.25: The boundary layer construction for a 2D example. (a) The subdivision result; (b) the first boundary layer for the subdivision boundary; and (c) the second boundary layer for the input boundary surface.

for the subdivision result is marked in blue, and the second boundary layer around the boundary is marked in yellow. These two boundary layers serve as a transition region to connect the input boundary spline surface with the subdivision result. It is easy to observe that the extraordinary nodes are inside the volume with two layers of elements away from the boundary. Some T-junctions are generated between the boundary layers for the transition purpose, with the input boundary representation preserved exactly. In other words, the resulting solid T-spline is conformal to the input NURBS or T-spline surface parameterization. The presented method can be extended to high-genus objects using polycubes.

7.4 Eigenfunction-Based Surface Parameterization

In Section 7.3, we discussed how to preserve topological features of the input objects during polycube construction using Boolean operations [245] and skeletons [246]. In addition to topological features, it is also important to preserve geometric features on the surface. In this section, we will discuss how to align parametric lines with major structure features represented by eigenfunctions [241].

Surface parameterization has been used in many geometric modeling applications, including quadrilateral meshing [56], texture mapping and synthesis [225, 388]. In surface parameterization, it is important to align parametric lines with surface feature directions like curvatures. Simplification with a good user control was used to build coarse domain meshes [266, 299]. People also used the harmonic field to capture features [209, 374], but feature alignment is limited because it is difficult to generate a good field and place singularities

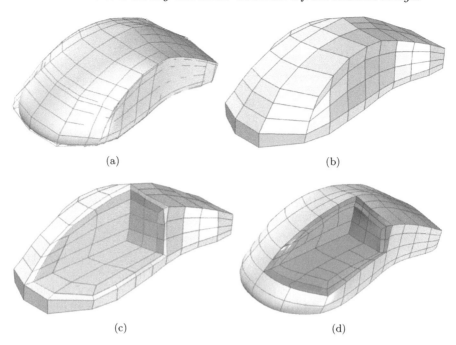

(a)

(b)

(c)

(d)

FIGURE 7.26: The mouse model (pictures from [439]). (a) The input T-spline surface; (b) the T-mesh after constructing two boundary layers; (c) the interior of the T-mesh with two layers; and (d) the extracted solid Bézier elements with some elements removed to show the interior structure.

in proper locations. Recently, the cross field-based methods were developed to capture features using the principal curvatures [57, 191, 209, 309].

As we all know, eigenfunctions [224, 311, 349] reflect the shape behavior and structural features at different scales. They vary along the object surface and are invariant to different poses or deformation, which make them superior for many applications such as surface segmentation and reconstruction [311], shape matching [224], surface quadrangulation or parameterization [116, 176, 443], and pose-invariant Reeb graph [349].

In this section, we describe an approach [241] to define a guidance for cross field generation using eigenfunctions, and generate a structure-aligned surface parameterization for the input triangular mesh. Given a smooth surface, we compute the eigenfunctions of the Laplacian−Beltrami operator (LBO) by solving $-\triangle f = \lambda f$ using the cotangent scheme [224], where $\triangle$ is the LBO and λ is the eigenvalue.

7.4.1 Guidance Estimation Using Eigenfunctions

Various eigenfunctions reflect surface features at different scales. Compared to high modes, the low-mode eigenfunctions tend to capture the major structure of the object and are less sensitive to the detailed surface features. In the following, we choose several low-mode eigenfunctions to design a direction guidance for building a cross field, from which we can then generate a structure-aligned surface parameterization.

We compute the gradient of the eigenfunctions and use them to represent the structural feature directions. Each eigenfunction plays a dominant role in certain regions of the surface, where its gradient reflects the structural features at a certain scale. Such region is called a *feature region* associated to that eigenfunction. By combining the gradient in the feature regions associated to several low-mode eigenfunctions, we can build a structure-aligned guidance and construct a cross field.

People usually use isocontours to represent the feature region. Instead, here we divide the surface into patches with each patch representing a feature region. Then, we define a positive characteristic value $C_{i,k}$ for each triangle T_i of the kth mode. For a set of mode indices K, triangle T_i is assigned to the kth mode if the kth mode is dominant on it or $k = arg\ max_{j \in K} C_{i,j}$, where $C_{i,j} = ||\nabla f_{i,j}||$. Triangles assigned to the same mode form a feature patch. Compared to the feature bands defined by the isocontours, this method yields separated feature bands with nonconsistent guidance directions avoided automatically. Figure 7.27(b) shows the obtained guidance directions for the hand model using the characteristic value.

7.4.2 Cross Field Generation and Surface Parameterization

With the guidance directions created from eigenfunctions, we compute a cross field via a smoothing process and also generate a surface parameterization using the mixed integer method [57]. Basically we solve two optimization problems. Let T^g denote the set of guidance triangles, and T^f denote the set of free triangles on the surface. We represent the guidance direction and the cross field in triangle T_i using (θ_i, e_i) and (γ_i, e_i), where θ_i and γ_i are a module of $\pi/2$, and e_i is the reference edge in triangle T_i. A smoothness energy [191] of the cross field is defined as

$$\Gamma^s = \sum_{e_{ij} \in E} \left(\theta_i + \kappa_{ij} + \frac{\pi}{2} p_{ij} - \theta_j \right)^2 + \bar{\lambda} \sum_{T_i \in T^g} (\theta_i - \gamma_i)^2, \qquad (7.35)$$

where e_{ij} is the edge shared by triangles T_i and T_j, E is the set of all the edges in the mesh, κ_{ij} is the angle between the reference edges of triangle T_i and T_j, and p_{ij} is the integer-valued period jump of the cross field across e_{ij}. A Lagrange multiplier $\bar{\lambda}$ is used to balance the alignment to the guidance directions and the field smoothness. We obtain a smooth cross field (U_i, V_i) by minimizing the energy function; see Figure 7.27(c).

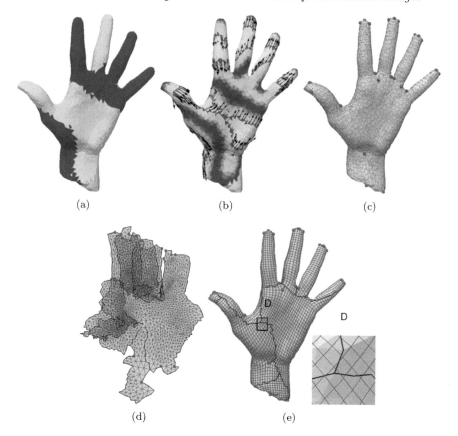

FIGURE 7.27: Define the guidance directions for the hand model using the characteristic value (pictures from [241]). (a) The feature patches for Modes 1 (green) and 2 (blue); (b) the Dijkstra distance distribution on the surface with guidance triangles (red triangles) and guidance directions (black arrows); (c) the constructed cross field; (d) the disk-like planar region from the original surface; and (e) the resulting parametric lines.

To obtain a surface parameterization, we cut the surface M into a disk-like region, and obtain the parametric coordinates (u_i, v_i) for each vertex by minimizing an orientation energy

$$\Gamma^o = \sum_{T_i \in M} A_i \cdot \left(||h\nabla_T u_i - U_i||^2 + ||h\nabla_T v_i - V_i||^2 \right), \qquad (7.36)$$

where A_i is the area of triangle T_i, h is a parameter controlling the spacing of the resulting parametric lines, and $\nabla_T u_i$ and $\nabla_T v_i$ are two gradients. We set integer constraints on the parametric coordinates for the other vertices on the planar region boundary to make the integer-value parametric lines meet

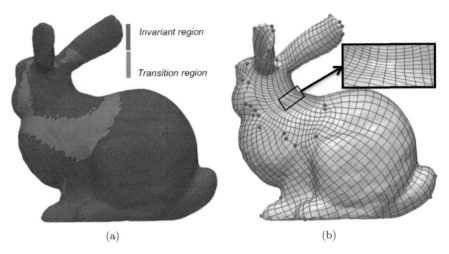

FIGURE 7.28: Anisotropic parameterization of the bunny model using Mode 1 eigenfunction (pictures from [241]). (a) The invariant and transition regions; and (b) the anisotropic parameterization result.

seamlessly at the boundary. As a result, we obtain a quadrangulation of the surface directly from the parametric lines; see Figure 7.27(d, e).

The resulting parametric lines follow the gradient directions of the cross field, and their spacings are determined by the parameter h. A smaller h yields denser parametric lines, and each parametric line corresponds to the integer value of the parametric coordinate u or v. Uniform parametric lines can be obtained when a constant h is used, while a non-uniform h tends to yield adaptive and anisotropic parameterization with various local spacing of the parametric lines. We define the invariant and transition regions and use them to adapt the original cross field to the spacing variation of the parametric lines; see Figure 7.28. Note that we may introduce new extraordinary nodes in the transition region.

7.5 Truncated Hierarchical Catmull–Clark Subdivision Modeling

Local refinement is an important issue for efficient computation in isogeometric analysis [104, 179]. In recent years, several techniques have been developed to break the regular tensor product structure of NURBS and support local refinement, including T-splines [332, 334], PHT-splines [110], hierarchical B-splines [59, 380] and locally refined splines [115, 185].

As an extension of NURBS, T-splines support local refinement with T-junctions and have been widely used in geometric modeling and isogeometric analysis [42, 329]. In the initial definition of T-splines, linear independence [66, 235] of the basis functions was not guaranteed, which is an important prerequisite for isogeometric analysis. Analysis-suitable T-splines [330] were then investigated to emphasize this property along with several others. Hierarchical B-splines [141] also support local refinement, and a global selection mechanism [210] can be used to select linearly independent basis functions from different B-spline hierarchical levels. To satisfy partition of unity and to decrease the overlapping of basis functions at different levels for better numerical conditioning, a truncation mechanism [153] was developed recently for hierarchical B-splines. However, hierarchical B-splines are restricted to a global rectangular parametric domain, and they cannot handle complex topology with extraordinary nodes.

On the other hand, subdivision surfaces are superior in handling extraordinary nodes [77, 289, 310]. As surveyed by Zorin [448] and Sabin [316], the popular Catmull–Clark subdivision [79] is based on bicubic B-splines, and its explicit basis functions [357] were derived with coarse patches simply overlaid with fine ones. The linear independence property is not guaranteed. Using subdivision basis functions, Cirak *et al.* [96, 97] developed a unified framework to integrate geometric design and analysis for thin shell structures. In addition, Catmull–Clark subdivision solids were recently applied in isogeometric analysis [69]. Grinspun *et al.* [160] developed adaptive simulations with basis refinement rather than the traditional element refinement for subdivision surfaces. Due to the loss of partition of unity, the convex hull property cannot be guaranteed and basis refinement may produce ill-shaped control mesh that is not suitable for geometric design.

In this section, we present a new method named Truncated Hierarchical Catmull–Clark Subdivision (THCCS) [399], which incorporates extraordinary nodes and supports local refinement using a similar hierarchical structure as in truncated hierarchical B-splines. In other words, we introduce the truncation concept into Catmull–Clark subdivision basis functions to guarantee partition of unity, which is essential to the convex hull property in geometric design. In addition, the local selection mechanism is adopted to ensure the linear independence of basis functions. THCCS preserves the exact geometry during adaptive h-refinement and it inherits the continuity of Catmull–Clark subdivision surface, that is, C^2-continuity everywhere except C^1-continuity at extraordinary nodes. As a result, the THCCS basis functions satisfy several nice properties that are important for both geometric design and isogeometric analysis, including partition of unity, linear independence, convex hull at each hierarchical level, supporting local refinement and extraordinary nodes, as well as smooth surface continuity. In the following, let us briefly review truncated hierarchical B-splines, and then discuss how to construct THCCS.

7.5.1 A Brief Review of Truncated Hierarchical B-Splines

Developed by Giannelli *et al.* [153], the truncation mechanism aims at enabling partition of unity for hierarchical B-spline basis functions and reducing the overlapping of basis functions for better numerical conditioning. As shown in Figure 7.29, the construction of truncated hierarchical B-splines is similar to hierarchical B-splines, which consists of three steps. First, the to-be-refined basis functions at Level l are identified such as the green dashed curve in (a). Due to the refinability, each Level-l basis function can be represented by a summation of all its children basis functions at Level $(l + 1)$. Therefore, we find out all the children of the to-be-refined basis function and set them to be active; see the solid blue curves in (b), and also set all the other Level-$(l + 1)$ basis functions to be passive. From Level-l basis functions, we identify those basis functions that have any shared children with the to-be-refined basis function like the four red curves in (a), and then truncate them using

$$trunN_i^l = \sum_{passive\ children} C_{ij} N_j^{l+1}, \tag{7.37}$$

where C_{ij} are the refinement coefficients from midknot insertions. In the end, we collect all the active basis functions to form the truncated hierarchical B-spline basis functions as shown in Figure 7.29(c).

Note that truncated hierarchical B-splines are limited to a rectangular parametric domain only, and they do not work for complex topology with extraordinary nodes. In the following, we extend this truncation idea to work for Catmull−Clark subdivision basis functions.

7.5.2 Construction of THCCS

We couple the truncation mechanism with the refinability relationship of Catmull−Clark basis functions to develop the THCCS. Given an unstructured quadrilateral control mesh, the regular regions are actually the same as truncated hierarchical B-splines. Here we mainly discuss how to construct the THCCS for unstructured control meshes with extraordinary nodes, which also consists of three main steps: identification and truncation (Step 1), refinement (Step 2) and collection (Step 3).

1. Identification and Truncation − In this step, we identify all the to-be-refined basis functions and elements, define active and passive basis functions and construct truncated basis functions;

2. Refinement − We subdivide the identified elements and define active basis functions at the finer level; and

3. Collection − We collect all the active basis functions and elements to obtain the THCCS basis functions and control elements.

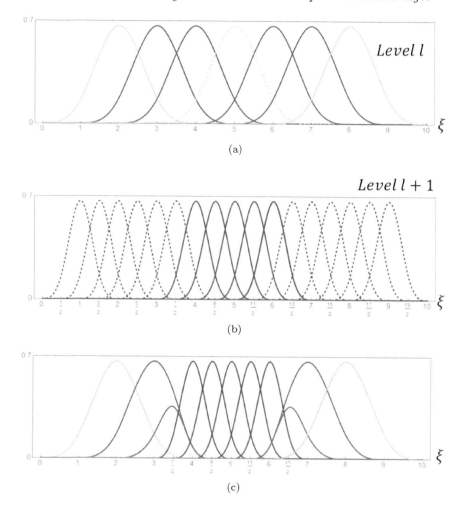

FIGURE 7.29: Truncated hierarchical B-splines. (a) Level-l basis functions and the green dashed curve represents the to-be-refined basis function; (b) Level-$(l+1)$ basis functions; and (c) four red basis functions are truncated.

Figure 7.30 shows an example of THCCS construction for a local region around a valence-3 extraordinary node using the above three steps. In (a), the red dot associates to the to-be-refined basis function, and we tend to compute the truncated basis function associated to the green dot, whose children basis functions are shown in (b) in green and blue dots. All the green children lie within the support area of the to-be-refined basis function, so we call them active children and each blue dot associates to a passive child. All the passive children are used to construct the truncated basis function at the green dot

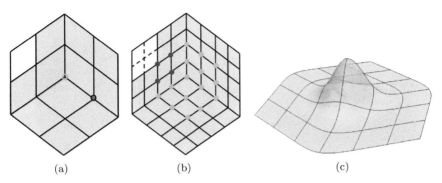

(a)   (b)   (c)

FIGURE 7.30: An example of THCCS construction around a valence-3 extraordinary node. (a) Level-l basis functions with the to-be-refined basis function (the red dot) and the to-be-truncated basis function (the green dot); (b) Level-$(l+1)$ basis functions, all the green dots are active children and all the blue dots are passive children; and (c) the truncated basis function compared with the nontruncated one (in transparent).

in (a). (c) shows a comparison between the truncated basis function with the nontruncated one (in transparent).

The THCCS basis functions have several very nice properties. The truncation mechanism ensures the property of partition of unity and convex hull, and the selection scheme ensures a global linear independence. In addition, local refinement does not change the original geometry, and the THCCS inherits the same surface continuity as Catmull–Clark subdivision, that is, C^2-continuous everywhere except C^1-continuous at extraordinary nodes. Furthermore, we also improved the efficiency of local refinement in the THCCS and developed the extended THCCS (eTHCCS) with a new basis-function-insertion scheme [400]. The Catmull–Clark subdivision basis functions are generalized to work for elements with more than one extraordinary node. Figure 7.31 shows the eTHCCS results of a bunny model and a human head model.

7.6 Weighted T-Spline and Trimmed Surfaces

Constructing spline models is important for isogeometric analysis to integrate design with analysis. In the literature, various schemes have been developed to construct surface and volumetric spline models. However, not all the T-splines are suitable for analysis, and the requirements to define an analysis-suitable T-spline was investigated in [330]. Among those requirements, partition of unity and linear independence are two important prerequisites for isogeometric analysis.

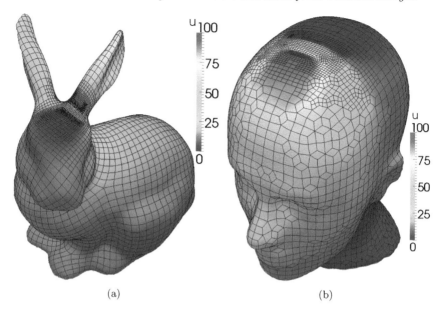

(a)   (b)

FIGURE 7.31: The constructed eTHCCS surface of a bunny model and a human head model. The color shows an isogeometric analysis result by solving the Laplace equation.

To ensure the constructed T-splines analysis-suitable, we can apply topological constraints to the control mesh, and require the T-junction extensions along different parametric directions not intersect with each other [330]. These constraints guarantee the defined T-spine basis functions satisfy partition of unity. As we discussed earlier for hierarchical B-splines [153] and subdivision surfaces [400, 399, 447], the partition of unity property can also be satisfied via a truncation mechanism. The truncated B-spline basis functions reduce the overlapping of basis functions at different hierarchical levels, but they cannot handle extraordinary nodes. Rational T-spline basis functions satisfy partition of unity for control meshes with arbitrary topology by definition [394], but they are no longer polynomials and therefore it is troublesome to evaluate their high order derivatives.

In the following, we describe a new type of T-spline named the weighted T-spline [248]. Instead of performing T-junction extensions, we calculate a new weight for each basis function in order to achieve the partition of unity property. We can also prove that the obtained weighted T-spline basis functions are linearly independent. In addition, the weighted T-splines can be used to reparameterize trimmed surfaces, handle extraordinary nodes and build volumetric models.

7.6.1 Weighted T-Spline

T-spline basis functions are defined based on local knot vectors satisfying the property of refinability. A T-spline basis function $N_i(\xi)$ can be represented by a linear combination of its children basis functions $N_{i,p}^c(\xi)$. By inserting k knots into the local knot vector, we have

$$N_i(\xi) = \sum_{p=1}^{k+1} c_{i,p} N_{i,p}^c(\xi), \tag{7.38}$$

where the positive $c_{i,p}$ are the refinement coefficients obtained from knot insertion. These basis functions satisfy partition of unity, and we have $\sum_{i=1}^{m} N_i(\xi, \eta) = 1$. In 2D, Eq. (7.38) can be rewritten as

$$N_i(\xi, \eta) = \sum_{k=1}^{n_c} R_{i,k} N_k^c(\xi, \eta), \tag{7.39}$$

where

$$R_{i,k} = \begin{cases} c_{i,p} & if \ N_k^c(\xi, \eta) = N_{i,p}^c(\xi, \eta) \\ 0 & otherwise \end{cases} \tag{7.40}$$

and n_c is the number of all children basis functions $N_k^c(\xi, \eta)$.

For the refinement of a basis function, if it is defined on uniform knot intervals, we bisect all the intervals. Otherwise, we refine it by inserting knots to make all the intervals equal to the existing highest level knot interval. Given the jth basis function $N_j^r(\xi, \eta)$ defined on the locally-refined T-mesh, refinability indicates

$$N_j^r(\xi, \eta) = \sum_q c_{j,q}^r N_{j,q}^c(\xi, \eta), \tag{7.41}$$

where $N_{j,q}^c(\xi, \eta)$ is the qth children basis function of $N_j^r(\xi, \eta)$. These basis functions do not satisfy partition of unity, and they can be rewritten as

$$N_j^r(\xi, \eta) = \sum_{k=1}^{n_c} R_{j,k}^r N_k^c(\xi, \eta), \tag{7.42}$$

where

$$R_{j,k}^r = \begin{cases} c_{j,q}^r & if \ N_k^c(\xi, \eta) = N_{j,q}^c(\xi, \eta), \\ 0 & otherwise. \end{cases} \tag{7.43}$$

We define the weighted T-spline basis function as

$$N_j^w(\xi, \eta) = \sum_q h_{j,q} N_{j,q}^c(\xi, \eta) = \sum_{k=1}^{n_c} s_k R_{j,k}^r N_k^c(\xi, \eta), \tag{7.44}$$

where $h_{j,q}$ is the weighting coefficient and s_k is a factor. To ensure the basis functions satisfy partition of unity, we require

$$
\begin{aligned}
0 &= \sum_{j=1}^{n} N_j^w(\xi,\eta) - 1 = \sum_{j=1}^{n} N_j^w(\xi,\eta) - \sum_{i=1}^{m} N_i(\xi,\eta) \\
&= \sum_{j=1}^{n} \sum_{k=1}^{n_c} s_k R_{j,k}^r N_k^c(\xi,\eta) - \sum_{i=1}^{m} \sum_{k=1}^{n_c} R_{i,k} N_k^c(\xi,\eta), \\
&= \sum_{k=1}^{n_c} N_k^c(\xi,\eta) \left(s_k \sum_{j=1}^{n} R_{j,k}^r - \sum_{i=1}^{m} R_{i,k} \right).
\end{aligned}
\tag{7.45}
$$

Then we can obtain

$$
s_k = \frac{\sum_{i=1}^{m} R_{i,k}}{\sum_{j=1}^{n} R_{j,k}^r}.
\tag{7.46}
$$

After obtaining the weighted T-spline basis functions, we compute control points in the physical space by solving a linear system.

Figure 7.32 shows a weighted T-spline surface patch. The weighted T-spline has several nice properties. For example, less topological constrain is needed, and T-junction extension is no longer necessary. Compared to standard T-splines, the weighted T-splines reduce the number of control points for the same level of refinement (about 25% to 30%).

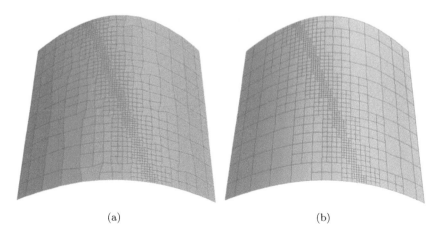

(a)   (b)

FIGURE 7.32: (a) T-mesh with four levels of refinement; and (b) the constructed weighted T-spline. (Pictures from [248].)

7.6.2 Trimmed Surfaces

Although most CAD models are represented by boundary NURBS surface patches, unfortunately they are not watertight which limits their usage in analysis. In CAD, many trimmed NURBS patches created by Boolean operations lose the tensor product structure, therefore we need to reparameterize them before applying them in simulations. Various schemes have been developed to create watertight T-splines and volumetric T-splines from T-spline surfaces [333, 439], while reparameterizing trimmed NURBS surfaces with T-splines still needs a lot of further investigation. The resulting trimmed surfaces should be analysis-suitable with bounded surface error.

We have developed an edge interval extension algorithm [248] to reparameterize the trimmed NURBS surface using weighted T-spline basis functions. The basic idea is to use the trimming curves as a guidance for the local refinement of the input NURBS surface patch, and then we group the resulting T-mesh elements into two categories: preserved elements and removed elements. As shown in Figure 7.33, the boundary of preserved elements like the red zip-zag edges in (a) is aligned to one parametric direction by extending edge intervals and modifying the surrounding T-mesh connectivity; see (c, d).

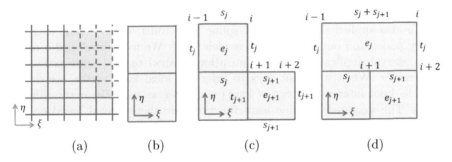

(a)   (b)   (c)   (d)

FIGURE 7.33: (a) Preserved elements (green) and removed elements (pink); (b) the first configuration of preserved elements that does not need a configuration modification; (c) the second configuration of preserved elements that needs a configuration modification; and (d) the connectivity modification result of (c).

Figure 7.34 shows a surface patch with two corners trimmed off. The trimming curve is preserved exactly, and the introduced surface error is bounded (<0.5%) within three-ring neighboring elements around the trimming curve. Compared to the standard T-spline, the weighted T-spline reduces the number of control points by 19% to 31%, and reduces the number of T-mesh elements by 14% to 33% in the reparameterization results.

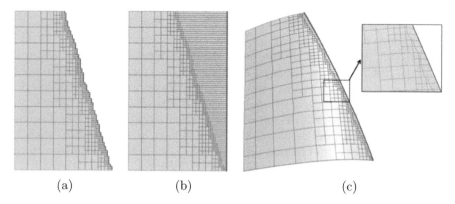

(a)   (b)   (c)

FIGURE 7.34: Reparameterizing trimmed surface with two corners trimmed off (pictures from [248]). (a) Deleting removed elements; (b) modifying the topology of preserved elements and extending edge intervals to align with the boundary of the rectangular parametric domain; and (c) reparameterized trimmed surface using the weighted T-spline basis functions.

7.6.3 Extraordinary Nodes and Volumetric Modeling

We also applied the weighted T-spline basis functions to handle extraordinary nodes and build volumetric models [247]. We have developed a novel knot interval duplication and optimization method to deal with extraordinary nodes. When we shoot rays to infer the local knot vector, we duplicate the knot interval (instead of knot) whenever we meet the extraordinary nodes. The weighted T-spline basis functions are then derived based on the obtained local knot vectors. With this method, the resulting T-spline surface is C^0-continuous across the edge shared by two first-ring neighboring Bézier elements, C^1-continuous across the edge shared by the first-ring and second-ring Bézier elements, and C^2-continuous everywhere else. We then perform degree elevation to compute the biquartic Bézier coefficients within the first-ring neighborhood. In addition, an optimization procedure is adopted to recompute these coefficients, yielding G^1-continuity for the first-ring Bézier elements.

Both the sweeping and parametric mapping methods can be used to construct volumetric weighted T-splines. Starting from a designed T-spline cross-section patch, we sweep it along one certain direction to create the T-spline volume. We support partial extraordinary nodes in the resulting T-mesh. In the parametric mapping method, we take a water-tight T-spline surface as the input to build a volumetric model. The input model is topologically equivalent to a unit cube and we do not introduce any new extraordinary nodes during the volumetric T-spline construction. Figure 7.35 shows two models: the wrench model was created using the sweeping method and the bolt model was created using parametric mapping. They were incorporated into Abaqus to perform mechanics study using isogeometric analysis.

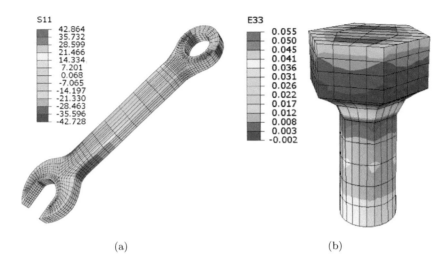

(a)   (b)

FIGURE 7.35: Two volumetric weighted T-spline models with extraordinary nodes. (a) A wrench model created using the sweeping method; and (b) a bolt model created using parametric mapping.

7.7 Incorporating T-Splines into Commercial Software

We have linked our T-spline modeling techniques with two commercial software packages, Rhinoceros 3D (Rhino) and SIMULIA Abaqus Unified FEA (Abaqus). In this section, we present a CAD-CAE software platform for T-spline based IGA [215]. Developing an IGA software from scratch is difficult and takes years of effort. It is well known that Rhino has been built specifically for design, and Abaqus has been built specifically for engineering analysis. Instead of building a standalone IGA design-through-analysis software, we create a T-spline based IGA software platform built upon Rhino and Abaqus to allow users to benefit from these two software packages.

As shown in Figure 7.36, this integrated software framework has three primary steps: creating CAD surface in Rhino with T-spline Plugin, converting surface T-spline representations into volumetric T-spline, and performing IGA of 2D T-spline surface or 3D T-spline volume in Abaqus through its User Element Subroutine (UEL/UELMAT). Unique aspects of this platform include:

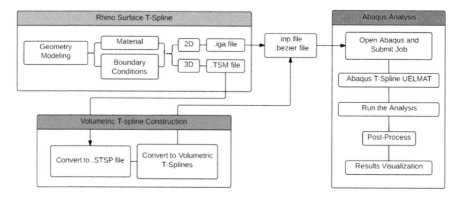

FIGURE 7.36: A software pipeline overview from Rhino 3D to Abaqus.

- Boundary value problem (BVP) definition interface for T-spline geometry;

- Surface-to-volume T-spline conversion;

- Efficient and compact trivariate T-spline data structure; and

- Abaqus UEL subroutine for T-spline based IGA based on Bézier extraction.

The framework contains two self-developed plugins, the Rhino plugin in the gray and blue blocks and the Abaqus plugin in the red block. We first build surface geometry in Rhino. With the Rhino plugin, we specify the boundary value problem and preprocess the geometry to generate Abaqus input files. When a 3D geometry is needed, users can construct volumetric T-splines. For a 2D problem, we directly convert a Rhino .iga file into the Abaqus .inp and .bezier files. For a 3D problem, we save .TSM files and convert it to our .STSP data structure for volumetric T-spline construction, which is later converted to the Abaqus .inp and .bezier file for analysis. Once the analysis is done through the Abaqus plugin, we call a postprocessing function in the Abaqus plugin to generate the .odb file for visualization. Based on the simulation results, the user can go back to Rhino to optimize the geometry. Figure 7.35 shows two simulation results from Abaqus using our framework.

Homework

Problem 7.1. Perform a survey for NURBS and T-spline modeling; discuss pros and cons for each technique. What are the challenges and what are the future directions?

Problem 7.2. What is an extraordinary node? Perform a survey on various techniques to handle it.

Problem 7.3. What are the requirements for analysis-suitable T-splines for isogeometric analysis? Perform a literature study and discuss why.

Problem 7.4. Explain the meaning of refinability of hierarchical B-splines. What important property can the truncation scheme ensure?

Chapter 8

Finite Element and Isogeometric Analysis Applications

The finite element method (FEM) has very broad applications in a lot of research areas, and isogeometric analysis (IGA) is a new advancement based on FEM to integrate design with analysis. In this chapter, we first review the basic algorithm of finite element analysis (FEA) and its new developments, including IGA, extended FEM and immersed FEM. Then we overview the broad application areas and explain how the five topics in Chapter 1 are used in various applications through three real projects: multiscale biomolecular modeling, patient-specific cardiovascular model construction, and applications in material sciences like critical feature determination of polycrystalline materials and bondline composites.

8.1   Introduction to Finite Element Method and Its New Developments

As a popular and powerful numerical method to solve partial differential equations over complex domains, FEM has been developed rapidly and used in many research areas including computational medicine, biology and engineering. This section starts with an introduction of FEM using a simple one-dimensional problem, and then talks about three new developments based on the traditional FEM: IGA, extended FEM (XFEM) and immersed FEM (IFEM).

8.1.1 Finite Element Method

The FEM is a general technique to solve boundary value problems (BVP) with uniformly and non-uniformly spaced grids or meshes [178, 282]. This method requires to discretize the domain of the solution into a finite number of simple subdomains called the finite elements, and then use the variational concepts to construct an approximation solution over the discretized elements. The domain discretization process is called *mesh generation*, which generally takes 80% of the total analysis effort in many engineering applications. Due to the generality and richness of the ideas underlying this method, FEM has been used with great success in solving complex problems in many research areas, including mechanical engineering, civil engineering, aerospace, biomedical engineering, material sciences, petroleum and so on.

Formulation of FEM is complicated, but in the end it will reduce to solving a set of linear algebraic equations. Let us first take a look at a general 1D "two-point" BVP characterized by a linear second-order ordinary differential equation (ODE) problem,

$$-\frac{d}{dx}\left(k(x)\frac{du}{dx}\right) + b(x)\frac{du}{dx} + c(x)u(x) = F(x) \tag{8.1}$$

with boundary conditions at $x = 0$ and $x = l$, where $k(x)$, $b(x)$, $c(x)$ and $F(x)$ are known. We want to solve for $u(x)$ over $0 \leq x \leq l$. We rewrite Eq. (8.1) in a compact notation and obtain

$$-(ku')' + bu' + cu = F. \tag{8.2}$$

When $k=1$, $b=0$ and $c=1$, the ODE becomes $-u'' + u = F$, which is the governing equation of the deflection of a string on an elastic foundation or the temperature distribution in a rod. In FEM, we assume the solution is $\hat{u}$, for which we would like to minimize the residual

$$r(\hat{u}) = -(k\hat{u}')' + b\hat{u}' + c\hat{u} - F. \tag{8.3}$$

Note that when $\hat{u} = u(x)$, we have $r(u(x)) = 0$. The requirement that a solution satisfies the differential equation at every point is too strong. To relax the requirement from this strong form, we use weak or variational formulation to reformulate the BVP to admit weaker conditions on the solution and its derivatives. We choose the weighting functions $w(x)$ (also called the *test functions*) such that $\int r(\hat{u}(x))w_i(x)dx = 0$ over the solution domain with $w(0) = w(l) = 0$, and then we have

$$\int_0^l w(x)\left(-(k\hat{u}')' + b\hat{u}' + c\hat{u} - F\right)dx = 0, \tag{8.4}$$

where the first term can be simplified by integration by parts:

$$-\int_0^l w(k\hat{u}')'dx = -wk\hat{u}'\big|_0^l + \int_0^l k\hat{u}'w'dx$$

$$= \int_0^l k\hat{u}'w'dx,$$

leading to the "weak" form

$$\int_0^l (k\hat{u}'w' + b\hat{u}'w + c\hat{u}w - Fw)dx = 0. \tag{8.5}$$

Given an infinite set of basis functions $\phi_i(x)$ in the space, both the test function and the trial function can be represented as a linear combination of these basis functions. The Galerkin method takes only a finite number of terms or N terms, and we have $\hat{u}(x) = \sum_{j=1}^N \alpha_j\phi_j(x)$ and $w(x) = \sum_{i=1}^N \beta_i\phi_i(x)$. Then, we plug them into the weak form and obtain

$$\int_0^l \left\{ k(x)\left[\sum_{j=1}^N \alpha_j\phi_j'(x)\right]\left[\sum_{i=1}^N \beta_i\phi_i'(x)\right] + b(x)\left[\sum_{j=1}^N \alpha_j\phi_j'(x)\right]\left[\sum_{i=1}^N \beta_i\phi_i(x)\right] \right.$$

$$\left. + c(x)\left[\sum_{j=1}^N \alpha_j\phi_j(x)\right]\left[\sum_{i=1}^N \beta_i\phi_i(x)\right] - F(x)\left[\sum_{i=1}^N \beta_i\phi_i(x)\right] \right\}dx = 0,$$

which can be reorganized into

$$\sum_{i=1}^N \beta_i\left\{\sum_{j=1}^N \alpha_j\left[\int_0^l k(x)\phi_j'(x)\phi_i'(x)dx + \int_0^l b(x)\phi_j'(x)\phi_i(x)dx\right.\right.$$

$$\left.\left. + \int_0^l c(x)\phi_j(x)\phi_i(x)dx\right] - \int_0^l F(x)\phi_i(x)dx\right\} = 0.$$

We can write the above equation into the compact form

$$\sum_{i=1}^N \beta_i\left(\sum_{j=1}^N k_{ij}\alpha_j - F_i\right) = 0, \tag{8.6}$$

where

$$k_{ij} = \int_0^l k(x)\phi_j'(x)\phi_i'(x)dx + \int_0^l b(x)\phi_j'(x)\phi_i(x)dx$$

$$+ \int_0^l c(x)\phi_j(x)\phi_i(x)dx, \qquad (8.7)$$

$$F_i = \int_0^l F(x)\phi_i(x)dx. \qquad (8.8)$$

We need this expression to be true for all the weighting functions and thus for all β_i, therefore we obtain

$$\sum_{j=1}^N k_{ij}\alpha_j - F_i = 0, \qquad (8.9)$$

which is a system of linear algebraic equations in α_j. Here α_j are referred to as the degrees of freedom in the approximation. Note that the stiffness matrix is symmetric, which helps reduce the computational cost. The spaces of trial and test functions coincide, therefore we only need to construct one set of basis functions.

Next, let us specify the basis functions $\phi_i(x)$. The basis function can be defined piecewise over finite elements, for example, it can be piecewise polynomials of low degree. First, we partition the solution domain into segments (mesh generation), and define one $\phi_i(x)$ function (the "hat" function) at each node,

$$\phi_i(x) = \begin{cases} \frac{x - x_{i-1}}{h_{i-1}} & x_{i-1} \le x \le x_i, \\ \frac{x_{i+1} - x}{h_i} & x_i \le x \le x_{i+1}, \\ 0 & otherwise. \end{cases} \qquad (8.10)$$

We will build the solution and the weighting function using these basis functions. We have $\hat{u}(x) = \sum_{i=1}^N \alpha_i\phi_i(x)$ and $\hat{u}(x_i) = \alpha_i$. There are three fundamental criteria to construct a set of basis functions:

- The basis functions are piecewise (or element by element) defined by simple low-degree functions over the finite element mesh;

- The basis functions and their derivatives are smooth enough to represent the test and trial functions; and

- The basis functions are chosen such that the parameters α_j define the approximation solution precisely at the nodal points.

In the implementation, the element stiffness matrix and element load vector are computed element by element, and then assembled together into the global stiffness matrix and global load vector. If we number nodes properly, the constructed stiffness matrix can be banded. Finally, the approximation solution can be interpreted and understood together with the specific application background. With analytical solution known, we can compare our approximation solution with it in terms of the energy norm, mean-square norm or maximum norm. People also use a-priori estimates and a-posteriori estimates for error analysis. The former requires no information from the actual finite element solution, which is known before the construction of solution. The latter uses the finite element solution to give more detailed estimates.

8.1.2 IGA, XFEM and IFEM

In recent years, there have been many developments and advances based on the traditional FEM, including IGA [104, 179], XFEM [106, 361, 362] and IFEM [250, 424]. Let us briefly review each of them as follows.

IGA. Invented by Dr. Thomas J.R. Hughes in The University of Texas at Austin in 2005, IGA has gained a rapid development in recent years with the exponential growth speed. The root idea of IGA is to integrate design with analysis by introducing the isoparametric concept into the FEM. It utilizes the same basis functions like NURBS, T-spline, subdivision and generalized B-spline basis functions to construct the geometry as well as the solution space. Inspired by CAD, IGA shares some common features with FEM and meshfree methods. It also has many unique advantages over them in solving problems of linear elasticity [113, 179, 331], fluid-structure interaction [43, 44], structure vibrations [105], shell analysis [46] and electromagnetics [67]. In addition to h-refinement, p-refinement and hp-refinement, IGA also supports a new local refinement strategy called the k-refinement. A detailed comparison between IGA and FEM can be found in [104, 179].

XFEM. The initial development of the XFEM took place in Northwestern University by Dr. Ted Belytschko and his co-workers in 2000 [106, 361, 362], which was developed to solve crack problems in fracture mechanics. Unlike the FEM, the XFEM does not require the finite element mesh to conform to the internal boundaries such as cracks, material interfaces and voids. In the XFEM, a single mesh is sufficient in capturing the evolution of material interfaces and cracks in 2D and 3D, which allows the modeling of crack growth without remeshing. The basic idea of the XFEM is to use the framework of partition of unity to enrich the standard FEM. It is a special case of the partition of unity FEM or the generalized FEM. A crack can be modeled by enriching the approximation by step functions and asymptotic near-tip fields. Multiple branched cracks, voids and cracks emanating from holes are also modeled by introducing corresponding enrichment functions. XFEM has shown overriding advantages over existing FE-based technology for modeling cracks [258], together with various mesh-free methods developed in recent years [83, 307, 409].

IFEM. In the 1970s, Peskin [295] developed the immersed boundary method to study flow patterns around heart valves, using Eulerian and Lagrangian descriptions for the fluid and solid domains, respectively. The fluid domain is represented by uniform background grids and solved by finite difference methods, while the submerged solid structure is represented by a fiber network. Nodal forces and velocities are coupled between these two domains. The fiber-like 1D immersed structure has one limitation: it may carry mass, but occupies no volume in the fluid domain. Following the poineering work, the IFEM [424] was developed by Professor Wing Kam Liu at Northwestern University, which uses finite element formulations for both fluid and solid domains and can handle non-uniform grids accurately. This method has been successfully applied to many biological systems [250]. Following the similar idea to handle boundaries, in recent years people developed finite cell methods [324], immersed boundary methods [323] and cartesian grid-based methods [264].

FEM and its variants have very broad applications. They all need quality geometric models and finite element meshes. In the following sections, we will talk about several application projects in computational biology, medicine and material sciences.

8.2   Multiscale Biomolecular Modeling

FEA has been applied broadly to study many biological and physical phenomena of biomolecular complexes such as simulating electrostatic potential distributions [149, 166, 410], diffusion-based reaction rate constants [89, 354, 355, 422] and calcium signaling [198]. Mesh generation is a key step in FEM for numerical computation. Many times, geometry only exists in a density map or scanned images, and researchers have to construct finite element meshes from these inputs to enable mechanics computation. For large biomolecular complexes, we need to deal with huge amounts of data and computation, which brings a great challenge for modeling, meshing and simulation. To efficiently represent complicated biomolecular complexes, a multiscale modeling method with a local resolution control is required, which can help reduce the mesh size and computational cost significantly. In addition, multicore CPU and GPU-based parallel computation provides us powerful means to accelerate the entire modeling process.

There are three important and commonly used biomolecular surfaces [101, 102], including the van der Waals surface (VdW), the solvent-accessible surface (SAS) and the solvent-excluded surface (SES) or sometimes called the

Lee-Richards surface. In the VdW, each atom is represented as a hard sphere and the biomolecular surface is defined as the envelope of the union of these hard spheres. The SAS and SES can be defined by rolling a probe ball around the biomolecule. The SAS is basically the trajectory of the probe center, while the topological boundary of the union of all possible probes is the SES. A lot of research has been conducted in approximating the SES, including the alpha-shapes [20, 124], the beta-shapes [201, 315], the MSMS [318], the advancing front and generalized Delaunay approaches [217], NURBS approximation [39] and PDE-based methods [175, 398, 442]. In addition, biomolecules can also be represented implicitly. The Gaussian kernel functions were applied in constructing density maps for the biomolecules [54, 159, 219, 421, 441], which sometimes were filtered in order to obtain a smooth biomolecular surface [154]. Efficient computation is critical for large biomolecular modeling [376, 389], and several algorithms have been developed to improve the modeling efficiency, such as the Fast Fourier Transform (FFT) [175, 442] and the programmable GPU [200, 360].

In this section, we present an efficient and parallel multiscale modeling framework [242, 441] for biomolecules based on a multilevel summation of Gaussian kernel functions. The entire modeling process contains two main steps: (1) efficient Gaussian density map computation using neighboring search coupled with the KD-tree structure and the bounding volume hierarchy (BVH); and (2) parallel tetrahedral mesh generation and quality improvement. Both the multicore CPU and GPU-based parallel computation techniques are adopted in these two steps. In the following, let us take a look at each step in detail.

8.2.1 Density Map Construction for Multiscale Modeling

In the PDB/PQR format files provided by the protein data bank [47], all the major atoms making up the molecule are given for protein and nucleic acid structures. From the PDB/PQR data [54, 159], the summation of kernel functions created at each atom can be used to construct a smooth volumetric electron density map, which is often sampled at each rectilinear grid point and the molecular surface is approximated as a level set.

Biomolecules usually have a nice hierarchical structure, ranging from the atomic scale to residual and chain scales. Such hierarchical structure can be used to enable a local resolution control for multiscale biomolecular modeling. The molecular surface is usually an isocontour of the density map, which is a summation of the Gaussian kernel function

$$G_k(x) = e^{\kappa\left(||x-x_i||^2 - r_i^2\right)}, \tag{8.11}$$

where κ is the decay rate, controlling how fast the Gaussian kernel function decays. x_i and r_i are the center and radius of the ith atom, respectively. A multilevel summation of Gaussian kernel functions was applied to control the resolution of biomolecule models [242, 441]. Lower level structures are classified into groups according to higher level structures. As the basic unit in the biomolecules, atoms are represented by $N_A = \{N_A(0), \cdots, N_A(n)\}$. $N_R^{(i)}$ ($i = 1, \cdots, n_R$) are the sets of residues, which are also subsets of N_A. The elements of $N_R = \{N_R^{(i)}\}_{i=1}^{n_R}$ are further grouped into chains or peptide subsets $N_C^{(i)}$ ($i = 1, \cdots, n_C$). In other words, the density distribution of a higher level structure can be obtained through the summation of lower level density. For small proteins, we can use a two-level summation (residue and chain) of Gaussian kernel functions to build their density map and obtain

$$G(x) = \sum_{k=1}^{n_C} \left(\sum_{j=1}^{n_R} \left(\sum_{i=1}^{n_A} G_k(x) \right)^{P_R} \right)^{P_C}, \tag{8.12}$$

where n_A, n_R and n_C are the number of atoms, residues and peptides, respectively. R_R and P_C are constant coefficients controlling the local resolution at the residual and peptide scales. For complex proteins, we can easily extend the above equation to a three-level summation (residue, chain and domain) and even more. To have a full control of the multiscale biomolecular models, the number of levels in the summation of Gaussian kernel functions should be the same as the number of hierarchical levels of biomolecular structures. Figure 8.1 shows one protein Ribosome 30S with one chain (Chain B) modeled at three different resolutions.

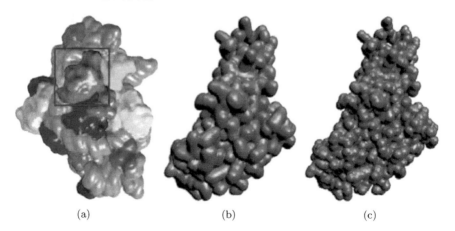

(a)   (b)   (c)

FIGURE 8.1: Multiscale models of Ribosome 30S (pictures from [441]). (a) Low level with the blue box showing one chain (Chain B); (b) Chain B at the residue level; and (c) Chain B at the atomic level.

During density map construction, for each grid point we need to compute the Gaussian density considering the contribution from all the atoms in the biomolecule. Given M atoms and N grid points, the time complexity is

$O(MN)$. Biomolecular surface can be extracted from the constructed density map as an isosurface, therefore we only need to compute the density for grid points around the biomolecular surface. To improve the computational efficiency, we introduce lower and upper thresholds and apply the neighboring search algorithm [29] to reduce the number of to-be-analyzed grid points. The time complexity can be improved from $O(MN)$ to $O(k_A MN)$, where k_A is the ratio of the analyzed grid points over the total grid points. Due to the decay property of the Gaussian kernel function, faraway atoms have little contribution to a specific grid point and they can be ignored. We use the KD-tree structure [200] and a BVH [140] to quickly find the contributing atoms for a grid point, with a time complexity of $O(logM)$. Therefore, the total time complexity for density map construction is further reduced to $O(k_A NlogM)$.

Moreover, the multicore CPU and GPU-assisted parallel computation are employed to further accelerate the computational speed. During the neighboring search, the to-be-analyzed grid points are distributed to various threads, and the obtained Gaussian density value is adopted as the input of the next step. The workload for each grid point can vary a lot due to the different number of contributing atoms. For the multicore CPU computation, we randomly distribute the grid points to different cores, which helps keep the computation at full load. In contrast, for the GPU computation we record the contributing atom number of the previously analyzed grids, and use it to estimate the workload of the to-be-analyzed grid points. In this way, we distribute grid points with similar estimated workload to the same GPU blocks and thus achieve a better balance.

8.2.2 Parallel Mesh Generation and Quality Improvement

Here we use the dual contouring method [428, 441] to generate tetrahedral meshes from the above constructed density map for biomoelcules. A strongly-balanced octree structure and a feature-sensitive error function are adopted to control the mesh adaptation. To resolve topology ambiguities, ambiguous cells are identified using a trilinear function, and then are split into tetrahedral cells [437]. For each leaf cell, we compute a dual vertex and tetrahedral elements are generated by analyzing sign change edges and interior edges. For boundary cells intersecting with the biomolecular surface, the dual vertex is chosen as the average of all the intersection points between the biomolecular surface and the grid edges. For interior cells, the dual vertex is simply the cell center.

We also apply multicore CPU and GPU-based techniques to the dual contouring method for adaptive tetrahedral mesh generation. Each leaf cell has 12 edges, we divide them into four groups and analyze one group in each step. Each octree cell is distributed to a CPU core or GPU thread, and the connectivity can be built in the parallel manner. To balance the workload, we distribute octree cells at the same or similar levels to the same GPU blocks.

As discussed in previous chapters, we choose the edge ratio, the Jue-Liu parameter [244] and the dihedral angle to measure mesh quality [428, 441].

Then, we adopt face swapping, edge contraction and geometric flow [223] to improve the mesh quality. Both face swapping and edge contraction change the local connectivity or topology, while geometric flow relocates vertices without changing the topology and it is the most time-consuming part for quality improvement. We apply the CPU-based parallel computation to the smoothing step and use METIS [196] to partition the mesh. For GPU-based computation, we relocate vertices after each step to avoid any conflict. To avoid serious workload imbalance, we group vertices based on their valence numbers.

8.2.3 Numerical Results and Applications

We have applied our multiscale modeling techniques to many proteins. All the steps, including Gaussian density computation, mesh generation and quality improvement, can be run in three different modes: CPU sequential, CPU 8-core parallel and GPU parallel. Some of our created quality meshes have been used in finite element applications. For example, our constructed monomer and tetramer mouse acetylcholinesterase models have been used to compute the diffusion-based reaction rate constants in the neuromuscular junction system [89, 354, 355, 422]. The SERCA bump and myofibril lattice models shown in Figure 8.2 have been used in simulating the human cardiac calcium signaling system [198].

8.3 Patient-Specific Geometric Modeling for Cardiovascular Systems

The human cardiovascular system consists of the heart and blood vessels, which supply each organ with blood. The amount of the blood provided to each organ and the flow pattern may vary depending on physiologic conditions and organ demands. Patient-specific modeling and blood flow simulations have been proposed as new paradigms in simulation-based treatment planning and prediction [23, 45, 369, 397]. In the following, we will discuss two application projects: vascular modeling for blood flow simulation and cardiac Hermite model construction.

8.3.1 Vascular Modeling for Blood Flow Simulations

Both FEM and IGA have been used broadly in simulating blood flow of vascular structures. We will talk about how to build patient-specific geometric models and finite element meshes for these two techniques.

Solid NURBS Modeling for Isogeometric Analysis. We have developed a four-step pipeline to construct solid NURBS vascular models directly

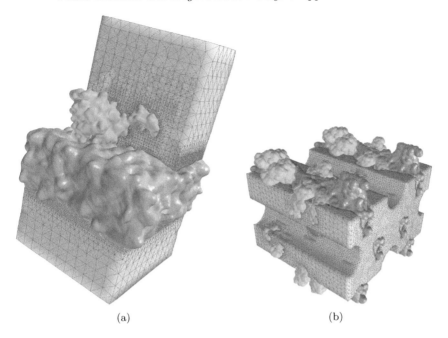

(a)   (b)

FIGURE 8.2: Two biomolecualr complexes in human cardiac calcium signaling system (pictures from [242]). (a) The SERCA pump; and (b) the myofibril lattice (ML).

from patient-specific medical images [430]. In scanned computed tomography (CT) or magnetic resonance imaging (MRI) data, we first pass these raw data into a preprocessing, including contrast enhancement, noise removal, classification and segmentation, to obtain the areas of interest. Then, isocontouring is used together with geometry editing to build the vascular surface models, from which the 1D skeleton is generated using Voronoi and Delaunay diagrams [158]. Next, we use a skeleton-based sweeping method to construct hexahedral NURBS control meshes. Templates for bifurcation, trifurcation and *n*-branching configurations are designed to match multiple branches together smoothly. Finally, we construct solid NURBS which have been applied to simulate fluid-structure interaction [43, 44] and nanoparticle delivery [168, 169, 170, 171] using IGA. Figure 8.3(a) shows the solid NURBS model for a local trifurcation structure.

Tetrahedral Finite Element Meshing. In addition to solid NURBS modeling, we also developed comprehensive and high-fidelity finite element meshing approaches [440] for patient-specific arterial geometries (e.g., cerebral aneurysms) from medical images. There are four main steps in our meshing method: (1) image segmentation and surface model extraction; (2) tetrahedral mesh generation for the fluid volume using the octree-based dual contouring

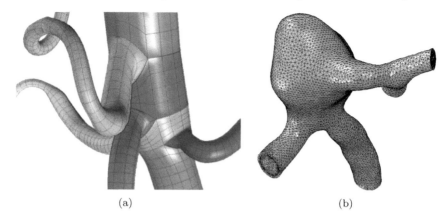

(a)   (b)

FIGURE 8.3: (a) Solid NURBS model of a local trifurcation vascular structure; and (b) tetrahedral mesh of an aneurysm.

method; (3) mesh quality improvement using edge-contraction, pillowing, optimization, geometric flow smoothing and mesh cutting; and finally (4) mesh generation for the blood vessel wall based on the boundary layer generation technique. The constructed meshes have been employed in fluid-structure interaction analysis of several cerebral aneurysms. Figure 8.3(b) shows the tetrahedral mesh of an aneurysm.

Note that due to the limitation of imaging resolution, it is hard to obtain the wall thickness and material property information directly from CT and MRI images. To construct an accurate computational model for weakened wall structures, we deform a healthy vessel onto an aneurysm through surface parameterization and a nonlinear spring system [186]. By comparing the original surface and the deformed mesh, the material strength and anisotropy can be estimated (e.g., the wall thickness and the Youngs modulus) [423]. From several tested models, we have observed that compared with using a uniform wall thickness, using the equivalent wall thickness yields a more accurate prediction of the rupture site.

8.3.2 Cardiac Hermite Model Construction

Many cardiac computational modeling and biomechanics techniques [62, 180, 199, 220] require smooth high-order cubic Hermite finite elements with much fewer degrees of freedom than piecewise-linear elements. They are important in simulation-based medical prediction in studying ventricular arrhythmia, atrial fibrillation and congestive heart failure. Constructing an accurate four-chamber cubic Hermite model is labor-intensive, requiring a tremendous amount of user input to create the hexahedral control mesh connectivity that represents the sophisticated topology correctly.

Here we describe an atlas-based geometry pipeline [434] for constructing

3D cubic Hermite finite element meshes of the human heart from medical images, taking into account four main chambers (the left and right atria, ventricles) and all major blood vessels. Given a patient's heart from segmented images, we first build a 1D centerline path tree to represent the complex cardiac topology and an atlas is created using the skeleton-based sweeping method. Then, we use optical flow based registration to deform the constructed atlas to match with new patients' images, as well as align diffusion tensor MRI data to include fiber and sheet orientation into finite element models. The constructed cubic Hermite meshes are suitable for cardiac mechanical and electrical function analysis. Extraordinary nodes were also studied in particular in such models [157].

In addition to finite element modeling, statistical shape matching is another important research area to characterize morphological variations between healthy and pathological anatomical structures. In particular, the eigenmodes of the Laplace–Beltrami Operator (LBO) have been shown to be effective for parametric representation of the overall shape and structural details of an object [148], providing a promising feature space for statistical shape analysis and potential biomedical applications.

8.4 Applications in Material Sciences

FEM and scanned images have also been used popularly over recent years to study composite material properties, fracture and fatigue. 3D datasets are usually quite large and complex. Incorporating them into predictive models requires advanced meshing techniques that can preserve essential features with as few elements as possible. In the following, we will talk about two applications of image-based modeling: polycrystalline materials [306] and adhesive bondline composites [130, 131].

8.4.1 Critical Feature Determination of Polycrystalline Materials

Scanning techniques and FEA have been employed a lot to study and predict the performance of microstructure materials [108, 165, 226, 227, 417]. For example, serial sectioning and optical microscopy with periodic electron backscatter diffraction (EBSD) [4, 313, 314] and high-energy X-ray diffraction microscopy are two commonly used scanning techniques. Due to the limited computational resource, an abbreviated representative volume element (RVE) is generally selected for modeling microstructures with certain criteria [71, 194]. In its simplest form, the RVE dataset consists of a 3D array of values representing a property or an identifier for each voxel. In FEA, hexahedral (hex) elements are preferred over tetrahedra due to their superior performance in terms of increased accuracy, fewer elements and improved reliability. A straightforward way for hex meshing is to simply convert each voxel to a 3D brick element as shown in Figure 8.4. The resulting mesh generally consists of a large number of elements with a stair-casing effect at the grain boundaries.

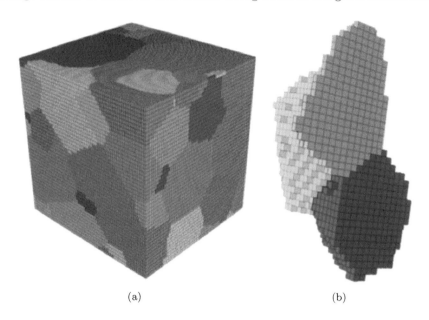

(a)   (b)

FIGURE 8.4: (a) The voxel-to-brick mesh for the 92-grain beta titanium data; and (b) a local mesh for three grains with stair-casing boundaries. (Pictures from [306].)

Here, we adopt an octree-based isocontouring algorithm to construct all-hex meshes [432]. The grain boundaries form non-manifold surfaces, which bring a lot of challenges for quality meshing. To improve the mesh quality and simultaneously preserve these non-manifold boundary features, we categorize all vertices into seven groups based on their relative location in the individual grain and the entire RVE dataset [306]. With this vertex classification, we apply proper algorithms to different groups of vertices to improve the mesh quality. A comprehensive method of pillowing, geometric flow and optimization techniques is developed. In the relaxation-based relocation process, non-manifold points are fixed. Planar boundary curves and interior spatial curves are distinguished and regularized via B-spline interpolation and re-sampling. Grain boundary surface patches and interior vertices are improved using geometric flow and the weighted averaging method. Finally, the optimization method eliminates all negative Jacobians and improves the worst aspect ratio of the mesh.

We have applied our quality improvement algorithms to several beta titanium datasets, and the generated all-hex meshes are validated via a statistics study [306]. To compare our meshing technique with the voxel-to-element technique, a 3D volume consisting of 92 grains is sampled from a reconstructed titanium microstructure [356]; see Figure 6.19. Our resulting volumetric mesh consists of 49,673 elements with conformal grain boundaries in contrast with 200,000 elements from the voxel-to-element technique with stair-casing grain boundaries. To compare these two meshes, a series of FEA with a user-material (UMAT) subroutine is carried out in ABAQUS [177]. The reduction of the number of elements (with only one quarter elements) results in a reduction of computation time by a factor of 10 in the plastic simulation, from approximately 4 days to 8 hours for the same parallel computing on 156 processors. While maintaining the accuracy of the global responses and increasing precision at the grain boundaries, such significant reduction in mesh size and computational time effectively demonstrates our meshing algorithm a superior method for meshing microstructures.

8.4.2 Damage Characterization of Adhesively Bonded Materials

Our meshing techniques have also been used in FEA of adhesive bondline composites recently, which is a composite structure used widely in aerospace, marine, automotive and many other industries with many attractive features. A bonded composite structure is sensitive to environmental conditions and the fabrication processing may introduce porosities, voids and inclusions. All these factors influence the integrity and durability of adhesively bonded composite structures. To test and optimize the design of a bonded composite structure, a 3D multiscale cohesive finite element model needs to be developed to characterize its damage, fracture and fatigue. Here we still use our octree-based dual contouring method [432] to generate all-hex meshes for a unit cell model embedded with glass beads and voids from scanning electron microscope images; see Figure 8.5. The resulting meshes have been successfully incorporated into ABAQUS for a multiscale damage characterization study [130, 131]. This is a collaborative project with Drs. Jim Lua and Eugene Fang in Global Engineering and Materials, Inc. sponsored by Naval Air Warfare Center, Aircraft Division.

In summary, image-based geometric modeling and mesh generation have very broad applications in FEA and IGA in many research areas, such as computational biology, medicine and material sciences. With the continuous development of scanning techniques and computer facilities, huge data and large scale of computation need more and more advanced modeling techniques to facilitate new simulation-based design, optimization and prediction. This is the most challenging but also interesting part of research!

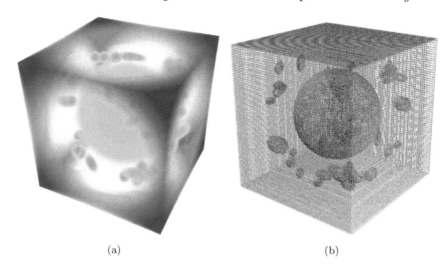

(a)   (b)

FIGURE 8.5: A unit cell model with embedded glass beads and voids. (a) Volume rendering result; and (b) an all-hex mesh visualized in wireframe.

Homework

Problem 8.1. The boundary value problem is defined as

$$-xu'' - u' + u = sinx, \qquad 0 < x < 1,$$

where $u(0) = u(1) = 0$.

1. Prove that its variational formulation is to find $u \in H_0^1$ such that

$$\int_0^1 (xu'v' + uv - vsinx)dx = 0 \qquad for \ \ v \in H_0^1.$$

2. The Galerkin approximation is to find $u_N \in H_0^N$ such that

$$\int_0^1 (xu_N'v_N' + u_Nv_N)dx = \int_0^1 v_N sinx dx,$$

for all $v_N \in H_0^N$. Suppose $u_N = \sum_{j=1}^N \alpha_j\phi_j(x)$ and $v_N = \sum_{i=1}^N \beta_i\phi_i(x)$. Derive the formulation for K_{ij} and F_i satisfying $\sum_{j=1}^N K_{ij}\alpha_j = F_i$.

Problem 8.2. Perform a survey and compare FEM, IGA, XFEM and IFEM. What are the challenges and what are the future directions?

Problem 8.3 (Project). Two students form a team. From the given list, each team picks one topic and finds at least six technical papers on this topic from conference proceedings and technical journals. If you are interested in other research topics related to this class, please discuss with me. You can choose one from the following two options to work on your topic.

- Option 1: Survey Project
  Read these papers, summarize problems, categorize previously proposed approaches, discuss the pros and cons of each approach, give discussion and observation, and wrap up with future directions that you think might be promising. Each team gives a presentation and submits the project proposal and final report together.

- Option 2: Programming Project
  From the six papers, you can choose one algorithm, implement it, and design some examples to test your code. Finally, summarize your results into one report and submit along with your codes.

Some example topics you can choose from:

1. Medical imaging filtering

2. Medical image segmentation

3. Medical image registration

4. Geometric modeling from imaging data (or isocontouring)

5. Delaunay triangulation and Voronoi diagram

6. Octree-based mesh generation

7. Tetrahedral mesh generation

8. Hexahedral mesh generation

9. Mesh quality improvement

10. NURBS and T-spline modeling

11. Implicit solvation model construction from atomic resolution data of protein data bank

12. Biomolecular electrostatic potential analysis by solving Poisson-Boltzman equations

13. Computational mechanics and its applications in biomedical engineering

14. Patient-specific cardiovascular blood flow simulation

15. Dynamic lung modeling and image registration

16. Bioheat transfer analysis for laser therapy of cancer treatment planning

Bibliography

[1] CUBIT Mesh Generation Toolkit. http://cubit.sandia.gov/.

[2] http://bishopw.loni.ucla.edu/air5.

[3] http://commons.wikimedia.org/wiki/File:X-ray_by_Wilhelm_Röntgen_of_Albert_von_Kölliker's_hand_-_18960123-01.jpg.

[4] http://en.wikipedia.org/wiki/Electron_backscatter_diffraction.

[5] http://fsl.fmrib.ox.ac.uk/fsl/fslwiki/flirt.

[6] http://jessicaz.me.cmu.edu.

[7] https://en.wikipedia.org/wiki/Ultrasound#/media/File:CRL_Crown_rump_lengh_12_weeks_ecografia_Dr._Wolfgang_Moroder.jpg.

[8] http://tetruss.larc.nasa.gov/.

[9] http://www.andrew.cmu.edu/user/suter/3dxdm/3dxdm.html.

[10] http://www.ansys.com.

[11] http://www.cmu.edu/me/xctf/facility/index.html.

[12] http://www.geom.uiuc.edu/software/cglist/welcome.html.

[13] http://www.itk.org.

[14] http://www.math.uwaterloo.ca/~vavasis/.

[15] http://www.scorec.rpi.edu.

[16] Real time interactive Voronoi and Delaunay diagrams, http://www.cs.cornell.edu/info/people/chew/delaunay.html.

[17] B. Adams and P. Dutré. Interactive Boolean operations on surfel-bounded solids. *ACM Transactions on Graphics*, 22(3):651–656, 2003.

[18] M. N. Ahmed, S. M. Yamany, N. Mohamed, A. A. Farag, and T. Moriarty. A modified fuzzy c-means algorithm for bias field estimation and segmentation of MRI data. *IEEE Transactions on Medical Imaging*, 21(3):193–199, 2002.

[19] M. Aigner, C. Heinrich, B. Jüttler, E. Pilgerstorfer, B. Simeon, and A. V. Vuong. Swept volume parameterization for isogeometric analysis. In *13th IMA International Conference on Mathematics of Surfaces XIII*, pages 19–44, 2009.

[20] L. Albou, B. Schwarz, O. Poch, J. Wurtz, and D. Moras. Defining and characterizing protein surface using alpha shapes. *Proteins*, 76(1):112, 2009.

[21] D. C. Alexander, C. Pierpaoli, P. J. Basser, and J. C. Gee. Spatial transformations of diffusion tensor magnetic resonance images. *IEEE Transactions on Medical Imaging*, 20(11):1131–1139, 2001.

[22] G. Amati. A multi-level filtering approach for fairing planar cubic B-spline curves. *Computer Aided Geometric Design*, 24:53–66, 2006.

[23] M. Anand, J. Kwack, and A. Masud. A new generalized Oldroyd-B model for blood flow in complex geometries. *International Journal of Engineering Science*, 72:78–88, 2013.

[24] C. W. Anderson and S. Crawford-Hines. Fast generation of NURBS surfaces from polygonal meshmodels of human anatomy. *Technical Report, Colorado State University*, 2000.

[25] P. R. Andresen and M. Nielsen. Non-rigid registration by geometry-constrained diffusion. *Medical Image Analysis*, 5(2):81–88, 2001.

[26] R. Araiza, M. G. Averill, G. R. Keller, and S. A. Starks. 3-D image registration using fast Fourier transform, with potential applications to geoinformatics and bioinformatics. In *International Conference on Information Processing and Management of Uncertainty in Knowledge-Based Systems, Paris, France*, pages 817–824, 2006.

[27] A. Z. Arifin and A. Asano. Image segmentation by histogram thresholding using hierarchical cluster analysis. *Pattern Recognition Letters*, 27(13):1515–1521, 2006.

[28] C. Armstrong, D. Robinson, R. McKeag, T. Li, S. Bridgett, R. Donaghy, and C. McGleenan. Medials for meshing and more. In *Proceedings of the 4th International Meshing Roundtable*, pages 277–288, 1995.

[29] S. Artemova, S. Grudinin, and S. Redon. A comparison of neighbor search algorithms for large rigid molecules. *Journal of Computational Chemistry*, 32(13):2865–2877, 2011.

[30] F. B. Atalay and S. Ramaswami. Quadrilateral meshes with bounded minimum angle. In *Proceedings of the 17th International Meshing Roundtable*, pages 73–91, 2008.

[31] O. K.-C. Au, C.-L. Tai, H.-K. Chu, D. Cohen-Or, and T.-Y. Lee. Skeleton extraction by mesh contraction. *ACM Transactions on Graphics*, 27(3):44:1–44:10, 2008.

[32] P. Baehmann, S. Wittchen, M. Shephard, K. Grice, and M. Yerry. Robust geometrically based, automatic two-dimensional mesh generation. *International Journal of Numerical Methods in Engineering*, 24:1043–1078, 1987.

[33] C. Bajaj, E. Coyle, and K.-N. Lin. Arbitrary topology shape reconstruction from planar cross sections. *Graphical Modeling and Image Processing*, 58(6):524–543, 1996.

[34] C. Bajaj, E. Coyle, and K.-N. Lin. Tetrahedral meshes from planar cross sections. *Computer Methods in Applied Mechanics and Engineering*, 179:31–52, 1999.

[35] C. Bajaj, V. Pascucci, and D. Schikore. Fast isocontouring for improved interactivity. In *Proceedings of the IEEE Symposium on Volume Visualization*, pages 39–46, 1996.

[36] C. Bajaj, V. Pascucci, and D. Schikore. The contour spectrum. In *IEEE Visualization*, pages 167–174, 1997.

[37] C. Bajaj, Q. Wu, and G. Xu. Level set based volume anisotropic diffusion. In *ICES Technical Report 301, The University of Texas at Austin*, 2002.

[38] C. Bajaj and G. Xu. Anisotropic diffusion of surfaces and functions on surfaces. *ACM Transactions on Graphics*, 22(1):4–32, 2003.

[39] C. L. Bajaj, V. Pascucci, A. Shamir, R. J. Holt, and A. N. Netravali. Dynamic maintenance and visualization of molecular surfaces. *Discrete Applied Mathematics*, 127(1):23–51, 2003.

[40] R. Bajcsy and S. Kovacic. Multiresolution elastic matching. *Computer Vision, Graphics, and Image Processing*, 46(1):1–21, 1989.

[41] J. L. Barron, D. J. Fleet, and S. S. Beauchemin. Systems and experiment: performance of optical flow techniques. *International Journal of Computer Vision*, 12:43–77, 1994.

[42] Y. Bazilevs, V. M. Calo, J. A. Cottrell, J. A. Evans, T. J. R. Hughes, S. Lipton, M. A. Scott, and T. W. Sederberg. Isogeometric analysis using T-splines. *Computer Methods in Applied Mechanics and Engineering*, 199(5-8):229–263, 2010.

[43] Y. Bazilevs, V. M. Calo, T. J. R. Hughes, and Y. Zhang. Isogeometric fluid-structure interaction: theory, algorithms, and computations. *Computational Mechanics*, 43:3–37, 2008.

[44] Y. Bazilevs, V. M. Calo, Y. Zhang, and T. J. R. Hughes. Isogeometric fluid-structure interaction analysis with applications to arterial blood flow. *Computational Mechanics*, 38:310–322, 2006.

[45] Y. Bazilevs, K. Takizawa, and T. E. Tezduyar. *Computational Fluid-Structure Interaction: Methods and Applications*. Wiley, New York, 2013.

[46] D. J. Benson, Y. Bazilevs, M.-C. Hsu, and T. J. R. Hughes. Isogeometric shell analysis: the Reissner-Mindlin shell. *Computer Methods in Applied Mechanics and Engineering*, 199(5-8):276–289, 2010.

[47] H. M. Berman, J. Westbrook, Z. Feng, G. Gilliland, T. N. Bhat, H. Weissig, I. N. Shindyalov, and P. E. Bourne. The protein data bank. *Nucleic Acids Research*, 28:235–242, 2000.

[48] M. Bern and D. Eppstein. Quadrilateral meshing by circle packing. *International Journal of Computational Geometry and Application*, 10(4):347–360, 2000.

[49] M. W. Bern, D. Eppstein, and J. R. Gilbert. Provably good mesh generation. In *IEEE Symposium on Foundations of Computer Science*, pages 231–241, 1990.

[50] C. J. Bishop. Quadrilateral meshes with no small angles. *Manuscript, www.math.sunysb.edu/ bishop/papers/quadmesh.ps*, 1991.

[51] T. Blacker. The cooper tool. In *Proceedings of the 5th International Meshing Roundtable*, pages 13–29, 1996.

[52] T. Blacker and M. Stephenson. Paving: a new approach to automated quadrilateral mesh generation. *International Journal of Numerical Methods in Engineering*, 32:811–847, 1991.

[53] T. D. Blacker and R. J. Myers. Seams and wedges in plastering: a 3D hexahedral mesh generation algorithm. *Engineering with Computers*, 2:83–93, 1993.

[54] J. F. Blinn. A generalization of algebraic surface drawing. *ACM Transactions on Graphics*, 1(3):235–256, 1982.

[55] W. Boehm. Inserting new knots into B-spline curves. *Computer Aided Design*, 12(4):199–201, 1983.

[56] D. Bommes, B. Lévy, N. Pietroni, E. Puppo, C. Silva, and D. Zorin. Quad-mesh generation and processing: a survey. *Computer Graphics Forum*, 32:51–76, 2013.

[57] D. Bommes, H. Zimmer, and L. Kobbelt. Mixed-integer quadrangulation. *ACM Transactions on Graphics*, 28:1–10, 2009.

[58] M. J. Borden, M. A. Scott, J. A. Evans, and T. J. R. Hughes. Isogeometric finite element data structures based on Bézier extraction of NURBS. *International Journal for Numerical Methods in Engineering*, 87(1-5):15–47, 2011.

[59] P. B. Bornermann and F. Cirak. A subdivision-based implementation of the hierarchical B-spline finite element method. *Computer Methods in Applied Mechanics and Engineering*, 253:584–598, 2013.

[60] F. J. Bossen and P. S. Heckbert. A pliant method for anisotropic mesh generation. In *5th International Meshing Roundtable*, pages 63–76, 1996.

[61] Y. Boykov and G. Funka-Lea. Graph cuts and efficient ND image segmentation. *International Journal of Computer Vision*, 70(2):109–131, 2006.

[62] C. Bradley, A. Pullan, and P. Hunter. Geometric modeling of the human torso using cubic Hermite elements. *Annals of Biomedical Engineering*, 25(1):96–111, 1997.

[63] D. Brodsky and B. Watson. Model simplification through refinement. *Graphics Interface*, pages 221–228, 2000.

[64] J. R. Bronson, J. A. Levine, and R. T. Whitaker. Lattice cleaving: a multimaterial tetrahedral meshing algorithm with guarantees. *IEEE Transactions on Visualization and Computer Graphics*, 20(2):223–237, 2014.

[65] L. G. Brown. A survey of image registration techniques. *ACM Computing Surveys*, 24(4):325–376, 1992.

[66] A. Buffa, D. Cho, and G. Sangalli. Linear independence of the T-spline blending functions associated with some particular T-meshes. *Computer Methods in Applied Mechanics and Engineering*, 199:1437–1445, 2010.

[67] A. Buffa, G. Sangalli, and R. Vázquez. Isogeometric analysis in electromagnetics: B-splines approximation. *Computer Methods in Applied Mechanics and Engineering*, 199:1143–1152, 2010.

[68] W. Burger and M. J. Burge. *Digital Image Processing: An Algorithmic Approach Using Java*. Springer. ISBN 1846283795 and ISBN 3540309403, 2007.

[69] D. Burkhart, B. Hamann, and G. Umlauf. Iso-geometric finite element analysis based on Catmull-Clark subdivision solids. *Computer Graphics Forum*, 29:1575–1584, 2010.

[70] W. Cai, S. Chen, and D. Zhang. Fast and robust fuzzy c-means clustering algorithms incorporating local information for image segmentation. *Pattern Recognition*, 40(3):825–838, 2007.

[71] G. Cailletaud, S. Forest, D. Jeulin, F. Feyel, I. Galliet, V. Mounoury, and S. Quilici. Some elements of microstructural mechanics. *Acta Materialia*, 27:351–374, 2003.

[72] S. A. Canann. Plastering and optismoothing: new approaches to automated, 3D hexahedral mesh generation and mesh smoothing. *Ph.D. Dissertation, Brigham Young University, Provo, UT*, 1991.

[73] S. A. Canann, J. R. Tristano, and M. L. Staten. An approach to combined laplacian and optimization-based smoothing for triangular, quadrilateral and quad-dominant meshes. In *7th International Meshing Roundtable*, pages 479–494, 1998.

[74] Y. Cao, L. Ju, Y. Zhou, and S. Wang. 3D superalloy grain segmentation using a multichannel edge-weighted centroidal Voronoi tessellation algorithm. *IEEE Transactions on Image Processing*, 22(9-10):4123–4135, 2013.

[75] Y. Cao, L. Ju, Q. Zou, C. Qu, and S. Wang. A multichannel edge-weighted centroidal Voronoi tessellation algorithm for 3D super-alloy image segmentation. In *IEEE Conference on Computer Vision and Pattern Recognition (CVPR)*, pages 17–24, 2011.

[76] H. Carr, J. Snoeyink, and U. Axen. Computing contour trees in all dimensions. *Computational Geometry: Theory and Applications*, 24(2):75–94, 2003.

[77] T. J. Cashman, U. H. Augsdörfer, N. A. Dodgson, and M. A. Sabin. NURBS with extraordinary points: high-degree, non-uniform, rational subdivision schemes. *ACM Transactions on Graphics*, 28(46):146, 2009.

[78] K. Castleman. *Digital Image Processing.* Prentice Hall, Englewood Cliffs, NJ, 1996.

[79] E. Catmull and J. Clark. Recursively generated B-spline surfaces on arbitrary topological meshes. *Computer Aided Design*, 10:350–355, 1978.

[80] T. F. Chan and L. A. Vese. Active contour and segmentation models using geometric PDE's for medical imaging. In *Geometric Methods in Bio-Medical Image Processing*, pages 63–75. Springer, 2002.

[81] S. C. Chapra. *Applied Numerical Methods with MATLAB for Engineers and Scientists, 3rd Edition.* McGraw-Hill, New York, 2012.

[82] C. Charalambous and A. Conn. An efficient method to solve the minimax problem directly. *SIAM Journal of Numerical Analysis*, 15(1):162–187, 1978.

[83] J. S. Chen, C. Marodon, and H. Y. Hu. Model order reduction for meshfree solution of Poisson singularity problems. *International Journal for Numerical Methods in Engineering*, 102(5):1211–1237, 2015.

[84] S. Chen and D. Zhang. Robust image segmentation using FCM with spatial constraints based on new kernel-induced distance measure. *IEEE Transactions on Systems, Man, and Cybernetics, Part B: Cybernetics*, 34(4):1907–1916, 2004.

[85] T. Chen and D. N. Metaxas. Markov random field models. *Insight into Images. A K Peters/CRC Press, Boca Raton, FL. ISBN 1-56881-217-5*, 2004.

[86] Y. T. Chen. A level set method based on the Bayesian risk for medical image segmentation. *Pattern Recognition*, 43(11):3699–3711, 2010.

[87] S.-W. Cheng and T. K. Dey. Quality meshing with weighted Delaunay refinement. In *Proc. 13th ACM-SIAM Sympos. Discrete Algorithms (SODA)*, pages 137–146, 2002.

[88] S.-W. Cheng, T. K. Dey, H. Edelsbrunner, M. A. Facello, and S. Teng. Sliver exudation. *Proc. Journal of ACM*, 47:883–904, 2000.

[89] Y. Cheng, C. A. Chang, Z. Yu, Y. Zhang, M. Sun, T. S. Leyh, M. J. Holst, and J. A. McCammon. Diffusional channeling in the sulfate activating complex: combined continuum modeling and coarse-grained Brownian dynamics studies. *Biophysical Journal*, 95(10):4659–4667, 2008.

[90] A. N. Chernikovb and N. P. Chrisochoides. Parallel 2D constrained Delaunay mesh generation. *ACM Transactions of Mathematical Software*, 34(1):6:1–6:20, 2008.

[91] A. N. Chernikovb and N. P. Chrisochoides. Multitissue tetrahedral image-to-mesh conversion with guaranteed quality and fidelity. *SIAM Journal on Scientific Computing*, 33(6):3491–3508, 2011.

[92] L. P. Chew. Guaranteed-quality Delaunay meshing in 3D (short version). In *13th ACM Symposium on Computational Geometry*, pages 391–393, 1997.

[93] J. Cho, T. Kim, and K. Lee. Surface fairing with boundary continuity based on the wavelet transform. *Electronics and Telecommunications Research Institute*, 23:85–95, 2001.

[94] G. Christensen, R. Rabbitt, and M. Miller. Deformable templates using large deformation kinematics. *IEEE Transactions on Image Processing*, 5(10):1435–1447, 1996.

[95] K. S. Chuang, H. L. Tzeng, S. Chen, J. Wu, and T. J. Chen. Fuzzy c-means clustering with spatial information for image segmentation. *Computerized Medical Imaging and Graphics*, 30(1):9–15, 2006.

[96] F. Cirak, M. Ortiz, and P. Schröder. Subdivision surfaces: a new paradigm for thin shell analysis. *International Journal for Numerical Methods in Engineering*, 47:2039–2072, 2000.

[97] F. Cirak, M. J. Scott, E. K. Antonsson, M. Ortiz, and P. Schröder. Integrated modeling, finite-element analysis, and engineering design for thin shell structures using subdivision. *Computer Aided Design*, 34:137–148, 2002.

[98] U. Clarenz, M. Droske, and M. Rumpf. Towards fast non-rigid registration. In *In Inverse Problems, Image Analysis and Medical Imaging, AMS Special Session Interaction of Inverse Problems and Image Analysis*, pages 67–84, 2002.

[99] B. Clark, N. Ray, and X. Jiao. Surface mesh optimization, adaption, and untangling with high-order accuracy. In *21st International Meshing Roundtable*, pages 385–402, 2013.

[100] E. Cohen, R. F. Riesenfeld, and G. Elber. *Geometric Modeling with Splines: An Introduction*. AK Peters, Natick, MA, 2001.

[101] M. L. Connolly. Analytical molecular surface calculation. *Journal of Applied Crystallography*, 16(5):548–558, 1983.

[102] M. L. Connolly. Molecular surface: a review. *Network Science*, 1996.

[103] N. D. Cornea, D. Silver, and P. Min. Curve-skeleton properties, applications, and algorithms. *IEEE Transactions on Visualization and Computer Graphics*, 13(3):530–548, 2007.

[104] J. A. Cottrell, T. J. R. Hughes, and Y. Bazilevs. *Isogeometric Analysis: Toward Integration of CAD and FEA*. Wiley, New York, 2012.

[105] J. A. Cottrell, A. Reali, Y. Bazilevs, and T. J. R. Hughes. Isogeometric analysis of structural vibrations. *Computer Methods in Applied Mechanics and Engineering*, 195:5257–5296, 2006.

[106] C. Daux, N. Moës, J. Dolbow, N. Sukumar, and T. Belytschko. Arbitrary branched and intersecting cracks with the extended finite element method. *International Journal for Numerical Methods in Engineering*, 48(12):1741–1760, 2000.

[107] C. Davatzikos, J. Prince, and R. Bryan. Image registration based on boundary mapping. *IEEE Transactions on Medical Imaging*, 15(1):112–115, 1996.

[108] P. Dawson, D. Mika, and N. Barton. Finite element modeling of lattice misorientations in aluminum polycrystals. *Scripta Materialia*, 47:713–717, 2002.

[109] B. N. Delaunay. Sur la sphére vide. *Izvestia Akademii Nauk SSSR, Otdelenie Matematicheskikh I Estestvennykh Nauk*, 7:793–800, 1934.

[110] J. Deng, F. Chen, X. Li, C. Hu, W. Tong, Z. Yang, and Y. Feng. Polynomial splines over hierarchical T-meshes. *Graphical Models*, 70:76–86, 2008.

[111] M. Desbrun, M. Meyer, P. Schroder, and A. H. Burr. Discrete differential-geometry operators in nD. *http://www. multires.caltech.edu/pubs/*, 2000.

[112] E.W. Dijkstra. A note on two problems in connexion with graphs. *Numerische Mathematlk*, 1:269–271, 1959.

[113] R. Dimitri, L. D. Lorenzis, M. A. Scott, P. Wriggers, R. Taylor, and G. Zavarise. Isogeometric large deformation frictionless contact using T-splines. *Computer Methods in Applied Mechanics and Engineering*, 269:394–414, 2014.

[114] G. L. Dirichlet. Über die reduktion der positiven quadratischen formen mit drei unbestimmten ganzen zahlen. *Journal für die Reine und Angewandte Mathematik*, 40:209–227, 1850.

[115] T. Dokken, T. Lyche, and K.F. Pettersen. Polynomial splines over locally refined box-partitions. *Computer Aided Geometric Design*, 30:331–356, 2013.

[116] S. Dong, P. Bremer, M. Garland, V. Pascucci, and J. C. Hart. Spectral surface quadrangulation. *ACM Transactions on Graphics*, 25:1057–1066, 2006.

[117] S. Dong, S. Kircher, and M. Garland. Harmonic functions for quadrilateral remeshing of arbitrary manifolds. *Computer Aided Geometric Design*, 22:392–423, 2005.

[118] M. Droske and W. Ring. A Mumford-Shah level-set approach for geometric image registration. *SIAM Journal on Applied Mathematics*, 66(6):1–19, 2006.

[119] Q. Du, V. Faber, and M. Gunzburger. Centroidal Voronoi tessellations: applications and algorithms. *SIAM Review*, 41(4):637–676, 1999.

[120] Q. Du, M. Gunzburger, L. Ju, and X. Wang. Centroidal Voronoi tessellation algorithms for image compression, segmentation, and multichannel restoration. *Journal of Mathematical Imaging and Vision*, 24(2):177–194, 2006.

[121] Q. Du, M. D. Gunzburger, and L. Ju. Constrained centroidal Voronoi tessellations for surfaces. *SIAM Journal on Scientific Computing*, 24(5):1488–1506, 2003.

[122] M. Ebeida, A. Patney, J. Owens, and E. Mestreau. Isotropic conforming refinement of quadrilateral and hexahedral meshes using two-refinement templates. *International Journal of Numerical Methods in Engineering*, 88(10):974–985, 2011.

[123] M. Eck and J. Hadenfeld. Local energy fairing of B-spline curves. *Computing Supplemenentum, vol. 10, Springer*, 1995.

[124] H. Edelsbrunner and E. P. Mücke. Three-dimensional alpha shapes. *ACM Transactions on Graphics*, 13(1):43–72, 1994.

[125] J. Edgel. An adaptive grid-based all hexahedral meshing algorithm based on 2-refinement. *Master Thesis, Brigham Young University*, 2010.

[126] G. Elber and E. Cohen. Offset approximation improvement by control points perturbation. *Mathematical Methods in Computer Aided Geometric Design II, Academic Press*, pages 229–237, 1992.

[127] D. Eppstein. Linear complexity hexahedral mesh generation. In *Symposium on Computational Geometry*, pages 58–67, 1996.

[128] J. Escobar, E. Rodriguez, R. Montenegro, G. Montero, and J. Gonzalez-Yuste. Simultaneous untangling and smoothing of tetrahedral meshes. *International Journal for Numerical Methods in Enginnering*, 192(25):2775–2787, 2003.

[129] J. M. Escobar, J. M. Cascón, E. Rodríguez, and R. Montenegro. A new approach to solid modeling with trivariate T-splines based on mesh optimization. *Computer Methods in Applied Mechanics and Engineering*, 200(45-46):3210–3222, 2011.

[130] E. Fang, J. Lua, J. Zhang, A. Rahman, and N. D. Phan. A multi-scale bondline damage characterization and hybrid analysis approach for adhesively bonded composite structures. In *AHS International 71st Annual Forum & Technology Display. Virginia Beach, VA. May 5-7*, 2015.

[131] E. Fang, J. Lua, Y. J. Zhang, Y. Lai, and W. Seneviratne. Multi-scale characterization of an adhesive bondline with fabrication induced defects. In *American Society for Composite 30th Conference. East Lansing, MI. Sept. 28-30*, 2015.

[132] G. Farin. Fairing cubic B-spline curves. *Computer Aided Design*, 4(1-2):91–103, 1987.

[133] G. Farin. *A History of Curves and Surfaces in CAGD, Handbook of CAGD.* Elsevier, New York, 2002.

[134] P. F. Felzenszwalb and D. P. Huttenlocher. Efficient graph-based image segmentation. *International Journal of Computer Vision*, 59(2):167–181, 2004.

[135] M. Ferrant, S. K. Warfield, C. R. G. Guttmann, R. V. Mulkern, F. A. Jolesz, and R. Kikinis. 3D image matching using a finite element based elastic deformation model. In *Medical Image Computing and Computer-Assisted Intervention-MICCAI'99, Lecture Notes in Computer Science*, volume 1679, pages 202–209, 1999.

[136] D. A. Field. Laplacian smoothing and Delaunay triangulations. *Communications in Applied Numerical Methods*, 4:709–712, 1988.

[137] M. S. Floater. Parametrization and smooth approximation of surface triangulations. *Computer Aided Geometric Design*, 14(3):231–250, 1997.

[138] M. S. Floater and K. Hormann. Surface parameterization: a tutorial and survey. *Advances in Multiresolution for Geometric Modelling, Mathematics and Visualization*, pages 157–186, 2005.

[139] N. Folwell and S. Mitchell. Reliable whisker weaving via curve contraction. In *Proceedings, 7th International Meshing Roundtable*, pages 365–378, 1998.

[140] R. Fonseca and P. Winter. Bounding volumes for proteins: a comparative study. *Journal of Computational Biology*, 19(10):1203–1213, 2012.

[141] D. R. Forsey and R. H. Bartels. Hierarchical B-spline refinement. *Computer Graphics*, 22:205–212, 1988.

[142] P. A. Foteinosa, A. N. Chernikovb, and N. P. Chrisochoides. Guaranteed quality tetrahedral Delaunay meshing for medical images. *Computational Geometry*, 47(4):539–562, 2014.

[143] L. Freitag and C. Ollivier-Gooch. Tetrahedral mesh improvement using swapping and smoothing. *International Journal for Numerical Methods in Engineering*, 40:3979–4002, 1997.

[144] L. A. Freitag. On combining Laplacian and optimization-based mesh smoothing techniques. *AMD-Vol. 220 Trends in Unstructured Mesh Generation*, pages 37–43, 1997.

[145] P. J. Frey, H. Borouchaki, and P.-L. George. Delaunay tetrahedralization using an advancing-front approach. In *5th International Meshing Roundtable*, pages 31–48, 1996.

[146] I. Fujishiro, Y. Maeda, H. Sato, and Y. Takeshima. Volumetric data exploration using interval volume. *IEEE Transactions on Visualization and Computer Graphics*, 2(2):144–155, 1996.

[147] Y. Furukawa and H. Masuda. Compression of NURBS surfaces with error evaluation. In *NICOGRAPH International*, 2002.

[148] Y. Gao, Y. Zhang, and P. Menon. 3D shape comparison using a Laplace spectral shape matching approach. *The Special Issue of CompIMAGE'14 in Computer Methods in Biomechanics and Biomedical Engineering: Imaging & Visualization, DOI: 10.1080/21681163.2015.1057867*, 2015.

[149] W. Geng and S. Zhao. Fully implicit ADI schemes for solving the nonlinear Poisson-Boltzmann equation. *Molecular Based Mathematical Biology*, 1:109–123, 2013.

[150] P. L. George. Tet meshing: construction, optimization and adaptation. In *8th International Meshing Roundtable*, pages 133–141, 1999.

[151] P. L. George and H. Borouchaki. *Delaunay Triangulation and Meshing, Application to Finite Elements*, pages 230–234, 1998.

[152] C. F. Gerald and P. O. Wheatly. *Applied Numerical Analysis, Fifth Edition*. Addison Wesley, Reading MA, 1994.

[153] C. Giannelli, B. Jüttler, and H. Speleers. THB-splines: the truncated basis for hierarchical splines. *Computer Aided Geometric Design*, 29:485–498, 2012.

[154] J. Giard and B. Macq. Molecular surface mesh generation by filtering electron density map. *International Journal of Biomedical Imaging*, pages 263–269, 2010.

[155] S. F. Gibson. Using distance maps for accurate surface representation in sampled volumes. In *IEEE Visualization*, pages 23–30, 1998.

[156] R. Goldman and T. Lyche. *Knot Insertion and Deletion Algorithms for B-Spline Curves and Surfaces*. Society for Industrial and Applied Mathematics, Philadelphia, 1993.

[157] M. J. Gonzales, G. Sturgeon, A. Krishnamurthy, J. Hake, R. Jonas, P. Stark, W.-J. Rappel, S. M. Narayan, Y. Zhang, W. P. Segars, and A. D. McCulloch. A three-dimensional finite element model of human atrial anatomy: new methods for cubic Hermite meshes with extraordinary vertices. *Medical Image Analysis*, 17(5):525–537, 2013.

[158] S. Goswami, T. K. Dey, and C. L. Bajaj. Identifying flat and tubular regions of a shape by unstable manifolds. In *11th Symposium of Solid and Physical Modeling*, pages 27–37, 2006.

[159] J. A. Grant and B. T. Pickup. A Gaussian description of molecular shape. *Journal of Physical Chemistry*, 99(11):3503–3510, 1995.

[160] E. Grinspun, P. Krysl, and P. Schröder. CHARMS: a simple framework for adaptive simulation. *ACM Transactions on Graphics*, 21:281–290, 2002.

[161] X. Gu, Y. Wang, and S. Yau. Volumetric harmonic map. *Communications in Information & Systems*, 3(3):191–202, 2003.

[162] P. Hansbo. Generalized Laplacian smoothing of unstructured grids. *Communications in Numerical Methods in Engineering*, 11:455–464, 1995.

[163] J. A. Hartigan and M. A. Wong. Algorithm AS 136: a k-means clustering algorithm. *Applied Statistics*, 28(1):100–108, 1979.

[164] A. Hatcher. Pants decomposition of surfaces. *arXiv:math/9906084*, 1999.

[165] J. D. Hochhalter, D. J. Littlewood, M. D. Veilleux, J. E. Bozek, A. M. Maniatty, A. D. Rollett, and A. R. Ingraffea. A geometric approach to modeling microstructurally small fatigue crack formation: III. development of a semi-empirical model for nucleation. *Modelling and Simulation in Materials Science and Engineering*, 19(3):035008, 2011.

[166] M. Holst, N. Baker, and F. Wang. Adaptive multilevel finite element solution of the Poisson-Boltzmann equation algorithms I: algorithms and examples. *Journal of Computational Chemistry*, 21:1319–1342, 2000.

[167] B. K. P. Horn and B. G. Schunck. Determining optical flow. *Artifical Intelligence*, 17:185–203, 1981.

[168] S. S. Hossain, A. M. Kopacz, Y. Zhang, S.-Y. Lee, T.-R. Lee, M. Ferrari, T. J. R. Hughes, W. K. Liu, and P. Decuzzi. Multiscale modeling for the vascular transport of nanoparticles. *Nano and Cell Mechanics. Wiley Series in Micro and Nano Technologies. Wiley, New York*, 2012.

[169] S. S. Hossain and Y. Zhang. Application of isogeometric analysis to simulate local nanoparticulate drug delivery in patient-specific coronary arteries. *Multiscale Simulations and Mechanics of Biological Materials: Wing Liu's 60 Anniversary Volume. John Wiley & Sons, New York*, 2012.

[170] S. S. Hossain, Y. Zhang, X. Fu, G. Brunner, J. Singh, T. J. R. Hughes, D. Shah, and P. Decuzzi. MRI-based computational modeling of blood flow and nanomedicine deposition in patients with peripheral arterial disease. *Journal of the Royal Society Interface*, 12(106):20150001, 2015.

[171] S. S. Hossain, Y. Zhang, X. Liang, F. Hussain, M. Ferrari, T. J. R. Hughes, and P. Decuzzi. In silico vascular modeling for personalized nanoparticle delivery. *Nanomedicine*, 8(3):343–357, 2013.

[172] K. Hu, J. Qian, and Y. Zhang. Adaptive all-hexahedral mesh generation based on a hybrid octree and bubble packing. In *22nd International Meshing Roundtable*, 2013.

[173] K. Hu and Y. Zhang. Extended edge-weighted centroidal Voronoi tessellation for image segmentation. *CompIMAGE (Computer Modeling of Objects Presented in Images: Fundamentals, Methods, and Applications). Lecture Notes in Computer Science*, 8461:164–175, 2014.

[174] K. Hu and Y. Zhang. Image segmentation and adaptive superpixel generation based on harmonic edge-weighted centroidal Voronoi tessellation. *Computer Methods in Biomechanics and Biomedical Engineering, accepted*, 2015.

[175] L. Hu, D. Chen, and G. Wei. High-order fractional partial differential equation transform for molecular surface construction. *Molecular Based Mathematical Biology*, 1:125, 2013.

[176] J. Huang, M. Zhang, J. Ma, X. Liu, L. Kobbelt, and H. Bao. Spectral quadrangulation with orientation and alignment control. *ACM Transactions on Graphics*, 27(147):1–9, 2008.

[177] Y. Huang. A user-material subroutine incorporating single crystal plasticity in the ABAQUS finite element program. *Division of Applied Sciences*, 23:3647–3679, 1991.

[178] T. J. R. Hughes. *The Finite Element Method: Linear Static and Dynamic Finite Element Analysis*. Prentice Hall, Englewood Cliffs, NJ, 1987.

[179] T. J. R. Hughes, J. A. Cottrell, and Y. Bazilevs. Isogeometric analysis: CAD, finite elements, NURBS, exact geometry, and mesh refinement. *Computer Methods in Applied Mechanics and Engineering*, 194:4135–4195, 2005.

[180] P. J. Hunter, A. J. Pullan, and B. H. Smaill. Modeling total heart function. *Annual Review of Biomedical Engineering*, 5:147–177, 2003.

[181] Y. Ito, A. Shih, and B. Soni. Octree-based reasonable-quality hexahedral mesh generation using a new set of refinement templates. *International Journal of Numerical Methods in Engineering*, 77(13):1809–1833, 2009.

[182] Y. Jia, Y. Zhang, and T. Rabczuk. A novel dynamic multilevel technique for image registration. *Computers and Mathematics with Applications*, 69(9):909–925, 2015.

[183] W. Jiang, M. Baker, Q. Wu, C. Bajaj, and W. Chiu. Applications of bilateral denoising filter in biological electron microscopy. *Journal of Structural Biology*, 144(1-2):114–122, 2003.

[184] X. Jiao, A. Colombi, X. Ni, and J. Hart. Anisotropic mesh adaptation for evolving triangulated surfaces. In *15th International Meshing Roundtable*, pages 173–190, 2006.

[185] K. A. Johannessen, T. Kvamsdal, and T. Dokken. Isogeometric analysis using LR B-splines. *Computer Methods in Applied Mechanics and Engineering*, 269:471–514, 2014.

[186] E. Johnson, Y. Zhang, and K. Shimada. Estimating an equivalent wall-thickness of a cerebral aneurysm through surface parameterization and a non-linear spring system. *International Journal for Numerical Methods in Biomedical Engineering*, 27(7):1054–1072, 2011.

[187] P. Joshi, M. Meyer, T. DeRose, B. Green, and T. Sanocki. Harmonic coordinates for character articulation. *ACM Transactions on Graphics*, 26(3):71:1–9, 2007.

[188] L. Ju, Q. Du, and M. Gunzburger. Probabilistic methods for centroidal Voronoi tessellations and their parallel implementations. *Parallel Computing*, 28(10):1477–1500, 2002.

[189] T. Ju, F. Losasso, S. Schaefer, and J. Warren. Dual contouring of hermite data. *SIGGRAPH*, 21(3):339–346, 2002.

[190] T. Ju and T. Udeshi. Intersection-free contouring on an octree grid. In *Proceedings of Pacific Graphics*, 2006.

[191] F. Kälberer, M. Nieser, and K. Polthier. QuadCover - surface parameterization using branched coverings. *Computer Graphics Forum*, 26:375–384, 2007.

[192] T. Kanai, H. Suzuki, and F. Kimura. 3D geometric metamorphosis based on harmonic maps. In *Pacific Graphics*, pages 97–104, 1997.

[193] T. Kanaya, Y. Teshima, K. Kobori, and K. Nishio. A topology-preserving polygonal simplification using vertex clustering. *Graphite*, pages 117–120, 2005.

[194] T. Kanit, S. Forest, I. Galliet, V. Mounoury, and D. Jeulin. Determination of the size of the representative volume element for random composites: statistical and numerical approach. *International Journal of Solids and Structures*, 40:3647–3679, 2003.

[195] T. Kanungo, D. M. Mount, N. S. Netanyahu, C. D. Piatko, R. Silverman, and A. Y. Wu. An efficient k-means clustering algorithm: analysis and implementation. *IEEE Transactions on Pattern Analysis and Machine Intelligence*, 24(7):881–892, 2002.

[196] G. Karypis and V. Kumar. A fast and high quality multilevel scheme for partitioning irregular graphs. *SIAM Journal on Scientific Computing*, 20(1):359–392, 1998.

[197] M. Kass, A. Witkin, and D. Terzopoulos. Snakes: active contour models. *International Journal of Computer Vision*, pages 321–331, 1998.

[198] P. M. Kekenes-Huskey, T. Liao, A. K. Gillette, J. E. Hake, Y. Zhang, A. P. Michailova, A. D. McCulloch, and J. A. McCammon. Molecular and subcellular-scale modeling of nucleotide diffusion in the cardiac myofilament lattice. *Biophysical Journal*, 105(9):2130–2140, 2013.

[199] R. Kerckhoffs, S. Healy, T. Usyk, and A. McCulloch. Computational methods for modeling cardiac electromechanics. *Proceedings of the IEEE*, 94(4):769–783, 2006.

[200] B. Kim, K. J. Kim, and J. K. Seong. GPU accelerated molecular surface computing. *Applied Mathematics & Information Sciences*, 6(1S):185S–194S, 2012.

[201] D.-S. Kim, J. Seo, D. Kim, J. Ryu, and C.-H. Cho. Three-dimensional beta shapes. *Computer-Aided Design*, 38(11):1179–1191, 2006.

[202] J. A. Kjellander. Smoothing of cubic parametric splines. *Computer Aided Design*, 15(3):175–179, 1983.

[203] P. Knupp. Next-generation sweep tool: a method for generating all-hex meshes on two-and-one-half dimensional geometries. In *Proceedings of the 7th International Meshing Roundtable*, pages 505–513, 1998.

[204] P. Knupp. Matrix norms and the condition number: a general framework to improve mesh quality via node movement. In *Proceedings of the 8th Meshing Roundtable*, pages 13–22, 1999.

[205] P. Knupp. Achieving finite element mesh quality via optimization of the Jacobian matrix norm and associated quantities. Part I - a framework for surface mesh optimization. *International Journal of Numerical Methods in Engineering*, 48:401–420, 2000.

[206] P. Knupp. Achieving finite element mesh quality via optimization of the Jacobian matrix norm and associated quantities. Part II - a framework for volume mesh optimization and the condition number of the Jacobian matrix. *International Journal of Numerical Methods in Engineering*, 48:1165–1185, 2000.

[207] L. Kobbelt, M. Botsch, U. Schwanecke, and H. Seidel. Feature sensitive surface extraction from volume data. In *Proceedings of SIGGRAPH*, pages 57–66, 2001.

[208] C. Kober and M. Matthias. Hexahedral mesh generation for the simulation of the human mandible. In *Proceedings of the 9th International Meshing Roundtable*, pages 423–434, 2000.

[209] D. Kovacs, A. Myles, and D. Zorin. Anisotropic quadrangulation. In *14th ACM Symposium on Solid and Physical Modeling*, pages 137–146, 2010.

[210] R. Kraft. Adaptive and linearly independent multilevel B-splines. *A. L. Méhauté, C. Rabut, L. L. Schumaker (Eds.), Surface Fitting and Multiresolution Methods, Vanderbilt University Press*, pages 209–218, 1997.

[211] A. Kumar, P. Mandal, Y. Zhang, and S. Litster. Image restoration of phase contrast nano scale X-ray CT images. *Lecture Notes in Computer Science*, 8461:280–285, 2014.

[212] G. V. V. Ravi Kumar, K. G. Shastry, and B. G. Prakash. Computing non-self-intersecting off-sets of NURBS surfaces. *Computer Aided Design*, 34:209–228, 2002.

[213] F. Labelle and J. R. Shewchuk. Isosurface stuffing: fast tetrahedral meshes with good dihedral angles. *ACM Transactions on Graphics*, 26(3):57.1–57.10, 2007.

[214] M. Lai, S. Benzley, G. Sjaardema, and T. Tautges. A multiple source and target sweeping method for generating all hexahedral finite element meshes. In *Proceedings of the 5th International Meshing Roundtable*, pages 217–225, 1996.

[215] Y. Lai, L. Liu, Y. Zhang, J. Chen, E. Fang, and J. Lua. Rhino 3D to Abaqus: a T-spline based isogeometric software platform. *The edited volume of the Modeling and Simulation in Science, Engineering and Technology Book Series devoted to AFSI 2014 - a birthday celebration conference for Tayfun Tezduyar. Springer Publisher*, 2015.

[216] R. Laramee and R. Bergeron. An isosurface continuity algorithm for super adaptive resolution data. *Advances in Modelling, Animation, and Rendering: Computer Graphics International (CGI)*, pages 215–237, 2002.

[217] P. Laug and H. Borouchaki. Molecular surface modeling and meshing. *Engineering with Computers*, 18:199–210, 2002.

[218] C. Lee and S. Lo. A new scheme for the generation of a graded quadrilateral mesh. *Computers and Structures*, 52(5):847–857, 1994.

[219] M. S. Lee, M. Feig, F. R. Salsbury, and C. L. Brooks. New analytic approximation to the standard molecular volume definition and its application to generalized born calculations. *Journal of Computational Chemistry*, 24(14):1348–1356, 2003.

[220] I. Legrice, P. Hunter, A. Young, and B. Smaill. The architecture of the heart: a data-based model. *Philosophical Transactions of the Royal Society A*, 359(1783):1217–1232, 2001.

[221] J. Leng, G. Xu, and Y. Zhang. Medical image interpolation based on multi-resolution registration. *Computers and Mathematics with Applications*, 66:1–18, 2013.

[222] J. Leng, G. Xu, Y. Zhang, and J. Qian. Quality improvement of segmented hexahedral meshes using geometric flows. *Image-Based Geometric Modeling and Mesh Generation. Springer Publisher*, 2013.

[223] J. Leng, Y. Zhang, and G. Xu. A novel geometric flow approach for quality improvement of multi-component tetrahedral meshes. *Computer-Aided Design*, 45(10):1182–1197, 2013.

[224] B. Lévy. Laplace-Beltrami eigenfunctions: towards an algorithm that understands geometry. In *IEEE International Conference on Shape Modeling and Applications*, pages 13–21, 2006.

[225] B. Lévy, S. Petitjean, N. Ray, and J. Maillot. Least squares conformal maps for automatic texture atlas generation. *ACM Transactions on Graphics*, 21:362–371, 2002.

[226] A. Lewis, J. Bingert, D. Rowenhorst, A. Gupta, A. Geltmacher, and G. Spanos. Two and three dimensional microstructural characterization of a super-austenitic stainless steel. *Materials Science and Engineering*, 418:11–18, 2005.

[227] A. Lewis, K. Jordan, and A. Geltmacher. Determination of critical microstructural features in an austenitic stainless steel using image-based finite element modeling. *Metallurgical and Materials Transactions A: Physical Metallurgy and Materials Science*, 39(5):1109–1117, 2008.

[228] B. Li, X. Li, K. Wang, and H. Qin. Generalized polycube trivariate splines. In *Shape Modeling International Conference*, pages 261–265, 2010.

[229] C. Li, C. Xu, C. Gui, and M. D. Fox. Distance regularized level set evolution and its application to image segmentation. *IEEE Transactions on Image Processing*, 19(12):3243–3254, 2010.

[230] Q. Li, N. Mitianoudis, and T. Stathaki. Spatial kernel k-harmonic means clustering for multi-spectral image segmentation. *IET Image Processing*, 1(2):156–167, 2007.

[231] S. Li. *Markov Random Field Modeling in Computer Vision.* Springer, Berlin, 1995.

[232] W. Li. Automatic mesh to spline surface conversion. *PhD Thesis, Institut National Polytechnique de Lorraine, France,* 2006.

[233] W. Li, N. Ray, and B. Lévy. Automatic and interactive mesh to T-spline conversion. In *The Fourth Eurographics Symposium on Geometry Processing,* pages 191–200, 2006.

[234] X. Li, X. Guo, H. Wang, Y. He, X. Gu, and H. Qin. Harmonic volumetric mapping for solid modeling applications. In *ACM Symposium on Solid and Physical Modeling,* pages 109–120, 2007.

[235] X. Li, J. Zheng, T. W. Sederberg, T. J. R. Hughes, and M. A. Scott. On linear independence of T-spline blending functions. *Computer Aided Geometric Design,* 29(1):6376, 2012.

[236] Y. Li, Y. Liu, W. Xu, W. Wang, and B. Guo. All-hex meshing using singularity-restricted field. *ACM Transactions on Graphics,* 31(6):177:1–177:11, 2012.

[237] X. Liang, M. Ebeida, and Y. Zhang. Guaranteed-quality all-quadrilateral mesh generation with feature preservation. *Computer Methods in Applied Mechanics and Engineering,* 199(29-32):2072–2083, 2010.

[238] X. Liang and Y. Zhang. Hexagon-based all-quadrilateral mesh generation with guaranteed angle bounds. *Computer Methods in Applied Mechanics and Engineering,* 200(23-24):2005–2020, 2011.

[239] X. Liang and Y. Zhang. Matching interior and exterior all-quadrilateral meshes with guaranteed angle bounds. *Engineering with Computers,* 28(4):375–389, 2012.

[240] X. Liang and Y. Zhang. An octree-based dual contouring method for triangular and tetrahedral mesh generation with guaranteed angle range. *Engineering with Computers,* 30(2):211–222, 2014.

[241] T. Liao, G. Xu, and Y. Zhang. Structure-aligned guidance estimation in surface parameterization using eigenfunction-based cross field. *Graphical Models,* 76(6):691–705, 2014.

[242] T. Liao, Y. Zhang, P. Kekenes-Huskey, Y. Cheng, A. Michailova, A. D. McCulloch, M. Holst, and J. A. McCammon. Multi-core CPU or GPU-accelerated multiscale modeling for biomolecular complexes. *Molecular Based Mathematical Biology,* 1:164–179, 2013.

[243] E. B. De l'Isle and P. L. George. Optimization of tetrahedral meshes. *IMA Volumes in Mathematics and Its Applications,* 75:97–128, 1995.

[244] A. Liu and B. Joe. Relationship between tetrahedron shape measures. *BIT*, 34(2):268–287, 1994.

[245] L. Liu, Y. Zhang, T. J. R. Hughes, M. A. Scott, and T. W. Sederberg. Volumetric T-spline construction using Boolean operations. *Engineering with Computers*, 30(4):425–439, 2014.

[246] L. Liu, Y. Zhang, Y. Liu, and W. Wang. Feature-preserving T-mesh construction using skeleton-based polycubes. *Computer Aided Design*, 58:162–172, 2015.

[247] L. Liu, Y. Zhang, and X. Wei. Handling extraordinary nodes with weighted T-spline basis functions. *24th International Meshing Roundtable. Austin, TX*, 2015.

[248] L. Liu, Y. Zhang, and X. Wei. Weighted T-spline and its application in reparameterizing trimmed NURBS surfaces. *Computer Methods in Applied Mechanics and Engineering*, 295:108–126, 2015.

[249] S. Liu and C. Wang. Fast intersection-free offset surface generation from freeform models with triangular meshes. *EEE Transactions Automation Science and Engineering*, 8:347–360, 2011.

[250] W. K. Liu, Y. Liu, D. Farrell, L. Zhang, X. Wang, Y. Fukui, N. Patankar, Y. Zhang, C. L. Bajaj, J. Lee, J. Hong, X. Chen, and H. Hsu. Immersed finite element method and its applications to biological systems. *Computer Methods in Applied Mechanics and Engineering*, 195(13-16):1722–1749, 2006.

[251] S. H. Lo. Volume discretization into tetrahedra II. 3D triangulation by advancing front approach. *Computers and Structures*, 39:501–511, 1991.

[252] R. Lohner. Progress in grid generation via the advancing front technique. *Engineering with Computers*, 12:186–210, 1996.

[253] R. Lohner, K. Morgan, and O. C. Zienkiewicz. Adaptive grid refinement for compressible Euler equations. *Accuracy Estimates and Adaptive Refinements in Finite Element Computations, I. Babuska et. al. eds. Wiley*, pages 281–297, 1986.

[254] R. Lohner and P. Parikh. Three dimensional grid generation by the advancing-front method. *International Journal for Numerical Methods in Fluids*, 8:1135–1149, 1988.

[255] A. Lopes and K. Brodlie. Improving the robustness and accuracy of the marching cubes algorithm for isosurfacing. In *IEEE Transactions on Visualization and Computer Graphics*, pages 19–26, 2000.

[256] W. E. Lorensen and H. E. Cline. Marching cubes: a high resolution 3D surface construction algorithm. *SIGGRAPH*, 21(4):163–169, 1987.

[257] C. Lu, X. Jiao, and N. Missirlis. A hybrid geometric + algebraic multigrid method with semi-iterative smoothers. *Numerical Linear Algebra with Applications*, 21(2):221–238, 2014.

[258] E. De Luycker, D. J. Benson, T. Belytschko, Y. Bazilevs, and M.-C. Hsu. X-FEM in isogeometric analysis for linear fracture mechanics. *International Journal for Numerical Methods in Engineering*, 87(6):541–565, 2011.

[259] T. Lyche and K. Morken. Making the Oslo algorithm more efficient. *SIAM Journal of Numerical Analysis*, 23:663–675, 1986.

[260] T. Lyche and K. Morken. A data reduction strategy for splines. *University of Oslo, Technical Report, No. 107*, 1987.

[261] T. Lyche, K. Morken, and E. Quak. Theory and algorithms for non-uniform spline wavelets. *Multivariate Approximation and Applications*, pages 152–187, 2001.

[262] W. Y. Ma and B. S. Manjunath. Edgeflow: a technique for boundary detection and image segmentation. *IEEE Transactions on Image Processing*, 9(8):1375–1388, 2000.

[263] J. Maintz and M. Viergever. A survey of medical image registration. *Medical Image Analysis*, 2(1):1–36, 1998.

[264] O. Marco, R. Sevilla, Y. Zhang, J. J. Rodenas, and M. Tur. Exact 3D boundary representation in finite element analysis based on cartesian grids independent of the geometry. *International Journal for Numerical Methods in Engineering*, 103(6):445–468, 2015.

[265] L. Maréchal. Advances in octree-based all-hexahedral mesh generation: handling sharp features. In *Proceedings of the 18th International Meshing Roundtable*, pages 65–84, 2009.

[266] M. Marinov and L. Kobbelt. Automatic generation of structure preserving multiresolution models. *Computer Graphics Forum*, 24:479–486, 2005.

[267] T. Martin, G. Chen, S. Musuvathy, E. Cohen, and C. Hansen. Generalized swept mid-structure for polygonal models. *Computer Graphics Forum*, 31:805–814, 2012.

[268] T. Martin, E. Cohen, and R. M. Kirby. Volumetric parameterization and trivariate B-spline fitting using harmonic functions. *Computer Aided Geometric Design*, 26(6):648–664, 2009.

[269] D. N. Metaxas and T. Chen. Deformable models. *Insight into Images*. A K Peters/CRC Press, Boca Raton, FL. ISBN 1-56881-217-5, 2004.

[270] M. Meyer, M. Desbrun, P. Schröder, and A. Burr. Discrete differential-geometry operators for triangulated 2-manifolds. *VisMath02, Berlin*, 2002.

[271] S. Mitchell and T. Tautges. Pillowing doublets: refining a mesh to ensure that faces share at most one edge. In *Proceedings of the 4th International Meshing Roundtable*, pages 231–240, 1995.

[272] S. A. Mitchell and S. A. Vavasis. Quality mesh generation in higher dimensions. *SIAM Journal on Computing*, 29(4):1334–1370, 2000.

[273] M. Moelich and T. Chan. Joint segmentation and registration using logic models. In *UCLA CAM Report 03-06*, February 2003.

[274] D. Moore. Compact isocontours from sampled data. *Graphics Gems III*, pages 23–28, 1992.

[275] D. Mumford and J. Shah. Optimal approximations by piecewise smooth functions and associated variational problems. *Communications on Pure and Applied Mathematics*, 42:577–685, 1989.

[276] B. K. Natarajan. On generating topologically consistent isosurfaces from uniform samples. *The Visual Computer*, 11(1):52–62, 1994.

[277] G. M. Nielson and J. Sung. Interval volume tetrahedrization. In *IEEE Visualization*, pages 221–228, 1997.

[278] M. Nieser, U. Reitebuch, and K. Polthier. CUBECOVER - parameterization of 3D volumes. *Computer Graphics Forum*, 30(5):1397–1406, 2011.

[279] R. Nock and F. Nielsen. On weighting clustering. *IEEE Transactions on Pattern Analysis and Machine Intelligence*, 28(8):1223–1235, 2006.

[280] M. Nygards. Number of grains necessary to homogenize elastic materials with cubic symmetry. *Mechanics of Materials*, 35:1049–1057, 2003.

[281] A. Oddy, J. Goldak, M. McDill, and M. Bibby. A distortion metric for isoparametric finite elements. *Transactions of CSME, No. 38-CSME-32, Accession No. 2161*, 1988.

[282] J. T. Oden and G. F. Carey. *Finite Elements: A Second Course*. Prentice Hall, Englewood Cliffs, NJ, 1983.

[283] A. Oppenheim and A. Willsky. *Signals and Systems (2nd Edition)*. Prentice Hall, Englewood Cliffs, NJ, 1996.

[284] S. Osher and R. Fedkiw. *Level Set Methods and Dynamic Implicit Surfaces*. Springer-Verlag, 2003.

[285] S. Owen. A survey of unstructured mesh generation technology. In *7th International Meshing Roundtable*, pages 26–28, 1998.

[286] S. Owen and J. Shepherd. Embedding features in a cartesian grid. In *18th International Meshing Roundtable*, pages 117–138, 2009.

[287] S. Owen, M. Staten, S. Canann, and S. Saigal. Q-Morph: an indirect approach to advancing front quad meshing. *Int. J. Numer. Meth. Engng*, 44:1317–1340, 1999.

[288] N. R. Pal and S. K. Pal. A review on image segmentation techniques. *Pattern Recognition*, 26(9):1277–1294, 1993.

[289] Q. Pan, G. Xu, and Y. Zhang. Unified method for hybrid subdivision surface design using geometric partial differential equations. *Computer Aided Design*, 46:110–119, 2014.

[290] N. Paragios and R. Deriche. Geodesic active regions and level set methods for supervised texture segmentation. *International Journal of Computer Vision*, 46(3):223–247, 2002.

[291] V. Pascucci and C. Bajaj. Time critical adaptive refinement and smoothing. In *Proceedings of the ACM/IEEE Volume Visualization and Graphics Symposium*, pages 33–42, 2000.

[292] W. Pedrycz and J. Waletzky. Fuzzy clustering with partial supervision. *IEEE Transactions on Systems, Man, and Cybernetics, Part B: Cybernetics*, 27(5):787–795, 1997.

[293] S. Periaswamy, J. B. Weaver, D. M. Healy, D. N. Rockmore, P. J. Kostelec, and H. Farid. Differential affine motion estimation for medical image registration. In *SPIE 45th Annual Meeting, San Diego, CA*, 2000.

[294] P. Perona and J. Malik. Scale-space and edge detection using anisotropic diffusion. *IEEE Transactions on Pattern Analysis Machine Intelligence*, 12:629–639, 1990.

[295] C. S. Peskin. Flow patterns around heart valves: a digital computer method for solving the equations of motion. *Ph.D. thesis, Albert Einstein College of Medicine of Yeshiva University*, 1972.

[296] J.-M. Peyrat, M. Sermesant, X. Pennec, H. Delingette, C. Xu, E. R. McVeigh, and N. Ayache. A computational framework for the statistical analysis of cardiac diffusion tensors: application to a small database of canine hearts. *IEEE Transactions on Medical Imaging*, 26(10):1–15, 2007.

[297] L. Piegl and W. Tiller. *The NURBS Book*. Springer, New York, 1997.

[298] L. A. Piegl and W. Tiller. Computing offsets of NURBS curves and surfaces. *Computer Aided Design*, 31:147–156, 1999.

[299] N. Pietroni, M. Tarini, and P. Cignoni. Almost isometric mesh parameterization through abstract domains. *Visualization and Computer Graphics*, 16:621–635, 2010.

[300] S. Pirzadeh. Unstructured viscous grid generation by advancing-layers method. *AIAA-93 3453-CP, AIAA*, pages 420–434, 1993.

[301] W. H. Press, S. A. Teukolsky, W. T. Vetterling, and B. P. Flannery. *Numerical Recipes in C: The Art of Scientific Computing, Second Edition*. Cambridge University Press, 1992.

[302] M. A. Price and C. G. Armstrong. Hexahedral mesh generation by medial surface subdivision: Part I. *International Journal for Numerical Methods in Engineering*, 38(19):3335–3359, 1995.

[303] M. A. Price and C. G. Armstrong. Hexahedral mesh generation by medial surface subdivision: Part II. *International Journal for Numerical Methods in Engineering*, 40:111–136, 1997.

[304] J. Qian and Y. Zhang. Automatic unstructured all-hexahedral mesh generation from B-Reps for non-manifold CAD assemblies. *Engineering with Computers*, 28(4):345–359, 2012.

[305] J. Qian, Y. Zhang, D. T. O'Connor, M. S. Greene, and W. K. Liu. Intersection-free tetrahedral meshing from volumetric images. *Computer Methods in Biomechanics and Biomedical Engineering: Imaging & Visualization*, 1(2):100–110, 2013.

[306] J. Qian, Y. Zhang, W. Wang, A. C. Lewis, M. A. S. Qidwai, and A. B. Geltmacher. Quality improvement of non-manifold hexahedral meshes for critical feature determination of microstructure materials. *International Journal for Numerical Methods in Engineering*, 82(11):1406–1423, 2010.

[307] T. Rabczuk, S. Bordas, and H. Askes. Meshfree discretization methods for solid mechanics. *Encyclopedia of Aerospace Engineering, Editor: R. De Borst, Wiley & Sons, New York*, 2010.

[308] L. Ramshaw. Blossoming: a connect-the-dots approach to splines. *Digital Systems Research Center, Palo Alto, CA, Technical Report, SRC-RR-19*, 1987.

[309] N. Ray, W. Li, B. Lévy, A. Sheffer, and P. Alliez. Periodic global parameterization. *ACM Transactions on Graphics*, 25:1460–1485, 2006.

[310] U. Reif. A unified approach to subdivision algorithms near extraordinary vertices. *Computer Aided Geometric Design*, 12:153–174, 1995.

[311] M. Reuter, S. Biasotti, and D. Giorgi. Discrete Laplace-Beltrami operators for shape analysis and segmentation. *Computer Graphics*, 33:381–390, 2009.

[312] P. Rogelj and S. Kovacic. Symmetric image registration. In *Proceedings of SPIE, Vol. 5032, Medical Imaging 2003: Image Processing*, pages 484–493, 2003.

[313] A. Rollett, S. Lee, R. Campman, and G. Rohrer. Three-dimensional characterization of microstructure by electron back-scatter diffraction. *Annual Review of Materials Research*, 37:627–658, 2007.

[314] D. Rowenhorst, A. Gupta, C. Feng, and G. Spanos. 3D crystallographic and morphological analysis of coarse martensite: combining EBSD and serial sectioning. *Scripta Materialia*, 55:11–16, 2006.

[315] J. Ryu, R. Park, and D.-S. Kim. Molecular surfaces on proteins via beta shapes. *Computer-Aided Design*, 39(12):1042–1057, 2007.

[316] M. Sabin. Recent progress in subdivision: a survey. *Advances in Multiresolution for Geometric Modelling*, pages 203–230, 2005.

[317] A. Z. I. Salem, S. A. Canann, and S. Saigal. Robust distortion metric for quadratic triangular 2D finite elements. *AMD-Vol. 220 Trends in Unstructured Mesh Generation*, pages 73–80, 1997.

[318] M. F. Sanner, A. J. Olson, and J. C. Spehner. Reduced surface: an efficient way to compute molecular surfaces. *Biopolymers*, 38(3):305–320, 1996.

[319] E. Sarioz. An optimization approach for fairing of ship hull forms. *Ocean Engineering*, 33:2105–2118, 2006.

[320] S. P. Sastry and S. M. Shontz. Performance characterization of nonlinear optimization methods for mesh quality improvement. *Engineering with Computers*, 28(3):269–286, 2012.

[321] S. P. Sastry, S. M. Shontz, and S. A. Vavasis. A log-barrier method for mesh quality improvement and untangling. *Engineering with Computers*, 30(3):315–329, 2014.

[322] S. Schaefer, T. Ju, and J. Warren. Manifold dual contouring. *IEEE Transaction on Visualization and Computer Graphics*, 13(3):610–619, 2007.

[323] D. Schillinger, L. Dedè, M. A. Scott, J. A. Evans, M. J. Borden, E. Rank, and T. J. R. Hughes. An isogeometric design-through-analysis methodology based on adaptive hierarchical refinement of NURBS, immersed boundary methods, and T-spline CAD surfaces. *Computer Methods in Applied Mechanics and Engineering*, 249-252:116–150, 2012.

[324] D. Schillinger, M. Ruess, N. Zander, Y. Bazilevs, A. Düster, and E. Rank. Small and large deformation analysis with the p- and B-spline versions of the finite cell method. *Computational Mechanics*, 50(4):445–478, 2012.

[325] R. Schneiders. A grid-based algorithm for the generation of hexahedral element meshes. *Engineering with Computers*, 12:168–177, 1996.

[326] R. Schneiders. Refining quadrilateral and hexahedral element meshes. *5th International Conference on Grid Generation in Computational Field Simulations*, pages 679–688, 1996.

[327] R. Schneiders. An algorithm for the generation of hexahedral element meshes based on an octree technique. In *6th International Meshing Roundtable*, pages 195–196, 1997.

[328] R. Schneiders, R. Schindler, and F. Weiler. Octree-based generation of hexahedral element meshes. In *Proceedings of 5th International Meshing Roundtable*, pages 205–216, 1996.

[329] M. A. Scott, M. J. Borden, C. V. Verhoosel, T. W. Sederberg, and T. J. R. Hughes. Isogeometric finite element data structures based on Bézier extraction of T-splines. *International Journal for Numerical Methods in Engineering*, 88(2):126–156, 2011.

[330] M. A. Scott, X. Li, T. W. Sederberg, and T. J. R. Hughes. Local refinement of analysis-suitable T-splines. *Computer Methods in Applied Mechanics and Engineering*, 213-216:206–222, 2012.

[331] M. A. Scott, R. N. Simpson, J. A. Evans, S. Lipton, S. P. A. Bordas, T. J. R. Hughes, and T. W. Sederberg. Isogeometric boundary element analysis using unstructured T-splines. *Computer Methods in Applied Mechanics and Engineering*, 254(0):197–221, 2013.

[332] T. W. Sederberg, D. L. Cardon, G. T. Finnigan, N. S. North, J. Zheng, and T. Lyche. T-spline simplification and local refinement. In *ACM SIGGRAPH*, pages 276–283, 2004.

[333] T. W. Sederberg, G. T. Finnigan, X. Li, H. Lin, and H. Ipson. Watertight trimmed NURBS. *ACM Transactions on Graphics*, 27(3):79:1–79:8, 2008.

[334] T. W. Sederberg, J. Zheng, A. Bakenov, and A. Nasri. T-splines and T-NURCCs. *ACM Transactions on Graphics*, 22(3):477–484, 2003.

[335] N. Senthilkumaran and R. Rajesh. Edge detection techniques for image segmentation - a survey of soft computing approaches. *International Journal of Recent Trends in Engineering*, 1(2):250–254, 2009.

[336] E. Seveno. Towards an adaptive advancing front method. In *6th International Meshing Roundtable*, pages 349–362, 1997.

[337] A. Sheffer, E. Praun, and K. Rose. Mesh parameterization methods and their applications. *Foundations and Trends in Computer Graphics and Vision*, 2(2):105–171, 2006.

[338] M. S. Shephard and M. K. Georges. Three-dimensional mesh generation by finite octree technique. *International Journal of Numerical Methods in Engineering*, 32:709–749, 1991.

[339] M. S. Shephard and M. K. Georges. Three-dimensional mesh generation by finite octree technique. *International Journal for Numerical Methods in Engineering*, 32:709–749, 1991.

[340] J. Shepherd. Conforming hexahedral mesh generation via geometric capture methods. In *18th International Meshing Roundtable*, pages 85–102, 2009.

[341] J. Shepherd, S. Mitchell, P. Knupp, and D. White. Methods for multisweep automation. In *Proceedings of the 9th International Meshing Roundtable*, pages 77–87, 2000.

[342] J. Shepherd, Y. Zhang, C. Tuttle, and C. Silva. Quality improvement and Boolean-like cutting operations in hexahedral meshes. In *The 10th ISGG Conference on Numerical Grid Generation. FORTH, Crete, Greece*, 2007.

[343] M. E. Sherif and S. Oudom. Non-rigid image registration using a hierarchical partition of unity finite element method. In *IEEE 11th International Conference on Computer Vision*, pages 1–8, 2007.

[344] J. R. Shewchuk. Tetrahedral mesh generation by Delaunay refinement. In *Proceedings of the 4th Annual Symposium on Computational Geometry, Association for Computational Machinery*, pages 86–95, 1998.

[345] J. R. Shewchuk. Constrained Delaunay tetrahedrizations and provably good boundary recovery. In *11th International Meshing Roundtable*, pages 193–204, 2002.

[346] J. R. Shewchuk. Two discrete optimization algorithms for the topological improvement of tetrahedral meshes. In *Unpublished manuscript*, 2002.

[347] J. R. Shewchuk. What is a good linear element? Interpolation, conditioning, and quality measures. In *11th International Meshing Roundtable*, pages 115–126, 2002.

[348] J. Shi and J. Malik. Normalized cuts and image segmentation. *IEEE Transactions on Pattern Analysis and Machine Intelligence*, 22(8):888–905, 2000.

[349] Y. Shi, R. Lai, S. Krishna, N. Sicotte, I. Dinov, and A. W. Toga. Anisotropic Laplace-Beltrami eigenmaps: bridging Reeb graphs and skeletons. *Computer Vision and Pattern Recognition Workshops*, pages 1–7, 2008.

[350] K. Shimada. Current issues and trends in meshing and geometric processing for computational engineering analyses. *Journal of Computing and Information Science in Engineering*, 11(2), 2011.

[351] K. Shimada, A. Yamada, and T. Itoh. Anisotropic triangular meshing of parametric surfaces via close packing of ellipsoidal bubbles. In *6th International Meshing Roundtable*, pages 375–390, 1997.

[352] R. Siegel, E. Ward, O. Brawley, and A. Jemal. Cancer statistics, 2011. *CA: A Cancer Journal for Clinicians*, 61(4):212–236, 2011.

[353] J. M. Smith and N. A. Dodgson. A topologically robust algorithm for Boolean operations on polyhedral shapes using approximate arithmetic. *Computer Aided Design*, 39(2):149–163, 2007.

[354] Y. Song, Y. Zhang, C. L. Bajaj, and N. A. Baker. Continuum diffusion reaction rate calculations of wild-type and mutant mouse acetylcholinesterase: adaptive finite element analysis. *Biophysical Journal*, 87(3):1558–1566, 2004.

[355] Y. Song, Y. Zhang, T. Shen, C. L. Bajaj, J. A. McCammon, and N. A. Baker. Finite element solution of the steady-state Smoluchowski equation for rate constant calculations. *Biophysical Journal*, 86(4):2017–2029, 2004.

[356] G. Spanos, D. Rowenhorst, A. Lewis, and A. Geltmacher. Combining serial sectioning, EBSD analysis, and image-based finite element modeling. *MRS Bulletin*, 33:597–602, 2008.

[357] J. Stam. Exact evaluation of Catmull-Clark subdivision surfaces at arbitrary parameter values. *25th Annual Conference on Computer Graphics and Interactive Techniques*, pages 395–404, 1998.

[358] M. Staten, S. Canaan, and S. Owen. BMSweep: locating interior nodes during sweeping. In *Proceedings of the 7th International Meshing Roundtable*, pages 7–18, 1998.

[359] M. Staten, R. Kerr, S. Owen, and T. Blacker. Unconstrained paving and plastering: progress update. In *15th International Meshing Roundtable*, pages 469–486, 2006.

[360] J. E. Stone, D. J. Hardy, I. S. Ufimtsev, and K. Schulten. GPU-accelerated molecular modeling coming of age. *Journal of Molecular Graphics and Modelling*, 29(2):116–125, 2010.

[361] N. Sukumar, D. L. Chopp, N. Moës, and T. Belytschko. Modeling holes and inclusions by level sets in the extended finite-element method. *Computer Methods in Applied Mechanics and Engineering*, 190(46-47):6183–6200, 2001.

[362] N. Sukumar, N. Moës, B. Moran, and T. Belytschko. Extended finite element method for three-dimensional crack modelling. *International Journal for Numerical Methods in Engineering*, 48(11):1549–1570, 2000.

[363] G. Sußner, C. Dachsbacher, and G. Greiner. Hexagonal LOD for interactive terrain rendering. In *Vision Modeling and Visualization*, pages 437–444, 2005.

[364] G. Sußner and G. Greiner. Hexagonal Delaunay triangulation. In *18th International Meshing Roundtable*, pages 519–538, 2009.

[365] A. Tagliasacchi, I. Alhashim, M. Olson, and H. Zhang. Mean curvature skeletons. *Computer Graphics Forum*, 31:1735–1744, 2012.

[366] J. Talbert and A. Parkinson. Development of an automatic, two dimensional finite element mesh generator using quadrilateral elements and Bézier curve boundary definitions. *International Journal of Numerical Methods in Engineering*, 29:1551–1567, 1991.

[367] T. Tam and C. Armstrong. 2D finite element mesh generation by medial axis subdivision. *Advances in Engineering Software*, 13:313–324, 1991.

[368] T. J. Tautges, T. Blacker, and S. Mitchell. The whisker-weaving algorithm: a connectivity based method for constructing all-hexahedral finite element meshes. *International Journal for Numerical Methods in Engineering*, 39:3327–3349, 1996.

[369] C. A. Taylor, T. J. R. Hughes, and C. K. Zarins. Finite element modeling of blood flow in arteries. *Computer Methods in Applied Mechanics and Engineering*, 158:158–196, 1998.

[370] S.-H. Teng and C. W. Wong. Unstructured mesh generation: theory, practice, and perspectives. *International Journal of Computational Geometry and Applications*, 10(3):227–266, 2000.

[371] J. P. Thirion. Image matching as a diffusion process: an analogy with Maxwell's demons. *Medical Image Analysis*, 2(3):243–260, 1998.

[372] J. F. Thompson, B. K. Soni, and N. P. Weatherill. *Handbook of Grid Generation*. CRC Press, Boca Raton, FL, 1999.

[373] O. J. Tobias and R. Seara. Image segmentation by histogram thresholding using fuzzy sets. *IEEE Transactions on Image Processing*, 11(12):1457–1465, 2002.

[374] Y. Tong, P. Alliez, D. Cohen-Steiner, and M. Desbrun. Designing quadrangulations with discrete harmonic forms. In *4th Eurographics Symposium on Geometry*, pages 201–210, 2006.

[375] G. Varadhan, S. Krishnan, Y. Kim, and D. Manocha. Feature-sensitive subdivision and iso-surface reconstruction. In *IEEE Visualization*, pages 99–106, 2003.

[376] A. Varshney and F. P. Brooks Jr. Fast analytical computation of Richard's smooth molecular surface. In *IEEE Visualization*, pages 300–307, 1993.

[377] S. A. Vavasis. QMG website: http://www.cs.cornell.edu/home/vavasis/qmg-home.html.

[378] L. A. Vese and T. F. Chan. A multiphase level set framework for image segmentation using the Mumford and Shah model. *International Journal of Computer Vision*, 50(3):271–293, 2002.

[379] G. Voronoi. Nouvelles applications des paramtres continus á la théorie des formes quadratiques. *Journal für die Reine und Angewandte Mathematik*, 133:97–178, 1907.

[380] A.-V. Vuong, C. Giannelli, B. Jüttler, and B. Simeon. A hierarchical approach to adaptive local refinement in isogeometric analysis. *Computer Methods in Applied Mechanics and Engineering*, 200:3554–3567, 2011.

[381] H. Wang, L. Dong, J. O'Daniel, R. Mohan, A. Garden, K. Ang, D. Kuban, M. Bonnen, J. Chang, and R. Cheung. Validation of an accelerated demons' algorithm for deformable image registration in radiation therapy. *Physics in Medicine and Biology*, 50(12):2887–2905, 2005.

[382] H. Wang, Y. He, X. Li, X. Gu, and H. Qin. Polycube splines. In *Symposium on Solid and Physical Modeling*, pages 241–251, 2007.

[383] J. Wang, L. Ju, and X. Wang. An edge-weighted centroidal Voronoi tessellation model for image segmentation. *IEEE Transactions on Image Processing*, 18(8):1844–1858, 2009.

[384] J. Wang, L. Ju, and X. Wang. Image segmentation using local variation and edge-weighted centroidal Voronoi tessellations. *IEEE Transactions on Image Processing*, 20(11):3242–3256, 2011.

[385] J. Wang and X. Wang. VCells: simple and efficient superpixels using edge-weighted centroidal Voronoi tessellations. *IEEE Transactions on Pattern Analysis and Machine Intelligence*, 34(6):1241–1247, 2012.

[386] J. Wang and Z. Yu. Feature-sensitive tetrahedral mesh generation with guaranteed quality. *Computer Aided Design*, 44(5):400–412, 2012.

[387] K. Wang, X. Li, B. Li, H. Xu, and H. Qin. Restricted trivariate polycube splines for volumetric data modeling. *IEEE Transactions on Visualization and Computer Graphics*, 18:703–16, 2009.

[388] L. Wang, X. Gu, K. Mueller, and S. Yau. Uniform texture synthesis and texture mapping using global parameterization. *The Visual Computer*, 21:801–810, 2005.

[389] Q. Wang, J. JaJa, and A. Varshney. An efficient and scalable parallel algorithm for out-of-core isosurface extraction and rendering. *Journal of Parallel and Distributed Computing*, 67(5):592–603, 2007.

[390] W. Wang and Y. Zhang. Wavelets-based NURBS simplification and fairing. *Computer Methods in Applied Mechanics and Engineering*, 199(5-8):290–300, 2010.

[391] W. Wang, Y. Zhang, L. Liu, and T. J. R. Hughes. Trivariate solid T-spline construction from boundary triangulations with arbitrary genus topology. *Computer Aided Design*, 45(2):351–360, 2013.

[392] W. Wang, Y. Zhang, and J. Qian. Error-bounded solid NURBS construction for navy structures using offsets. In *Marine 2009. Trondheim, Norway*, 2009.

[393] W. Wang, Y. Zhang, M. A. Scott, and T. J. R. Hughes. Converting an unstructured quadrilateral mesh to a standard T-spline surface. *Computational Mechanics*, 48:477–498, 2011.

[394] W. Wang, Y. Zhang, G. Xu, and T. J. R. Hughes. Converting an unstructured quadrilateral/hexahedral mesh to a rational T-spline. *Computational Mechanics*, 50(1):65–84, 2012.

[395] Y. Wang and J. Zheng. Control point removal algorithms for T-spline surfaces. *Lecture Notes of Computer Sciences*, 4077:385–396, 2006.

[396] N. P. Weatherill and O. Hassan. Efficient three-dimensional delaunay triangulation with automatic point creation and imposed boundary constraints. *International Journal of Numerical Methods in Engineering*, 37:2005–2039, 1994.

[397] J. C. Weddell, J. Kwack, P. I. Imoukhuede, and A. Masud. Hemodynamic analysis in an idealized artery tree: differences in wall shear stress between Newtonian and non-Newtonian blood models. *PLoS ONE*, 10(4):e0124575, 2015.

[398] G. Wei, Y. Sun, Y. Zhou, and M. Feig. Molecular multiresolution surfaces. *arXiv math-ph/0511001*, 2005.

[399] X. Wei, Y. Zhang, T. J. R. Hughes, and M. A. Scott. Truncated hierarchical Catmull-Clark subdivision with local refinement. *Computer Methods in Applied Mechanics and Engineering*, 291:1–20, 2015.

[400] X. Wei, Y. Zhang, T. J. R. Hughes, and M. A. Scott. Extended truncated hierarchical Catmull-Clark subdivision. *Computer Methods in Applied Mechanics and Engineering, submitted*, 299:316–336, 2016.

[401] R. Westermann, L. Kobbelt, and T. Ertl. Real-time exploration of regular volume data by adaptive reconstruction of isosurfaces. *The Visual Computer*, 15(2):100–111, 1999.

[402] R. Whitaker. Isosurfaces and level sets. *Insight into Images. A K Peters/CRC Press, Boca Raton, FL. ISBN 1-56881-217-5*, 2004.

[403] R. Whitaker. Nonlinear image filtering with partial differential equations. *Insight into Images. A K Peters/CRC Press, Boca Raton, FL. ISBN 1-56881-217-5*, 2004.

[404] D. White, S. Saigal, and S. Owen. Automatic decomposition of multisweep volumes. *Engineering with Computers*, 20:222–236, 2004.

[405] D. R. White. Automated hexahedral mesh generation by virtual decomposition. In *Proceedings of the 4th International Meshing Roundtable*, pages 165–176, 1995.

[406] M. Wierse, J. Cabello, and Y. Mochizuki. Automatic grid generation with HEXAR. In *6th International Conference on Numerical Grid Generation in Computational Field Simulations*, pages 843–852, 1998.

[407] J. Wilhelm and A. Van Gelder. Octrees for faster isosurface generation. In *ACM Transactions on Graphics*, pages 57–62, 1992.

[408] Z. Wood, M. Desbrun, P. Schroder, and D. E. Breen. Semi-regular mesh extraction from volumes. In *Visualization Conference Proceedings*, pages 275–282, 2000.

[409] C. T. Wu, W. Hu, and J. S. Chen. A meshfree-enriched finite element method for compressible and near-incompressible elasticity. *International Journal for Numerical Methods in Engineering*, 90:882–914, 2012.

[410] Y. Xie, J. Cheng, B. Lu, and L. Zhang. Parallel adaptive finite element algorithms for solving the coupled electrodiffusion equations. *Molecular Based Mathematical Biology*, 1:90–108, 2013.

[411] G. Xu. Discrete Laplace-Beltrami operators and their convergence. *Computer Aided Geometric Design*, 21:767–784, 2004.

[412] G. Xu, Q. Pan, and C. Bajaj. Discrete surface modeling using partial differential equations. *Computer Aided Geometric Design*, 23(2):125–145, 2006.

[413] S. Yamakawa, I. Gentilini, and K. Shimada. Subdivision templates for converting a non-conformal hex-dominant mesh to a conformal hex-dominant mesh without pyramid elements. *Engineering with Computers*, 27(1):51–65, 2011.

[414] I. Yanovsky, P. Thompson, S. Osher, and A. Leow. Large deformation unbiased diffeomorphic nonlinear image registration: theory and implementation. *UCLA CAM Report 06-71*, 2006.

[415] T. S. Yoo. *Insight into Images*. A K Peters/CRC Press, Boca Raton, FL. ISBN 1-56881-217-5, 2004.

[416] T. S. Yoo, G. D. Stetten, and B. Lorensen. Basic image processing and linear operators. *Insight into Images. A K Peters/CRC Press, Boca Raton, FL. ISBN 1-56881-217-5*, 2004.

[417] S. Youssef, E. Maire, and R. Gaertner. Finite element modeling of the actual structure of cellular materials determined by x-ray tomography. *Acta Materialia*, 53:719–730, 2005.

[418] Z. Yu and C. Bajaj. Image segmentation using gradient vector diffusion and region merging. In *16th International Conference on Pattern Recognition*, volume 2, pages 941–944, 2002.

[419] Z. Yu and C. Bajaj. A fast and adaptive algorithm for image contrast enhancement. *IEEE International Conference on Image Processing*, 2:1001–1004, 2004.

[420] Z. Yu and C. Bajaj. Automatic ultra-structure segmentation of reconstructed Cryo-EM maps of icosahedral viruses. *IEEE Transactions on Image Processing: Special Issue on Molecular and Cellular Bioimaging*, 14(9):1324–1337, 2005.

[421] Z. Yu, M. J. Holst, Y. Cheng, and J. A. McCammon. Feature-preserving adaptive mesh generation for molecular shape modeling and simulation. *Journal of Molecular Graphics and Modeling*, 26(8):1370–1380, 2008.

[422] D. Zhang, J. Suen, Y. Zhang, Y. Song, Z. Radic, P. Taylor, M. J. Holst, C. L. Bajaj, N. A. Baker, and J. A. McCammon. Tetrameric mouse acetylcholinesterase: continuum diffusion rate calculations by solving the steady-state Smoluchowski equation using finite element methods. *Biophysical Journal*, 88(3):1659–1665, 2004.

[423] H. Zhang, Y. Jiao, E. Johnson, L. Zhan, Y. Zhang, and K. Shimada. Modeling anisotropic material property of cerebral aneurysms for fluid-structure interaction simulation. *Computer Methods in Biomechanics and Biomedical Engineering: Imaging & Visualization*, 1(3):164–174, 2013.

[424] L. Zhang, A. Gerstenberger, X. Wang, and W. K. Liu. Immersed finite element method. *Computer Methods in Applied Mechanics and Engineering*, 193(21-22):2051–2067, 2004.

[425] N. Zhang, W. Hong, and A. Kaufman. Dual contouring with topology preserving simplification using enhanced cell representation. In *IEEE Visualization*, pages 505–512, 2004.

[426] Y. Zhang. Challenges and advances in image-based geometric modeling and mesh generation. *Image-Based Geometric Modeling and Mesh Generation. Springer Publisher*, 2013.

[427] Y. Zhang and C. Bajaj. Adaptive and quality quadrilateral/hexahedral meshing from volumetric data. *Computer Methods in Applied Mechanics and Engineering (CMAME)*, 195(9-12):942–960, 2006.

[428] Y. Zhang, C. Bajaj, and B.-S. Sohn. 3D finite element meshing from imaging data. *Computer Methods in Applied Mechanics and Engineering*, 194(48-49):5083–5106, 2005.

[429] Y. Zhang, C. Bajaj, and G. Xu. Surface smoothing and quality improvement of quadrilateral/hexahedral meshes with geometric flow. *Communications in Numerical Methods in Engineering*, 25(1):1–18, 2009.

[430] Y. Zhang, Y. Bazilevs, S. Goswami, C. Bajaj, and T. J. R. Hughes. Patient-specific vascular NURBS modeling for isogeometric analysis of blood flow. *Computer Methods in Applied Mechanics and Engineering*, 196(29-30):2943–2959, 2007.

[431] Y. Zhang, B. C. Cheng, C. Oh, J. L. Spehar, and J. Burgess. Dynamic neural foramina cross section measurement and kinematic analysis of lumbar spine undergoing extension. *Computer Modeling in Engineering and Sciences*, 29(2):55–62, 2008.

[432] Y. Zhang, T. J. R. Hughes, and C. L. Bajaj. An automatic 3D mesh generation method for domains with multiple materials. *Computer Methods in Applied Mechanics and Engineering*, 199(5-8):405–415, 2010.

[433] Y. Zhang, Y. Jing, X. Liang, G. Xu, and L. Dong. Dynamic lung modeling and tumor tracking using deformable image registration and geometric smoothing. *Molecular & Cellular Biomechanics*, 9(3):213–226, 2012.

[434] Y. Zhang, X. Liang, J. Ma, Y. Jing, M. J. Gonzales, C. Villongco, A. Krishnamurthy, L. R. Frank, P. Stark, S. M. Narayan, and A. McCulloch. An atlas-based geometry pipeline for cardiac hermite model construction and diffusion tensor reorientation. *Medical Image Analysis*, 16(6):1130–1141, 2012.

[435] Y. Zhang, X. Liang, and G. Xu. A robust 2-refinement algorithm in octree and rhombic dodecahedral tree based all-hexahedral mesh generation. *Computer Methods in Applied Mechanics and Engineering*, 256:88–100, 2013.

[436] Y. Zhang and J. Qian. Dual contouring for domains with topology ambiguity. *Computer Methods in Applied Mechanics and Engineering*, 217-220:34–45, 2012.

[437] Y. Zhang and J. Qian. Resolving topology ambiguity for multiple-material domains. *Computer Methods in Applied Mechanics and Engineering*, 247–248:166–178, 2012.

[438] Y. Zhang, W. Wang, and T. J. R. Hughes. Solid T-spline construction from boundary representations for genus-zero geometry. *Computer Methods in Applied Mechanics and Engineering*, 249-252:185–197, 2012.

[439] Y. Zhang, W. Wang, and T. J. R. Hughes. Conformal solid T-spline construction from boundary T-spline representations. *Computational Mechanics*, 51(6):1051–1059, 2013.

[440] Y. Zhang, W. Wang, X. Liang, Y. Bazilevs, M.-C. Hsu, T. Kvamsdal, R. Brekken, and J. Isaksen. High fidelity tetrahedral mesh generation from medical imaging data for fluid-structure interaction analysis of cerebral aneurysms. *Computer Modeling in Engineering & Sciences*, 42(2):131–149, 2009.

[441] Y. Zhang, G. Xu, and C. Bajaj. Quality meshing of implicit solvation models of biomolecular structures. *Computer Aided Geometric Design*, 23(6):510–530, 2006.

[442] Q. Zheng, S. Yang, and G. Wei. Biomolecular surface construction by PDE transform. *International Journal for Numerical Methods in Biomedical Engineering*, 28(3):291–316, 2012.

[443] K. Zhou, H. Bao, and J. Shi. 3D surface filtering using spherical harmonics. *Computer Aided Design*, 36:363–375, 2004.

[444] Y. Zhou, B. Chen, and A. Kaufman. Multiresolution tetrahedral framework for visualizing regular volume data. In *Proceedings of the 8th IEEE Visualization*, pages 135–142, 1997.

[445] J. Zhu, O. Zienkiewicz, E. Hinton, and J. Wu. A new approach to the development of automatic quadrilateral mesh generation. *International Journal of Numerical Methods in Engineering*, 32:849–866, 1991.

[446] B. Zitová and J. Flusser. Image registration methods: a survey. *Image and Vision Computing*, 21:977–1000, 2003.

[447] U. Zore, B. Jüttler, and J. Kosinka. On the linear independence of (truncated) hierarchical subdivision splines. *Johannes Kepler University of Linz, no. 17*, 2014.

[448] D. Zorin and P. Schröder. Subdivision for modeling and animation. *ACM Siggraph Course Notes*, 2000.

Index